HISTORICAL DICTIONARY

The historical dictionaries present essential information on a broad range of subjects, including American and world history, art, business, cities, countries, cultures, customs, film, global conflicts, international relations, literature, music, philosophy, religion, sports, and theater. Written by experts, all contain highly informative introductory essays on the topic and detailed chronologies that, in some cases, cover vast historical time periods but still manage to heavily feature more recent events.

Brief A–Z entries describe the main people, events, politics, social issues, institutions, and policies that make the topic unique, and entries are cross-referenced for ease of browsing. Extensive bibliographies are divided into several general subject areas, providing excellent access points for students, researchers, and anyone wanting to know more. Additionally, maps, photographs, and appendixes of supplemental information aid high school and college students doing term papers or introductory research projects. In short, the historical dictionaries are the perfect starting point for anyone looking to research in these fields.

HISTORICAL DICTIONARIES OF WAR, REVOLUTION, AND CIVIL UNREST

Civil Wars in Africa, Second Edition, by Guy Arnold. 2008.
The Anglo-Boer War, by Fransjohan Pretorius. 2009.
The Zulu Wars, by John Laband. 2009.
Terrorism, Third Edition, by Sean K. Anderson with Stephen Sloan. 2009.
Korean War, Second Edition, by Paul M. Edwards. 2010.
United States Navy, Second Edition, by James M. Morris and Patricia M. Kearns. 2011.
War in Vietnam, by Ronald B. Frankum Jr. 2011.
The Civil War, Second Edition, by Terry L. Jones. 2011.
Spanish Civil War, by Francisco J. Romero Salvadó. 2013.
The Crusades, Second Edition, by Corliss K. Slack. 2013.
The Chinese Civil War, Second Edition, by Christopher R. Lew and Edwin Pak-wah Leung. 2013.
World War II: The War against Germany and Italy, by Anne Sharp Wells. 2014.
The French Revolution, Second Edition, by Paul R. Hanson. 2015.
Chechen Conflict, by Ali Askerov. 2015.
Chinese Cultural Revolution, Second Edition, by Guo Jian, Yongyi Song, and Yuan Zhou. 2015.
Russian Civil Wars, 1916–1926, by Jonathan Smele. 2015.
The Arab–Israeli Conflict, Second Edition, by P. R. Kumaraswamy. 2015.
The British and Irish Civil Wars 1637–1660, Second Edition, by Martyn Bennett. 2016.
Dirty Wars, Third Edition, by David Kohut and Olga Vilella. 2017.
The Northern Ireland Conflict, Second Edition, by Gordon Gillespie. 2017.
The Cold War, Second Edition, by Joseph Smith and Simon Davis. 2017.
The Russo-Japanese War, Second Edition, by Rotem Kowner. 2017.
Nuclear, Biological, and Chemical Warfare, Second Edition, by Benjamin C. Garrett. 2017.
American Revolution, Third Edition, by Terry M. Mays. 2019.
Arab Uprisings, by Aomar Boum and Mohamed Daadaoui. 2020.
Holocaust, Third Edition, by Jack R. Fischel. 2020.
Russian Revolution, by Jonathan Davis. 2020.
Syrian Uprising and Civil War, by Asaad Alsaleh. 2021.
Modern Coups D'état, by John J. Chin, Joseph Wright, and David B. Carter. 2022.
World War II: The War Against Japan, Second Edition, by Anne Sharp Wells. 2023.
Explosives, by Benjamin C. Garrett. 2026.

Historical Dictionary of Explosives

Benjamin C. Garrett

BLOOMSBURY ACADEMIC
NEW YORK • LONDON • OXFORD • NEW DELHI • SYDNEY

BLOOMSBURY ACADEMIC
Bloomsbury Publishing Inc, 1359 Broadway, New York, NY 10018, USA
Bloomsbury Publishing Plc, 50 Bedford Square, London, WC1B 3DP, UK
Bloomsbury Publishing Ireland, 29 Earlsfort Terrace, Dublin 2, D02 AY28, Ireland

First published in the United States of America 2026

Library of Congress Cataloging-in-Publication Data

Names: Garrett, Benjamin C., 1949- author
Title: Historical dictionary of explosives / Benjamin C. Garrett.
Description: New York : Bloomsbury Academic, 2026. | Series: Historical dictionaries of war, revolution, and civil unrest | Includes bibliographical references and index.
Identifiers: LCCN 2025035864 (print) | LCCN 2025035865 (ebook) | ISBN 9798765145715 hardback | ISBN 9798765145739 pdf | ISBN 9798765145722 epub
Subjects: LCSH: Explosives–History–Dictionaries | Explosives, Military–History–Dictionaries
Classification: LCC TP270.A2 G37 2026 (print) | LCC TP270.A2 (ebook)
LC record available at https://lccn.loc.gov/2025035864
LC ebook record available at https://lccn.loc.gov/2025035865

ISBN: HB: 979-8-7651-4571-5
ePDF: 979-8-7651-4573-9
eBook: 979-8-7651-4572-2

Series: Historical Dictionaries of War, Revolution, and Civil Unrest

Typeset by Deanta Global Publishing Services, Chennai, India

For product safety related questions contact productsafety@bloomsbury.com.

To find out more about our authors and books visit www.bloomsbury.com and sign up for our newsletters.

Contents

Preface

In selecting the entries for this volume, I have attempted to provide an overview of historical and technical aspects of explosives, of major events in which explosives were used intentionally or accidentally, and of persons whose work contributed to the development of explosives. My goal has been to maintain a balance within the entries in terms of coverage devoted to noteworthy events, notable individuals, major domestic and international conflicts, advances in the basic science of explosives, and the manufacture and use of explosives for commercial, criminal, or military purposes. I believe that the dictionary contains information that those who follow technical or historical developments regarding explosives will find unusual as well as useful.

I hope this work will assist readers to gain an enhanced understanding of how the explosives field has evolved from the trial-and-error approach of alchemists to the sophisticated laboratory work of nuclear physicists, physical chemists, and other scientists and engineers. The evolution is illustrated dramatically by the growth from gunpowder to nuclear weapons. I hope also that it can help to inform consideration of issues of continuing international concern, especially those issues associated with dual-use/dual-purpose materials, technology, and equipment that might support weapon development and production programs. Such issues have included, for example, information about the main technological stages necessary to develop a nuclear weapon.

I have had to make choices regarding content; consequently, certain gaps in coverage exist. For example, disasters and tragedies involving explosives are numerous and, unfortunately, continue to accumulate, with their associated costs in loss of life and destruction of property. Similarly, details regarding various inventions of explosives and their associated inventors can be fascinating. I realize I might disappoint a reader searching for a particular event or a favorite invention, but I hope that sufficient attention is paid to the bulk of these events, inventions, and inventors.

The information contained in this work is based on unclassified primary or other authoritative sources. It is hoped that this work will also assist to correct the occasional error in fact that may have found its way into the literature.

While some of the information or phrasing used here might be considered politically sensitive in some circles, this work contains no sensitive technical information. In fact, given the popularity with some individuals and groups of employing explosives for political assassinations, acts of terrorism, and other criminal activities, I have avoided providing what might be taken as recipes, stepwise instructions, or other similar how-to guidance on making and using explosives. Such guidance does exist and is accessible but is absent from this work. It is hoped that this work will also assist to correct the occasional error in fact that may have found its way into the literature.

Finally, I would like to thank April Snider, Acquisitions Editor of Bloomsbury Publications, for her patience, assistance, and encouragement. I am solely responsible, however, for any errors or omissions.

Benjamin C. Garrett

Acronyms and Abbreviations

AEC	Atomic Energy Commission
ANFO	ammonium nitrate-fuel oil mixture
ANNIE	ammonium nitrate-nitrobenzene improvised explosive
ANNM	ammonium nitrate-nitromethane mixture
ASEAN	Association of Southeast Asian Nations
AWE	Atomic Weapons Establishment
BASF	Baden Aniline and Soda Factory, a corporation headquartered in Germany (Badische Anilin- und Soda-Fabrik AG)
BL	breech-loading
BN	French designation for Novel White Powder (Blanc Nouvelle)
BSX	1,7-Diacetoxytetramethylene-2,4,6-trinitramine
C1; C-1	Composition C
C2; C-2	Composition C2
C3; C-3	Composition C3
C4; C-4	Composition C4
CAN	calcium ammonium nitrate
CANWFZ	Central America Nuclear Weapon-Free Zone
CE	composition exploding
CEA	Atomic Energy Commission of France (Commissariat à l'énergie atomique)
CFR	Code of Federal Regulations
CHP	concentrated hydrogen pyroxide
CIA	Central Intelligence Agency
CSB	Chemical Safety and Hazard Investigation Board
cwt	centum weight *or* central weight (hundredweight)
DADP	diacetone diperoxide
DBX	Depth Bomb Explosive
DLA	dextrinated lead azide
DMDB	2,3-dimethyl-2,3-dinitrobutane
DNG	dinitroglycerin
DNI	Director of National Intelligence
DNP	dinitrophenol

DoD	Department of Defense
DoE	Department of Energy
DSTL	Defence Science and Technology Laboratory (UK)
EBW	exploding bridgewire
EDNA	ethylene dinitramine
EFP	explosively formed projectile
EGDN	ethylene glycol dinitrate
EIF	entry into force
EOD	Explosive Ordnance Disposal
EU	European Union
FAE; FAX	fuel-air explosive
FBI	Federal Bureau of Investigation
FOAB	Father Of All Bombs
FofI	Figure of Insensitivity
g	gram
GBU	Guided Bomb Unit
GMT	Greenwich Mean Time
HBX	high blast explosive
HE	high explosive; high velocity explosive
HEAT	high explosive anti-tank
HEU	highly enriched uranium
HME	homemade explosive
HMTD	hexamethylene triperoxide diamine
HMX	high-melting unknown *or* high-melting explosive (1,3,5,7-tetranitro-1,3,5,7-tetrazocane)
HTP	high-test peroxide
IAEA	International Atomic Energy Agency
ICAO	International Civil Aviation Organization
IED	improvised explosive device
IHE	insensitive high explosive
IND	improvised nuclear device
IS	icing sugar
ISIL	Islamic State of Syria and the Levant
ISIS	Institute for Science and International Security
IUPAC	International Union of Pure and Applied Chemistry
kg	kilogram
KRL	A. Q. Khan Research Laboratories (Pakistan)
LEU	low-enriched uranium
LOX	liquid oxygen explosive
MD	modified cordite
MED	Manhattan Engineering District

Mk	Mark
mL	milliliter
μm	micrometer
mm	millimeter
MOAB	Massive Ordnance Air Blast Bomb; Mother Of All Bombs
MON	mixed oxides of nitrogen
MOP	Massive Ordnance Penetrator
mt	megatons
NATO	North Atlantic Treaty Organization
NFPA	National Fire Protection Association
NG	nitroglycerin
NM	nitromethane
NNSA	National Nuclear Security Administration
NNWS	Non-nuclear Weapon State
NPT	Nonproliferation Treaty
NSG	Nuclear Suppliers Group
NTO	nitrogen tetraoxide
NWFZ	Nuclear Weapon-Free Zone
NWS	Nuclear Weapon State
OAU	Organization for African Unity
o-MNT	*ortho*-mononitrotoluene
OSS	Office of Strategic Services
Pa	Pascals
PACT	The Sergeant First Class Heath Robinson Honoring our Promise to Address the Comprehensive Toxics Act of 2022
PBIED	person-borne improvised explosive device
PBX	plastic-bonded explosive *or* polymer-bonded explosive *or* pressed bonded explosive *or* pressed plastic-bonded explosive
PETN	pentaerythritol tetranitrate
PIRA	Provisional Irish Republican Army
p-MNT	*para*-mononitrotoluene
PNE	peaceful nuclear explosions
RCP	German designation for Tubular Gunpowder With Centralite (Röhrenpulver mit Centralit)
RDB	Research Department formula B
RDD	radiological dispersal device
RDS	Russian prefix designation for nuclear devices; literally "Special Jet Engine" (Reaktivnyi Dvigatel Spetsialnyi)
RDX	Research Department Composition X (1,3,5-Trinitro-1,3,5-triazinane)
RE	relative effectiveness

SADM	special atomic demolition munition
SC	solventless cordite
SD	French designation for solventless (Sans Dissolvent)
SHL	German designation for heavy shaped charge (Schwere Hohlladung)
SMP	sulfur-free meal powder
SNM	special nuclear material
SNPE	National Company of Powders and Explosives of France (Société Nationale des Poudres et Explosifs)
SQ	significant quantity
SPF	South Pacific Forum
TATB	2,4,6-triamino-1,3,5-trinitrobenzene
TATP	triacetone triperoxide
TNA	trinitroanisol
TNT	trinitrotoluene
TNX	trinitro-*m*-xylene
TSA	Transportation Security Administration
UN	United Nations
UNGA	United Nations General Assembly
UNSC	United Nations Security Council
UXO	unexploded ordnance
VBIED	vehicle-borne improvised explosive device
WASAG	German explosives manufacturing company headquartered in the Westphalian-Anhalt region (Westfälische-Anhaltische-Spre ngstoff-AG)

Chronology

142 Wei Boyang, a Taoist priest, discovers a mix of three solids that would "fly and dance violently," likely a crude explosive, perhaps black powder.

ca. 300 Ge Hong identifies a mix of sulfur, saltpeter, and a carbonaceous material as explosive.

808 An unknown Chinese alchemist writes *Taishang Shengzu Jindan Mijue* (*Secret Teachings of the Golden Elixir of the Supreme Holy Ancestor*) in which he identifies correctly the formula for black powder.

1040–1044 A Northern Song dynasty scholar produces *Wujing Zongyao* (*Collection of the Most Important Military Techniques*), describing weapons using gunpowder.

ca. 1260 English Franciscan friar and alchemist Roger Bacon publishes *De nullitate magiae* (*The nullity of magic*), containing a cipher thought to be a recipe for making black powder. This recipe, if correct, is likely the earliest description of black powder in a European text.

1347 26 August: English forces under King Edward III defeat a much larger French force under King Philip VI at the site of present-day Crécy, France. The English forces included some guns, marking one of the earliest uses of black powder by the English.

1743 26 August: Antoine-Laurent de Lavoisier is born in Paris. He will become known as the father of modern chemistry, where his contributions will include increasing the quality and quantity of explosives. Lavoisier becomes an important ally to the American Colonies during the Revolutionary War, and his teachings aid his assistant, Eleuthère Irénée du Pont, in forming what becomes a major producer of explosives and other chemicals.

1745 Sir William Watson invents the first detonator when he uses a friction machine to generate an electric spark that detonates black powder.

1775 Pierre Bayen discovers mercury fulminate.

1792 **March:** Franz Xaver von Baader publishes his concept of "shaped charge explosives" in an article titled "Versuch einer Theorie der Sprengbarkeit" ("Investigation of a theory of blasting"), published in *Bergmannisches Journal von Kohler und Hoffman.*

1831 **6 September:** Cornish merchant William Bickford patents the combination of gunpowder wound inside jute, then varnished to make it waterproof. He calls his invention a "safety rod," later changing the name to "safety fuse." It will become more commonly known as the Bickford fuse, used to ignite black powder, replacing less reliable methods previously used. His invention will see wide application for blasting rock in Cornwall, especially, and will be widely imitated.

1839 **23 October:** British Army major general Charles Pasley successfully uses an electric detonator to remove a submerged obstacle to navigation in Spithead, southeast coast of England.

1846 **17 May:** German-Swiss chemist Schönbein announces his discovery of guncotton, which will enjoy wide use in explosives.

1847 Ascanio Sobrero announces his discovery of pyroglycerine, which will become better known as nitroglycerin. He calls attention to its explosive nature and warns of difficulties in handling it.

1856 British East India Company major John Jacobs creates an explosive bullet, filled with mercury fulminate. He proposes to use such bullets against artillery caissons, the goal being that the explosive bullet would cause munitions within the caisson to explode. In Jacobs' words, the explosive bullet "would revolutionize the art of war."

1858 **14 January:** French emperor Napoléon III and his wife survive an attempted assassination in Paris when mercury fulminate-filled bombs are thrown at their carriage, although eight others will die of injuries received from the blasts. This attempt on their lives marks the first terrorist attack using a high explosive.

1859 5 April: Martha Jane Coston receives a patent for her invention of pyrotechnic night signal flares. These flares will come into routine use by the US Life-Saving Service (predecessor to the US Coast Guard) from its founding in 1878 for communicating with ships in distress or to warn ships of danger. Coston-type flares continue to be used internationally. In 2006, she was inducted into the National Inventors Hall of Fame.

1863 German chemist Julius Wilbrand reports the original synthesis of trinitrotoluene. He fails to note, though, the explosive nature of what will become better known as TNT.

1864 28 November: Alfred Nobel founds Nitroglycerin AB, the first industrial-scale producer of nitroglycerin.

1865 26 April: Edward A. L. Roberts receives US patent 59,936 for improvements in methods of increasing the capacity of oil wells. His methods use explosives, lowered into the borehole and detonated, increasing oil flow without igniting the oil. This same year, Roberts forms the Roberts Petroleum Torpedo Company, specializing in shooting gas, oil, and water wells.

1866 Alfred Nobel discovers dynamite. He receives his first patent on it the following year.

1868 11 December: Nineteen nations attending a conference in Saint Petersburg, Russia, sign an agreement banning the use in war of explosive bullets weighing 400 grams or less. This agreement marks the first international arms control treaty and becomes a model for other efforts to limit various types of weapons of war.

1871 6 April: German chemist Hermann Sprengel receives British patent 921 for a novel class of explosives in which the fuel or oxidizer is a liquid. These liquid explosives will see wide use and become known as "Sprengel explosives" in his honor.

1875 Alfred Nobel invents blasting gelatin, the first plastic explosive.

1878 12 March: Henry Julius Smith receives US patent 201,296 for an electric detonator. Marketed as the "Magneto No. 3," its plunger design will become iconic.

1884 French physicist Paul-Marie-Eugène Vieille invents the first successful smokeless powder. Vieille prepares this powder by mixing insoluble nitrocellulose and soluble nitrocellulose, giving a white colloidal mass. It becomes known as "Poudre B." This same year, François Eugène Turpin invents mélinite, a smokeless powder based on picric acid. Both will be widely used by the French military and others.

1885 **10 October:** The US Army Corps of Engineers detonates over 150 tons of explosives to demolish Flood Rock, a 9-acre obstacle to ship traffic in New York's East River. The detonation succeeds, eliminating Flood Rock as a hazard to navigation. The blast is one of the largest non-nuclear explosions ever recorded.

1891 German chemist Theodor Curtius discovers lead azide, which will see wide use as a detonator.

1895 German chemist Richard Wolffenstein reports his discovery of acetone peroxides and notes their explosive nature.

1898 **1 September:** The British use Lyddite for the first time in combat during the opening operations of the Battle of Omdurman, Sudan. A mix of picric acid, nitrobenzene, and petroleum jelly, Lyddite becomes the standard smokeless powder used in British artillery rounds. **15 July:** German pharmaceutical chemist Georg Friedrich Henning receives German patent 104,280 for a process of making a nitro compound from hexamethylenetetramine. He suggests that the compound will be useful in making medicinal products. Decades later, it is recognized that what Henning made is RDX, making him the discoverer of this powerful explosive.

1899 German chemist Ernst Richard Escales receives German and Belgian patents covering the basic concept of ammonium nitrate and aluminum powder mixtures for use as explosives. This concept will be exploited widely, especially with the Amatol family of explosives, and also in homemade explosives.

1905 *La Salute è in Voi!* appears in print. The title is variously translated as *Health is in you!* or *Salvation is within You!* The book itself is a manual for making, handling, and using explosives. Authorship is uncertain, but the audience is anarchists willing to use violence in the face of capitalists' oppression. This audience includes those who engage in a US bombing campaign during the first decades of the twentieth century. **27–28 May:** The Russian Baltic fleet suffers heavy losses during the two-day Battle of Tsushima. Seven battleships and 14 other ships are sunk, and four battleships are captured. The use of smokeless powder by the Japanese Imperial Navy contributes mightily to this victory, as it ensures a clear line-of-sight for spotting targets as well as the impact of its artillery.

1908 A patent is granted in France for a lead tube filled with molten TNT. The resultant product, cordeau detonant, becomes the model for modern-day detonator cord.

1911 1 November: An Italian military aviator drops ammonium picrate-filled bombs on a Turkish camp, Ain Zara, Libya, marking the first delivery of explosives from an aircraft.

1917 6 December: At 9:08 a.m., local time, the French cargo ship, *Mont-Blanc*, detonates in the harbor of Halifax, Nova Scotia, Canada. Around 25 minutes earlier, the *Mont-Blanc* and the *Imo*, a Norwegian ship, collided at sea, touching off a fire aboard the Mont-Blanc, which carried a cargo that included nearly 3,000 tons of various explosives and other flammable materials, chiefly TNT, picric acid, and nitrocellulose. Halifax and surrounding communities suffer extensive damage from the blast itself, the ensuing fires, and a blast-induced tsunami that swept inland from Halifax harbor. Estimates of the dead and injured will be placed at approximately 2,000 and 9,000, respectively, but an accurate count proved impossible. The blast is estimated to have been equal to 3 kilotons of TNT-equivalent, making it the largest accidental explosion then recorded.

1921 21 September: A silo containing ammonium nitrate and ammonium sulfate explodes at a BASF facility in Oppau, Germany, killing at least 561. The devastation is vast, leaving more than 6,500 homeless. The proximate cause of the explosion remains uncertain.

1926 10 July: Temporary Magazine Number 8, Lake Denmark Naval Ammunition Depot, New Jersey, catches fire from a lightning strike. The fire causes some 800 tons of stored munitions and two other magazines to explode. Most of the depot and much of the adjoining Picatinny Arsenal are destroyed or damaged in the ensuing fires and explosions, and 19 men are killed.

1928 Partly in response to the events at Lake Denmark Naval Ammunition Depot two years earlier, the US Army adopts the American Table of Distances for Storage of Explosives. It also establishes the Armed Forces Safety Board.

1936 Ensign-Bickford Company, Simsbury, Connecticut, develops Primacord, a detonator cord modeled on an earlier French invention but replacing the lead tube with plastic and the filling of TNT with a high explosive, commonly PETN.

1939 11 February: The journal *Nature* features an article by Lise Meitner and Otto Robert Frisch providing mathematical proof that the Hahn-Strassmann experiments conducted the previous December resulted in a process that Frisch names "nuclear fission." **2 August:** Physicist Albert Einstein addresses the first of several letters to US president Franklin D. Roosevelt in which he describes how uranium, through a chain fission reaction, could be used as a powerful energy source and expresses concern that Germany might be able to harness this energy as a weapon. This letter and subsequent ones prompt Roosevelt to establish what becomes known as the Manhattan Project.

1945 16 July: "Gadget" detonates successfully at the Trinity Test Site near Alamogordo, New Mexico, becoming the first man-made nuclear explosion. **6 August:** US Army Air Forces plane *Enola Gay* drops the "Little Boy" nuclear bomb over Hiroshima, Japan. **9 August:** The United States drops the "Fat Man" nuclear bomb over Nagasaki, Japan.

1947 17 April: The SS *Grandcamp* explodes in the Port of Texas City, Galveston Bay, Texas, when a fire causes the ship's cargo of 2,300 tons of ammonium nitrate to detonate, obliterating the port. This explosion touches off a cascade of events, leading to 581 deaths, more than 5,000 injuries, and the destruction of 500 homes.

1949 29 August: The Soviet Union conducts its first successful detonation of a nuclear weapon, dubbed "Joe-1" by US intelligence but known officially as "RDS-1." Its design is similar to that of Fat Man.

1952 3 October: The United Kingdom conducts its first successful detonation of a nuclear weapon, "Hurricane," in the Montebello Islands off the west coast of Australia.

1954 12 September: The US Joint Chiefs of Staff recommend to US president Dwight D. Eisenhower that nuclear weapons be used against military targets on the Asian mainland as part of the ongoing Korean Conflict.

1956 Phillips Petroleum Company launches pilot plant production of Composition C4.

1958 **December:** The "Baby Tooth Survey" begins in St. Louis, Missouri. Its aim is to assess whether children were being exposed to radioactivity attendant to fallout from atmospheric testing of nuclear weapons. The survey focuses on strontium-90, a radioisotope known to be carcinogenic.

1960 **13 February:** France conducts its first successful detonation of a nuclear weapon, "Gerboise Bleue," at the Saharan Military Experiments Center, near Reggane, Algeria.

1961 **30 October:** The Soviet Union detonates RDS-220, a hydrogen bomb, over the Mityushikha Bay test range north of the Arctic Circle. The bomb becomes known as "Tsar Bomba" and is the most powerful nuclear weapon ever detonated, with an estimated yield of more than 50 megatons. **24 November:** The journal *Science* releases preliminary results of the baby tooth survey documenting the deleterious effects of nuclear fallout on humans.

1962 **26 June:** Lawrence H. Johnson receives US patent 3,040,660 for the exploding bridgewire detonator, which functions 10,000 times faster than does an ordinary electric detonator, typically taking 0.1 microsecond or less to go from initiation to detonation. Johnson and his mentor, Luis Alvarez, invented it 18 years earlier during the Manhattan Project and oversaw its use in the successful detonations of Gadget and Fat Man.

1964 **16 October:** China detonates its first successful detonation of a nuclear device, "596," at the Lop Nur test site.

1966 **17 January:** Two US Air Force aircraft, a B-52 bomber and a K-135 tanker, collide over Palomares, Spain, resulting in the breakup of both. Four nuclear weapons aboard the bomber are ejected, with three landing on the ground and one in the Mediterranean Sea. Two of the weapons undergo non-nuclear detonations on land, contaminating the countryside with nuclear materials. **28 December:** China detonates its first successful detonation of a thermonuclear weapon at the Lop Nur test site.

1974 **18 May:** India conducts its first successful detonation of a nuclear device, which will be given the code name "Smiling Buddha." India characterizes the detonation as a peaceful nuclear explosion test.

1981 30 March: John Hinckley, Jr., fires explosive bullets at US president Ronald Reagan. None strikes Reagan directly, but three people are struck, including Press Secretary James Brady. The lead azide charge within the bullet detonates in Brady's brain, leaving him permanently disabled.

1988 21 December: Pan American Airlines flight 103 departs London for New York. While still in British airspace, the aircraft explodes, raining debris over the Scottish countryside. All 259 passengers and crew were killed. A part of a wing lands with explosive force in Lockerbie, killing 11 and bringing the death toll to 270. Subsequent investigations show residues of plastic explosives, most likely Semtex, ruling out mechanical failure or other accidental causes for this disaster; instead, it was a deliberate act of terrorism.

1993 24 March: F. W. de Klerk, President of the Republic of South Africa, announces in a speech to the South African Parliament that the country had built and dismantled six nuclear weapons. He states that all of this work was done in secret. He goes on to relate that the program to build these weapons dated to 1979 but had its roots in the late 1940s. He had ordered the dismantlement of these six and one partially assembled weapon in 1990, shortly after becoming president.

1994 19 August: The IAEA confirms that six nuclear weapons plus one partially assembled weapon have been dismantled in South Africa, making that nation the only one thus far to have eliminated its nuclear weapon stockpile voluntarily.

1995 19 April: A Ryder truck laden with explosives explodes at the Alfred P. Murrah Federal Building, Oklahoma City, killing 168, injuring hundreds, and destroying many buildings. The explosion will be shown to be the work of domestic terrorists.

1996 29 July: China conducts its 45th and final nuclear weapon test to date.

1998 11 May: India conducts test of nuclear devices, exploding three. **13 May:** India explodes two more nuclear devices. **28 May:** Pakistan explodes five nuclear devices. **30 May:** Pakistan explodes one more nuclear device, its sixth. **21 June:** The Convention on the Marking of Plastic Explosive for the Purpose of Detection enters into force. This international agreement is a direct response to the destruction of Pan Am Flight 103 10 years earlier. Among its provisions, the convention defines plastic explosives and designates detection agents, also called taggants, whose use will satisfy the requirements for marking. **17 September:** Swedish chemists Nikolai V. Latypov and four colleagues report the synthesis and properties of 1,1-diamino-2,2-dinitroethylene, soon to find use in nuclear weapons as an insensitive high explosive under the name "FOX-7."

2001 22 December: Richard Reid fails in his attempt to set off a bomb hidden within his running shoes while aboard a flight from Paris to Miami. His use of his sneakers to hide explosives prompts him to become known as "The Shoe Bomber." It also prompts aviation security procedures to be revised to include the removal and scanning of shoes prior to boarding.

2002 12 October: Around 11:00 p.m., a suitcase bomb explodes outside the US consulate, Denpasar, Bali, Indonesia. Damage is slight. Minutes later, a suicide bomber detonates an explosives-laden vest inside Paddy's Pub, Kata district. Twenty seconds later, fleeing patrons as well as passers-by are caught when a driver detonates a vehicle-borne improvised explosive device on the street opposite the pub. More than 200 are killed, including the two bombers. Investigation will link the bombers to Jemaah Islamiyah, a terrorist group dedicated to establishing an Islamic state in Southeast Asia. The bomb materials are a rare combination of homemade explosives, TNT, PETN, and electrically operated RDX detonators, indicating a sophisticated bomb-making operation.

2003 11 March: The US Air Force successfully detonates the GBU/43-B bomb at Eglin Air Force Base, Florida. It is believed to be the largest non-nuclear explosive device ever detonated up to that point, with a yield of 11 tons TNT-equivalent or about 0.1 percent the yield of Little Boy. The bomb is also referred to as the Massive Ordnance Air Blast Bomb and, more colloquially, as the Mother of All Bombs. **15 August:** The Libyan ambassador to the United Nations submits a letter to the UN Security Council (UNSC) in which he formally accepts responsibility for the actions of its citizens in relation to the 1988 bombing of Pan Am Flight 103 over Lockerbie, Scotland. Eventually, the Libyan government will pay over $2 billion in compensation to the families of victims. **19 December:** Libya renounces its pursuit of biological, chemical, and nuclear weapons and agrees to work with the UK, the United States, and the IAEA on dismantling its nuclear program.

2004 31 January: The Pakistani government arrests Abdul Qadeer Khan for illegally transferring nuclear technology. **4 February:** Khan appears on Pakistani television and admits his role as head of the so-called A. Q. Khan nuclear smuggling network that aided Iran, Libya, and North Korea, among others. He absolves the government and military of any involvement. He is sentenced to house arrest.

2005 **10 February:** North Korea declares publicly that it possesses nuclear weapons. **7 July:** Four terrorist attacks involving homemade explosives take place in London, targeting the public transit system during the morning rush hour. Three explosions hit the London underground, while the fourth explodes on a double-decker bus. The attackers leave videotaped testimonies, making clear that they are Islamic radicals. The explosions claim 52 lives, plus those of the four attackers, who commit suicide when they detonate bombs in their backpacks, and injure more than 800. **1 October:** Three bombs explode at sites frequented by tourists in and near the Jimbaran Beach Resort, Bali, Indonesia, in attacks similar to those on the island on 12 October 2002. Twenty people are killed, and more than 100 are injured. Recovered remains point to suicide bombers. Three unexploded bombs are recovered and rendered safe.

2006 **9 October:** North Korea announces that it has exploded a nuclear weapon. The yield is approximately 500–1,000 tons of TNT, and this low yield fuels speculation that the detonation was a fizzle. **14 October:** The UNSC unanimously adopts Resolution 1718, condemning North Korea's nuclear test and imposing sanctions on the country.

2007 **11 September:** Russia successfully detonates a large-yield, non-nuclear explosive device, containing 7.8 tons of explosive. The force of the explosion is placed at 44 tons TNT-equivalent, making it the largest non-nuclear device detonated thus far and earning it the nickname "Father of All Bombs." **3 November:** Iran admits it used the A. Q. Khan network to obtain equipment, materials, and specifications of nuclear technology, whose acquisition had been denied through legitimate channels.

2009 **25 December:** Umar Farouk Abdulmutallab attempts to detonate explosives hidden in his underwear. The 24-year-old native of Nigeria is aboard Northwest Airlines flight 253, en route from Amsterdam to Detroit, Michigan. His attempted suicide bombing fails when flames are seen on his trouser leg and are extinguished by the crew and passengers. He tells the attendants he has a packet of explosive sewn into his underwear. The location of this packet caused him to be referred to in the news as the "Underwear Bomber." On landing, he is arrested and volunteers that he is working on behalf of Al Qaeda.

2012 **20 January:** The Babine Sawmill explodes in Burns Lake, British Columbia, Canada, killing two men and injuring 20 others. Three months later, on 23 April, a sawmill at the Lakeland Mills, Ltd., plant in nearby Prince George, British Columbia, explodes, killing 2 men and injuring 22. In both cases, the explosion is attributed to dust from milling dry, beetle-killed pine trees, illustrating the explosive nature of sawdust.

2013 30 December: UN General Assembly adopts resolution 68/32 calling for 26 September to be observed annually as International Day for Total Elimination of Nuclear Weapons. According to this resolution, the day "provides an occasion for the world community to reaffirm its commitment to global nuclear disarmament through enhancing public awareness and education about the threat posed to humanity by nuclear weapons."

2014 March: Spanish authorities announce that Laura Vega has died. She was a victim of the 11 March 2004 Madrid subway bombings and had been in a coma since that event. Her death brings the official death toll from those bombings to 193. **30 June:** Iraq informs the UNSC that Islamic State terrorists had seized control of the University of Mosul, including its stockpile of nearly 90 pounds of nuclear material, believed to be low-enriched uranium.

2015 15 March: During an appearance on state television, Russian president Vladimir Putin claims he was so fearful at the height of the Ukraine crisis in February 2014 of an attack on Russia by the United States that Russia was preparing to arm its nuclear weapons. **19 October:** Spain and the United States sign a letter of intent to negotiate a binding agreement for the cleaning and restoration of the Palomares area, contaminated by an incident on 17 January 1966 in which two US nuclear weapons underwent non-nuclear detonations on Spanish territory. **13 November:** The first of six terrorist attacks happens around 9:00 p.m. when a suicide bomber detonates his vest outside the Stade de France, Paris, as a soccer match between France and Germany is getting underway. Two explosions follow shortly in the same area. In the midst of these attacks, another suicide bomber strikes at a café on the Boulevard Voltaire. Terrorists also conduct shootings at four Parisian locations, including the Bataclan theater, where they stage a hostage-taking. In all, these attacks and shootings kill 130 people plus seven of the terrorists. The bombers use a type of acetone peroxide in their vests and are linked to the Islamic State.

2016 6 January: A test of a nuclear device in North Korea is detected and is believed to be its fourth test. North Korea will declare later that the test was of a hydrogen bomb, although this claim is met with skepticism. **22 March:** Two explosions hit the check-in area of the Brussels Airport shortly before 8:00 a.m. An hour later, another explosion takes place in the middle carriage of the three-car train near Brussels' Maelbeek Metro Station. In all three cases, the explosions are the work of suicide bombers affiliated with the Islamic State and use a type of acetone peroxide explosive. Two other would-be bombers flee, one from each location, and their explosion-laden backpacks are recovered. All five terrorists are later linked to the November 2015 bombings in Paris. The attacks claim the lives of 32, plus the three bombers. **9 September:** North Korea conducts its fifth detonation of a nuclear device. With an estimated yield of 9–11 kilotons, it is the most powerful test to date.

2017 3 September: North Korea conducts its sixth and, to date, final detonation of a nuclear weapon.

2019 21 April: Eight suicide bombings rock Sri Lanka, killing at least 269, including the bombers. The first bomb strikes during Easter Sunday morning services in Colombo. Two more churches are hit that morning, and then three hotels are targeted in rapid succession. After a pause of several hours, a seventh bomb hits a hotel in Deliwala, followed by a bombing at a Dematagoda housing complex. A van loaded with explosives is found three days later and explodes without harming anyone. The explosive is identified as a type of acetone peroxide, and the bombs were laden with nails and ball bearings to maximize injuries. The bombings are attributed to a terrorist group aligned with the Islamic State. The tactics and materials are identical to those used in bombings in Paris in November 2015 and in Brussels in March 2016. **6 December:** The US Court of Appeals for Veterans Claims rules by 6–3 in favor of veterans exposed to radiation during the 1966 cleanup of debris from the non-nuclear detonation of two US nuclear weapons on the ground near Palomares, Spain. Their disability claims had been denied consistently for more than a half-century, but today's ruling reverses that situation. About 1,600 US Air Force personnel were detailed to this cleanup on site in Spain during January through April 1966 and forced to work without personal protective gear to avoid alarming the public.

2020 4 August: A fire erupts at a warehouse in the Port of Beirut, Lebanon. The fire causes around 3,000 tons of ammonium nitrate to detonate. A nearby warehouse of fireworks adds to the conflagration. The ammonium nitrate was confiscated in 2013 from MV *Rhosus* for non-payment of port charges. The fire and explosion claim at least 218 lives, leave nearly 300,000 homeless, and cause an estimated $15 billion in damages. Seismic data analysis places the size of the explosion as the sixth-largest non-nuclear explosion to date.

2021 22 January: The Treaty on the Prohibition of Nuclear Weapons enters into force. None of the nine states known to or suspected to possess nuclear weapons is a party to the treaty. **10 October:** Abdul Qadeer Khan dies in Islamabad, Pakistan. He is the eponym of the A. Q. Khan network that engaged in the illegal transfer of technology and know-how critical to producing nuclear weapons and is the father of Pakistan's nuclear weapon program.

2022 10 August: The Sergeant First Class Heath Robinson Honoring our Promise to Address Comprehensive Toxics (PACT) Act of 2022 takes effect in the United States. The PACT Act expands eligibility for healthcare and disability benefits to veterans exposed to toxic substances during their military service. These veterans include the nearly 1,600 Air Force personnel who were assigned to collect radioactive soil and other debris onsite from the 1966 Palomares Incident in Spain, but were directed to neglect wearing protective gear to avoid alarming the public.

2023 6 March: Spain asks the United States to begin removing soil contaminated with radioactivity associated with the Palomares incident of 17 January 1966 in which non-nuclear detonations of two nuclear weapons scattered radioactive materials, including plutonium-239, across the Palomares landscape. Spanish officials estimate around 50,000 cubic meters of land are off-limits and in need of removal. **7 September:** The Court of Justice of the European Union issues a ruling condemning Spain for its failures to comply with obligations regarding basic safety standards for protecting against the dangers arising from exposure to ionizing radiation. This ruling arises in part from the lack of an inventory of contaminated soils related to the 1966 Palomares incident.

2024 July: The US Director of National Intelligence reports that Iran continues to increase the size of its uranium stockpile, increase its enrichment capability, and develop, manufacture, and operate advanced centrifuges. Further, it has the infrastructure and experience to quickly produce weapons-grade uranium at multiple facilities, if it chooses to do so. **17 July**: US secretary of the Navy Carlos Del Toro announces full exoneration of 256 men who had been wrongly convicted of assorted charges related to the 1944 Port Chicago Disaster. Del Toro notes that significant legal errors occurred during the courts-martial of these men, that the defendants were tried together despite conflicting interests, and that the defendants were denied a meaningful right to counsel. **6 December:** The Spanish government publishes a royal decree approving regulations requiring, among other provisions, an updated inventory of radiologically contaminated soils. This inventory is to include those soils contaminated by the 1966 Palomares incident. **17 December:** A bomb planted in an unmanned electric scooter detonates in Moscow outside an apartment building, killing Lt. Gen. Igor Kirillov and an aide. **Same day:** FBI searches an Isle of Wight County, Virginia, farm. A container of hexamethylene triacetone diamine (HMTD) is recovered from a refrigerator, and a cache of 155 improvised explosive devices (IEDs) is found. The IEDs are in the form of pipe bombs, with some loaded in wearable vests. This cache is labeled the largest cache of IEDs in terms of numbers in the history of the FBI.

2025 **22 January:** Rafael Grossi, IAEA Director-General, announces that Iran is "pressing the gas pedal" on its enrichment of uranium to near weapons grade. Speaking at the World Economic Forum, Davos, Switzerland, Grossi notes that Iran has increased its monthly uranium enrichment from around 7 kilograms per month to more than 30 kilograms. UN secretary-general Antonio Guterres adds that Iran must take the first step to make clear that it does not aim to produce nuclear weapons. **25 April:** Lieutenant General Yaroslav Y. Moskalik is killed by an IED near his home in the eastern Moscow suburb of Balashikha. The IED was placed in a vehicle and detonated as Moskalik walked past. At the time of his death, Moskalik was the Deputy Chief, Main Operations Directorate, General Staff, Russian Armed Forces. No individual or group claimed responsibility for the bombing, which was rumored to be the work of Ukrainian operatives. **22 June:** The US Air Force delivers 14 Massive Ordnance Penetrators against underground targets in Iran in an effort to hobble that nation's efforts to develop nuclear weapons. **1 July:** Shortly before 6:00 p.m., a fireworks storage warehouse situated near Esparto, Yolo County, California, explodes. The blast and ensuing fire claim seven lives. The warehouse was owned by Devastating Pyrotechnics, a licensed pyrotechnics operator. The exact cause of the explosion is not apparent, but likely causes include human error, non-compliance with regulations, and errant static electricity. **18 July:** Brad K. Spafford pleads guilty in Federal Court in Norfolk, Virginia, to two charges related to the 17 December 2024 seizure of 155 IEDs and firearms from his Isle of Wight County, Virginia, farm. This seizure had been labeled by the FBI as the largest cache of IEDs ever recovered. Evidence produced in court notes Spafford had used a photo of then-president Joseph Biden for target practice and had lost his thumb and parts of two fingers on his right hand in an explosion three years prior while working on an explosive device.

Introduction

The history of explosives is told through accidents, inventions, tragedies, and war. The arc of discoveries of explosives runs from trial-and-error to theoretical nuclear physics, with an occasional boost from serendipity. The use of explosives encompasses hunting game, opening waterways, removing obstacles to navigation, harvesting minerals, pumping oil, and destroying human life on a massive scale. Attempted use has altered how we board airplanes, enter arenas, and access public buildings. This history is at times complex, entertaining, and frightening.

The earliest explosive to have been discovered is most likely black powder. It can be made from relatively accessible ingredients: charcoal (from the fireplace), nitrates (from caves, dovecotes, and latrines), and sulfur (deposited around volcanos). This accessibility made it a candidate for ancient alchemists as they experimented in their quest to find an elixir of life or a means of transmuting base metals into gold. Because the final product is a mixture rather than a discrete substance, no sophisticated chemical industry is required for the production of black powder. It can be made simply by mixing these three ingredients together. The proportions are approximate rather than absolute. While the precise date and place of its discovery are uncertain, it is certain that black powder was known to alchemists who were active in China, as confirmed by writings from the second and third centuries AD. The earliest use of black powder as gunpowder is difficult to date or locate precisely, but Chinese writings from the eleventh century AD describe the use of black powder as a propellant in weapons. Consequently, the Chinese knew of gunpowder and were using it as such by this time.

It also seems likely that the discoveries of black powder as an explosive and its use as gunpowder migrated west, appearing in Europe no later than the fourteenth century. Whenever it occurred, the discovery of the use of black powder as gunpowder proved enduring. No other explosive came into routine use as a substitute in firearms or for blasting until the mid-nineteenth century. Prior to that time, a largely empirical search for substitutes was undertaken, in hopes of discovering a mixture that was less expensive to produce, safer to store and handle, and more effective than black powder. This search led to the incorporation of such curious ingredients as dried coffee grounds (Ricker's Powders, 1862); spermaceti, the waxlike substance from the head of a sperm whale (Hahn's Powder, 1867); and cascarilla bark, a purgative from the *Croton eluteria* tree (Rogers's Composition, 1870). Some of these seemingly curious ingredients proved enduring and saw prolonged use in many explosive powder formulations, with two examples being sawdust and spent tanner's bark.

Alchemists had discovered another explosive by the sixteenth century: fulminating gold. The first half of its name speaks to its explosiveness. The root of *fulminate* is *fulmen*, Latin for lightning or a thunderbolt. Unlike black powder, which burns if ignited in the open, fulminating gold explodes, even in an open container. Furthermore, it can do so without a flame, the shock of being shaken or dropped being sufficient to set it off. But fulminating gold proved impossible to handle safely, as well as expensive, making it unsuitable as a replacement for black powder. In 1775, a less expensive and marginally safer related substance was discovered: mercury fulminate. While it was not usable as a propellant in firearms, it proved useful as a detonator, becoming the first high explosive to be adopted and used widely, especially in percussion caps. Its discovery and adoption facilitated civilian works, such as digging canals and boring tunnels, and its use increased the safety of blasting in mining and the removal of obstacles on land and underwater. Its use in percussion caps enabled pistols and rifles to be fired without using a match or flame to ignite the cartridge within, enhancing their use in the rain and snow and benefitting both hunting game and fighting wars.

The emergence of synthetic organic chemistry in the nineteenth century brought with it discoveries of what proved to be novel explosives. Unlike black powder, which is a mixture, these chemicals are discrete substances. An early example was the work of Italian chemist Ascanio Sobrero, who prepared nitroglycerin in 1847 and reported his discovery in the scientific literature. Its explosive properties were soon recognized, but it would take two decades before a suitable method was discovered to render nitroglycerin safe for handling, transport, and storage. Alfred Nobel would be credited with discovering that absorbing nitroglycerin into an inert material such as diatomaceous earth would provide the desired safety in a product he named "dynamite." That discovery would earn him enormous wealth, considerable fame, and many legal challenges from others claiming priority for this discovery. It would also help to reshape the landscape, with its use in blasting through mountains and other landmasses, as an aid in building roads and canals, and its use in clearing tree stumps and boulders to free acreage for cultivation.

While the search was underway for a way to make nitroglycerin safer, William Henry Perkins was using synthetic organic chemistry to create the first synthetic dye: mauve. His creation set off a fashion craze—the Mauve Age—and brought him fame and fortune. It also spawned many imitators, hoping to find their fortune through synthetic organic chemistry. Joseph Wilbrand was on such a quest, searching for a synthetic yellow dye, when he discovered trinitrotoluene in 1863. While his discovery failed to gain acceptance as a yellow dye, it came into general use as an explosive under its better-known name of TNT. Smokeless powders, discovered in the early years of the nineteenth century, had been slow to gain acceptance by militaries, owing largely to the problems that plagued nitrocellulose-based explosives. Once militaries started using them, however, they proved a boon, eliminating the puff of smoke that might betray the position of shooters while leaving a clear line-of-sight for their own shots. Such discoveries as dynamite, TNT, and smokeless powders were consistent with the transition in developing novel explosives from a largely experimental approach to a more science-driven discipline.

Synthetic organic chemistry continued to contribute novel explosives, as did an improved understanding of the chemistry of the explosions that they created. A prime example of this improved understanding was the systematic study of the relationship between the caliber of a gun and the size, shape, and density of explosives used as propelling charges. The first four decades of the twentieth century witnessed stepwise enhancements in explosives, particularly those used by militaries. Novel formulations lessened sensitivity to fire, shock, temperature, and vibration, thereby improving safety in production, transport, and storage. Improved detonators and fuzes were found. Black powder largely disappeared except for limited commercial applications and an enduring use for small arms. Shaped charge explosives moved from having limited commercial use to being highly desired by militaries for piercing armor and reducing fortifications to rubble. Early in the 1940s, the explosives RDX and HMX were developed and soon found wide application in ammunition. But the defining events of the 1940s relative to explosives were the successful development of nuclear explosives—both fission-based and fusion-based—and the use of such explosives in war.

Developing nuclear explosives made a reality out of seemingly abstract theories about the fundamental nature of atoms. That reality came about through a uniting of diverse disciplines such as chemistry, engineering, mathematics, and nuclear physics on a massive scale and by the collective efforts of an international team, working largely in the United States on the Manhattan Project. The mushroom clouds that rose over the New Mexico desert and the cities of Hiroshima and Nagasaki testified to the destructiveness of this newest type of explosive. The detonation of nuclear bombs meant that the nature of warfare was changed dramatically and unalterably. The Nuclear Age had dawned, and with it the newest chapter in the history of explosives was being written. Nations lacking such weapons sought to develop them. Nations having them sought more of them and more powerful ones. The race for supremacy in nuclear arms had begun.

The nuclear arms race may have spurred modern attempts to control access to and use of nuclear explosives. Such attempts predate the nuclear era, however. In particular, concerns over the destructive potential of explosive bullets surfaced when Eugène Pertuiset toured Africa, Asia, and Europe in the late 1860s, demonstrating his novel explosive powder that he used as a filling for bullets. His explosive bullets were unlike anything seen previously, with a single bullet creating crater-like wounds in a tethered horse used for target practice. Russian tsar Alexander II was greatly impressed by a demonstration put on by Pertuiset, awarding him a diamond and emerald ring. However, the Russian foreign minister was horrified. The prospect that what he had witnessed might be used on humans prompted him to seek its ban. He invited representatives from 18 other nations to a conference that succeeded in crafting the first international arms control agreement. Known as the Saint Petersburg Declaration of 1868, it dealt exclusively with bullets under 400 grams (14 ounces) containing an explosive charge and used in war.

This agreement appears to have held, as no evidence of the use of explosive bullets emerged during the assorted armed conflicts from 1868 onwards. Efforts to ban expanding bullets, such as Minié balls and dum-dum bullets, were attempted later, but were less successful. The Saint Petersburg Declaration also served as a start for more ambitious agreements, particularly those that followed the development of nuclear weapons. In the 80 years following the first nuclear explosions in 1945, treaties have been ratified creating nuclear-weapon-free zones that ban such explosives from regions ranging from Antarctica to the moon and from the earth's seabed to outer space. An ambitious treaty establishing a total prohibition on nuclear weapons has garnered widespread international support but lacks endorsement by any of the nations that are known to have or are suspected of having nuclear weapons. Other arms control agreements include the Nonproliferation Treaty (NPT), obligating parties to take measures to avoid assisting, encouraging, or inducing non-nuclear weapon states from gaining such weapons. However, the history of the A. Q. Khan nuclear technology smuggling network in aiding attempts by nations such as Iran, Libya, and North Korea to become nuclear weapon states highlights the ease with which arms control measures can be frustrated.

Terrorists exploited the discovery of explosives in their unlawful use of violence in pursuit of political aims or to extract retribution for perceived grievances. Their use of explosives likely began with the production of black powder and gunpowder, although accounts of such use prior to 1800 are limited and often anecdotal. A cartload of gunpowder is known to have been used in an unsuccessful attempt in 1800 to assassinate Napoléon I. He survived, although the attempt killed several people, injured many others, and damaged the surroundings. Something similar happened in 1920 when a cartload of gunpowder detonated on Wall Street in the heart of New York City. This explosion caused more human deaths and injuries and greater property damage than the 1800 incident. The real damage, though, was the uncertainty that followed as officials investigated and wondered who was responsible and whether more bombings might take place.

Terrorists use of explosives took advantage, too, of new explosives or novel formulations of existing explosives as these became available. For example, the Fenian Dynamite Campaign in Britain peaked during 1881–1885, less than two decades after Nobel's invention of this explosive. Pyrotol was a reformulation of excess stocks of smokeless powder following the First World War, intended as a low-cost explosive for American farmers. Distribution began in mid-1924. Three years later, a lone terrorist, angry at his local government, used his supply to destroy a Michigan school, killing more than 40 students, teachers, and others. Semtex, a plastic explosive, became commercially available in the 1960s. Its use by terrorists in the December 1988 downing of Pan Am Flight 103 left many dead and sparked the 1991 Convention on the Marking of Plastic Explosives for the Purpose of Detection, intended to enhance aviation safety. The effectiveness of that Convention is difficult to measure. What is known, though, is that there have been other terrorist attempts at blowing up commercial aircraft by other, seemingly less conventional means. In December 2001, Richard Reid failed in his attempt to ignite explosives hidden in his sneakers. Eight years later, Umar Farouk Abdulmutallab failed in his attempt, where he had hidden explosives in his underwear. Both attempts occurred aboard commercial aircraft while in flight. And both attempts prompted enduring changes to procedures for passenger screening prior to boarding, where one must remove shoes and undergo a full body scan.

By the end of the twentieth century, terrorists were resorting to less exotic means than shoe bombs and explosive-laden underpants, as well as more accessible materials. The explosive powder formulators of the nineteenth century had demonstrated how ordinary materials, such as sawdust and sugar, might be used to create explosives. More recent events had drawn attention to homemade explosives using readily accessible substances. Raw materials for the explosives used in the tragic 1995 terrorist bombing in Oklahoma City included fertilizer, fuel oil, and a motorsports additive. The bombs used in the 7 July 2005 terrorist attacks in London were made from hydrogen peroxide—available from a pharmacy—and black pepper—a condiment carried on the grocer's shelves. In December 2024, the FBI seized a jar from the refrigerator of a suspected domestic terrorist. The contents were a powerful explosive based on vinegar, an energy drink, and camp stove fuel. That same month, the FBI arrested an individual whose possessions included an Islamic State-authored bomb-making manual with stepwise instructions on formulating acetone peroxide-type explosives and using ball bearings to maximize human injuries following detonation. This tactic was seen in the 2002 and 2005 Bali bombings and shows the continuing challenges facing governments in controlling access to and use of explosives. These and other terrorist events speak to a tragic side to the history of explosives.

A

ABEL, FREDERICK AUGUSTUS (1827–1902). British chemist and inventor, noted for his contributions to the study of **explosives**. Educated at the Royal College of Chemistry, London, Abel spent most of his professional career at the Royal Military Academy, Woolwich. He carried out a long collaboration with Sir **Andrew Noble** on the behavior of **black powder**. In 1865, Abel received British patent 1,102 for a process involving the manufacture of **guncotton**. The patent was the first to make explicit claims relative to the importance of pulping as a treatment of guncotton. He would be widely credited with developing a process that both provided enhanced safety over existing processes for making guncotton and increased the usefulness of the finished product by pressing it into blocks. In reality, though, the blocks burned too fast and failed to perform uniformly with respect to their ballistics. The process did allow most impurities to be washed out of the guncotton, and this step proved important to future developments with guncotton.

Abel credited his assistant, Edwin Ormond Brown (1828–1885), with the critical developments that both enhanced the safety of the manufacture of guncotton and prepared a useful finished product. In 1868, Brown conducted two experiments in which he made important observations. First, he found that dry, compressed guncotton could be made to detonate by use of a fulminate detonator, such as those recently developed by **Alfred Nobel**. This observation was important because dry guncotton had been deemed too difficult to detonate for practical purposes. Second, Brown found that wet guncotton could be made to detonate with a small quantity of dry guncotton. This latter observation would later become recognized as the principle of using a booster. Taken together, the work of Brown made it possible to use large blocks of wet guncotton with comparative safety, and he would receive British patent 3,115 in 1868 covering his contributions. Sir **James Dewar** and Abel are credited with inventing **Cordite**, which became a standard **smokeless powder** for the British military. Abel received many awards and honors for his contributions to science and the government, including being knighted in 1891 and being created Baronet of Whitehall Court in 1893.

"A" BLASTING POWDER. **Black powder** containing **potassium nitrate** rather than **sodium nitrate**. *See also* "B" BLASTING POWDER; DU PONT, LAMMOT (1831–1884).

ABSOLUTE DUD. A **nuclear explosive** that fails to **detonate**. *See also* DWARF DUD; PREINITIATION.

ACETONE PEROXIDES. Family of at least four structurally related chemicals sharing acetone (C_3H_6O) and a peroxide, most commonly hydrogen peroxide (H_2O_2), as precursors. All are extremely **shock-sensitive** and unstable, decomposing explosively. The acetone peroxides are easily prepared from precursor chemicals readily available commercially and in quantity for legitimate purposes, making them candidates for illicit use as homemade explosives. The acetone peroxides of greatest concern relative to illicit use are **diacetone diperoxide (DADP)**, **hexamethylene triperoxide diamine (HMTD)**, and **triacetone triperoxide (TATP)**.

The earliest record in the scientific literature of acetone peroxide preparation was in 1895 by German chemist Richard Wolffenstein (1864–1926). He was studying the oxidation of the naturally occurring poison coniine, using acetone and hydrogen peroxide as solvents, when he noted a white powder. Wolffenstein characterized this powder and determined it would make a useful explosive. He obtained a patent that same year, covering its preparation and use. At that time and for several decades thereafter, this powder—most likely TATP—was thought to be shock-resistant and stable. This thinking proved incorrect, and the acetone peroxides lack any legitimate commercial, industrial, or military uses. However, they have been involved in or suspected of assorted acts of terrorism in which individuals or small groups prepare them from readily obtained precursors chemicals.

ADMINISTRATION RÉVOLUTIONNAIRE DES POUDRES ET SALPÊTRES. *See* GUNPOWDER SERVICE (*SERVICE DES POUDRES*).

AGENCE NATIONALE DES POUDRES ET SALPÊTRES. *See* GUNPOWDER SERVICE (*SERVICE DES POUDRES*).

AKERS, WALLACE ALAN (1888–1954). British chemist, industrialist, and science administrator. In the fall of 1941, the British government established a secret program to develop a **nuclear weapon**. Known as the Directorate of Tube Alloys, it was headed by Akers. At the time, he was serving as director of research for a large chemical firm, was respected for his skill as a science administrator, and enjoyed powerful friends in government and industry. However, Akers's confrontational style made him unpopular with many of the leading British scientists, and he engendered an acrimonious relationship with his American counterparts on nuclear weapon development efforts, Vannevar Bush and Leslie Groves. Consequently, Akers was replaced by **James Chadwick** in October 1943 as the liaison for such efforts, although he continued to head Tube Alloys. By the time of Akers's dismissal, the work of Tube Alloys had been largely subsumed into the **Manhattan Project**, on which Bush and Groves were key figures. In 1946, Akers was knighted for his contributions to the war effort.

ALARM CLOCK. Name given in 1946 by Edward Teller to a single-stage design for a **thermonuclear device**. He dubbed the design the "alarm clock" because he thought it would awaken the world to the possibility of a new generation of **nuclear explosives**. Although Teller was working in the United States at the time that he advanced this design, the United States never tested the design. However, some analysts have concluded that the Soviet Union used a single-stage design, such as an "alarm clock" for its first successfully **detonated** thermonuclear device, which the United States dubbed "**Joe-4**."

ALCHEMISTS AND EXPLOSIVES. Prior to the emergence of chemistry as a scientific discipline, individuals were experimenting with assorted substances and mixtures in hopes of finding one that would grant immortality, that would cure all diseases, or that would turn base metals, such as lead, into gold. Their experiments form the basis of alchemistry, and these individuals were the alchemists. Although their quests for these elixirs failed, their work had important, if unintended, consequences for the development of **explosives**. In particular, writings attributed to some of these alchemists describe results in which something resembling **black powder** was discovered.

The earliest of these writings still extant are credited to Taoist priests active in modern-day China. These priests were among the earliest alchemists, searching for an elixir of life that would grant immortality. Their searches were methodical rather than random, and notable results were written to inform other alchemists as an aid to further research, enabling progress and avoiding hazards. The earliest of these writings is attributed to Wei Boyang in 142. In his *Cantong qi* (*The Kinship of Three*), he described three solids that, on being mixed, would "fly and dance violently." While precise identification of the three solids is lacking, his description does fit the properties of an explosive, particularly black powder, if the three solids contained sources of carbon, nitrate, and **sulfur**.

The earliest written account of the three principal ingredients of black powder appeared around 300 in a work credited to Ge Hong. Like Wei Boyang, he was a Taoist priest, scholar, and, most importantly for advancements in explosives, an alchemist who was active during the latter years of the third century and the early years of the fourth century. In his *Baopuzi* (*The Master Who Embraces Simplicity*), Ge Hong identified these three ingredients correctly as sulfur, **saltpeter**, and a carbonaceous material. Various Chinese alchemists would continue research into these ingredients. In the ninth century, the earliest written accounts of a formula for explosives appeared, giving relative amounts of each ingredient. The *Taishang Shengzu Jindan Mijue* (*Secret Teachings of the Golden Elixir of the Supreme Holy Ancestor*), written in 808 by an as-yet unidentified Chinese scholar, gave this formula a mixture of six parts sulfur, six parts saltpeter, and one part birthroot, an herbaceous plant that would serve as a source of carbon.

Within the next 200 years, alchemistry in China had advanced such that not only ingredients and formulas for explosives were known, but their destructive potential and military importance were recognized. In particular, *Wujing Zongyao* (*Collection of the Most Important Military Techniques*), a work attributed to the Northern Song dynasty and dating to 1040–1044, went beyond describing black powder by describing weapons using **gunpowder**.

This knowledge would migrate westward over the next several centuries, perhaps via the Silk Road trade networks or as a result of Mongol invasions. Around 1260, **Roger Bacon**, an English alchemist, published *De nullitate magiae* (*The nullity of magic*), containing a cipher thought to be a recipe for making black powder. This recipe, if correct, is likely the earliest description of black powder in a European text. This knowledge had taken root firmly in Europe by 26 August 1347, when English forces under King Edward III defeated a much larger French force under King Philip VI at the site of present-day Crécy, France. The English forces included some guns, marking one of the earliest uses of black powder as a **propellant** in Europe. Gunpowder would go on to play a significant role for the duration of the Hundred Years' War (1337–1453). *See also* SCHWARZ, BERTHOLD.

ALUMINUM. Silvery white, crystalline solid nontoxic metal. Finely divided aluminum powder burns in air. This property has been exploited successfully in various **explosives**, such as **ammonal** and **H-6**, and in **pyrotechnics**.

AMATEX 20. *See* TERNARY EXPLOSIVE.

AMATOL. Family of **binary explosives** containing **ammonium nitrate** and **TNT**. The percentages of ammonium nitrate and TNT vary from 80/20 (ammonium nitrate/TNT) to 40/60, and these percentages are occasionally used in the name of the amatol. For instance, "amatol 60/40" refers to the mix of 60 percent ammonium nitrate and 40 percent TNT. The amatols were developed by the British in 1915 as a means of extending their limited supply of TNT and were used as a bursting charge in munitions. Amatols are prepared by mixing solid ammonium nitrate into molten TNT. Ammonium nitrate is insoluble in TNT; however, the use of molten TNT in the mixing process ensures thorough incorporation of solid ammonium nitrate into the solid mass that results on cooling. Ammonium nitrate both absorbs water and dissolves in water. Therefore, the amatols are substantially less water-resistant than is TNT alone. Because the amatols offer no clear advantage over TNT alone, their popularity declined during the Second World War when inventories of TNT proved adequate for military needs.

AMERICAN CIVIL WAR (1861–1865). Black powder and **mercury fulminate** figured prominently in the war-fighting efforts of both combatants, the Union and the Confederacy, during the American Civil War. **DuPont** estimated that 3.5 million pounds of black powder were supplied to the Union during the war. The estimate for the Confederacy is only slightly lower at 3.3 million pounds. Estimated supplies of mercury fulminate on hand at the start of the war are unknown, although the raw materials used to produce mercury fulminate—alcohol, nitric acid, and elemental mercury—were available initially. As the war continued, supplies of elemental mercury ran low in the Confederacy.

Selected military forces on both sides suffered a shortage of black powder on occasion, creating a tactical disadvantage during a battle or a campaign. But neither the Union nor the Confederacy suffered such shortages as a matter of routine. While this situation might have been expected for the more industrialized states of the Union, it is remarkable for the Confederacy, where military logistics were often handled ineptly. Both sides displayed ingenuity in producing black powder and, especially, in securing the raw materials needed for its manufacture. In the Confederacy, this ingenuity extended to public appeals to collect human urine, which might be turned over to the government for use in making **saltpeter**.

The Union. At the outbreak of hostilities in April 1861, the Union possessed demonstrated capabilities for manufacturing black powder and moderate stockpiles of raw materials to support this manufacturing. Supplies of **potassium nitrate**, however, were limited, and measures were needed to obtain supplies from outside the United States. In October 1861, **Lammot du Pont** was summoned to a meeting in Washington, D.C., where he agreed to undertake a business mission on behalf of the US government. That mission involved an attempt to corner the market in **India Saltpeter**, which was then the primary source for potassium nitrate. In November, he landed in

London and set about cornering the India Saltpeter market. His move largely succeeded, despite a setback over the *Trent* incident when the British government felt its neutrality had been violated by the US Navy.

In that incident, the commander of the US Navy vessel USS *San Jacinto* ordered his men to board the British mail ship *Trent* on 9 November 1861 while it was in the harbor off Havana, Cuba, well outside US territorial waters. Two passengers were taken prisoner: James Murray Mason and John Slidell, both commissioners for the Confederate government and bound for Europe to seek recognition for their government. The British government demanded the immediate release of Mason and Slidell as well as an apology, ordering 8,000 troops to Canada to back up its demands. In the meantime, the vast quantities of India Saltpeter that du Pont had purchased were embargoed. On 1 January 1862, Mason and Slidell were freed. Shortly thereafter, the embargo was lifted, and shipment of India Saltpeter to the United States resumed.

The initial fighting dashed Union hopes for a brief conflict; however, the government and manufacturers turned their attention to securing raw materials to support a prolonged conflict. They elected to import **Chile Saltpeter**, an impure form of **sodium nitrate**. Large quantities were available in South America and imported successfully. Once received, the sodium nitrate was mixed with potassium hydroxide (NaOH) or potassium chloride (KCl), and the resulting reaction converted it to potassium nitrate.

The Confederacy. When the Civil War began, the Confederacy had a less robust infrastructure to support manufacturing black powder and to provide the raw materials for its manufacture than did the Union. One measure of this lack of infrastructure was that, as of early 1861, there were merely two powder mills in the States that would form the Confederacy: a very small one in South Carolina and a larger one in Tennessee. This situation presented the Confederacy with three primary options for meeting its black powder needs: (1) Acquisition by conquest, typically when an arsenal was taken from the Union or in the form of abandoned supplies; (2) Purchase from abroad, including from merchants in the Union; and (3) Manufacture from ingredients from within the Confederacy.

At the beginning of the conflict, the Confederacy had on hand nearly 500,000 pounds of black powder either received prior to the Union embargo or seized from various sources when the Union forces surrendered or abandoned their positions. Another 300,000 pounds were seized by the Confederacy on 20 April 1861 when the Union forces abandoned the Norfolk Navy Yard in Virginia. But acquisition of black powder by conquest was episodic, and the Confederacy sought a more reliable option.

The second option was adopted and proved successful, but, like acquisition through conquest, it was episodic. Early in the conflict, the US government forced Union merchants having contracts with individuals, firms, or states in the Confederacy to cancel orders prior to fulfillment. Purchases of black powder and, especially, of sulfur and saltpeter from foreign sources continued throughout the Civil War. Fulfillment, though, proved problematic owing to the Union blockade of ports in the Confederacy and to both shortages of hard currency and lack of credit. Although the Confederacy would ultimately smuggle 2.8 million pounds of saltpeter during the course of the war, more reliable sources were sought.

The need for sufficient and predictable supplies of black powder prompted the Confederacy to establish a bureau with the mission of ensuring a reliable supply of sulfur and saltpeter and of various metals needed for other purposes. In April 1862, the Confederate Congress directed that the Niter Bureau (sometimes spelled "Nitre Bureau") be established within the Ordnance Department. A year later, an act of the Confederate Congress established the **Niter and Mining Bureau** as a standalone agency, although many of its personnel carried military rank and wore uniforms. By any measure, the bureau fulfilled its mission.

Similarly, the Confederacy succeeded in turning the raw materials of potassium nitrate, sulfur, and charcoal into black powder. George Washington Rains (1817–1898), West Point graduate and a career army officer prior to joining the Confederate military, was instrumental to this success. Shortly after the beginning of the conflict, Rains became part of the Ordnance Department, where he took up the task of enlarging existing powder-making facilities. By October 1861, he had overseen the return to operations of two idle powder mills near Nashville, Tennessee. During this same period, he wrote *Notes on making saltpeter from the earth of caves*, a booklet used by the Niter Bureau and later by the Niter and Mining Bureau in training its personnel. A second publication proved helpful as well in laying out the stepwise processes to make sodium and potassium nitrate. In 1862, Joseph Le Conte (1823–1901) released *Instructions for the Manufacture of Saltpeter*. At the time, Le Conte was a professor of chemistry and geology at South Carolina College, Columbia (later to become the University of South Carolina). His careful explanations of what is to be done, when to do it, and what to expect proved extraordinarily helpful to the otherwise inexperienced and poorly trained operators in the Niter and Mining Bureau.

Sometime after October 1861, Rains launched into his most ambitious project: the design, construction, and operation of the Confederate Powder Works. He selected Augusta, Georgia, as the site for building this facility, using an 1857 booklet of instructions for the Royal Gunpowder Factory,

Waltham Abbey, as an operating guide. The selection of this location proved fortuitous in that Augusta was never occupied by Union forces during the conflict. The Confederate Powder Works began operations on 10 April 1862. It operated continuously until the end of the conflict in April 1865, having produced over 2.7 million pounds in three years.

While George Washington Rains was producing explosives, his older brother Gabriel James Rains (1803–1881) was putting them to effective use. Gabriel Rains, also a West Point graduate and former career army officer, around 1862 was placed in command of the Confederate **Torpedo** Bureau. In this role, he oversaw the development of three novel munitions: the "sub-terra mine," a pressure-activated land mine; the keg torpedo; and the submarine mortar battery. All three proved effective, particularly in use against naval vessels. Rains related his designs for and experiences with torpedoes and mines in a text called, simply, *Torpedo Book*. In 1874, he sent a copy of his book to US president Ulysses Grant, along with a proposal that Rains teach a course on this subject at West Point. That proposal was denied. The interest that George and Gabriel Rains shared in explosives led to their being nicknamed the "bomb brothers."

The Confederacy experienced shortages beyond the ingredients for black powder. At that time, **primary explosives** used as detonators were often made with mercury fulminate. Producing and handling mercury fulminate was hazardous, as demonstrated early in the war in Richmond, Va., by two explosions—the first on 6 June 1861 that took the life of 38-year-old Edward Finch and the second on 3 July 1861 that killed 30-year-old Joseph Laidley. The noise of these explosions caused considerable excitement in the newly designated capital of the Confederacy, as residents feared the city was being bombarded by Union forces. Investigations promptly determined that no such attack was underway but that the victims were chemists, working in improvised laboratories and using drugstore supplies as raw materials.

Throughout the war, competition in the Confederacy for a key ingredient—elemental mercury—was keen. In addition to its use in making mercury fulminate, mercury was used directly to treat venereal diseases, especially syphilis, and as an ingredient in making calomel (Hg_2Cl_2, mercury (I) chloride), which had many medicinal uses, including as a purgative, a teething powder, and in douches. Because few natural sources of mercury lay within the Confederacy, shortages arose. This situation led Lt. Col. John W. Mallet (1832–1913), superintendent of the Confederate States Central Laboratory, Macon, Georgia, to issue an appeal to residents of the Confederate States to donate barometers and thermometers so that they might be broken apart and the mercury within might be collected. Mallet, an Irish-born chemist who had been serving on the faculty at the University of Alabama prior to the war, was also responsible for developing a means to ease the shortage of sulfur, which was needed for black powder. He directed that the naturally occurring ore iron pyrite (FeS_2; iron (II) sulfide) be harvested wherever possible. It was then roasted, creating sulfur oxide, which was processed further to make sulfuric acid or to extract the sulfur. *See also* BROWN'S ISLAND EXPLOSION; EXPLOSIVE BULLET; MULE-BORNE BOMBING; PETER MONKEY.

AMERICANITE. **Nitroglycerin**-based **high explosive** containing an alcohol to lessen sensitivity to **shock** and **friction**. Russian-born American chemist Sergey de Smolianinoff (1862–1901) selected Americanite as the name for mixtures containing **nitroglycerin** and an alcohol, commonly methanol. According to Smolianinoff, the mixtures were insensitive to shock and friction but exploded with greater force than did nitroglycerin alone. Beginning in the late 1880s, he toured the United States, demonstrating Americanite to executives of various firms that manufactured or used explosives, primarily for **blasting** and drilling. A newspaper account of one such demonstration appeared in the *Lima (Ohio) Daily Times* on 20 July 1889. According to that account, Smolianinoff filled a tin tube with Americanite and then threw the tube against a tree without any **detonation** of the contents. He then hammered the tube with stones, denting the tube but again without any detonation. Next, he placed the tube in an empty can, added **black powder** until it surrounded the tube, and detonated the powder. Again, the contents of the tube failed to explode. Finally, Smolianinoff placed equal quantities of nitroglycerin and Americanite on the ground and detonated both. Americanite produced a larger hole. After these demonstrations, one observer noted that he had suffered from a headache, smelled some Americanite, and the headache was cured.

On 15 July 1890, Smolianinoff received US patent 432,336, for an explosive mixture "insensitive to mere agitation." The patent, titled "Explosive Compound," and assigned to the Americanite Manufacturing Company of Virginia, covers Americanite and related formulations of nitroglycerin, an alcohol, and, possibly, other additives. That same year, Smolianinoff demonstrated Americanite for the US War Department and the US Navy for use in military munitions. Despite the marketing efforts of Smolianinoff, there is no record that the US military or any other military adopted Americanite. The Americanite Manufacturing Company appears to have failed to achieve commercial success with its nitroglycerin/alcohol-based products, as it ceased operations around 1900. Smolianinoff barely outlived his company, dying in Washington, D.C., on 11 February 1901.

AMMONAL. Family of **explosives** containing **ammonium nitrate**, **aluminum powder**, and, occasionally, other substances. The concept of ammonal explosives arose independently with two German chemists. In 1899, Ernst Richard Escales (1863–1925) received German and Belgian patents covering the basic concept of ammonium nitrate and aluminum powder mixtures for use as explosives. In 1900, Hans Ritter von Dahmen received German patents for a similar concept. Both Escales and von Dahmen continued to pursue these concepts, with various formulations being developed in assorted nations. The inclusion of aluminum powder in the various ammonals resulted in both a bright flash and an extremely hot flame when the explosive **detonated**.

The original military ammonal explosive, designated Ammonal I and known outside Germany as German Ammonal, contained 54 percent ammonium nitrate, 30 percent **TNT**, and 16 percent aluminum powder. Ammonal II, also a German formulation, contained 72 percent ammonium nitrate, 12 percent TNT, and 16 percent aluminum powder. A third German formulation was Ammonal B, which contained 93.4 percent ammonium nitrate, 3.6 percent aluminum powder, and 3 percent **charcoal**.

Shortly after receiving his German patents in 1900, von Dahmen went to work in Austria, where he was instrumental in the development of T-Ammonal, containing 47 percent ammonium nitrate, 30 percent TNT, 22 percent aluminum powder, and 1 percent charcoal. This formulation, also known as Toluol-Ammonal, was used by the Austrians as an underwater explosive in depth charges and **torpedoes**.

The success of the German and Austrian ammonals led to extensive copying by other nations, each seeking a militarily useful formulation. "French Ammonal," as it came to be known, contained 86 percent ammonium nitrate, 8 percent aluminum powder, and 6 percent stearic acid, a fatty acid that served as both a **binder** and a **fuel**. The United States fielded "Minol," containing 40 percent ammonium nitrate, 40 percent TNT, and 20 percent aluminum powder. All these ammonals proved successful, and many saw extensive use during the First and Second World Wars.

AMMONIA DYNAMITE; AMMONIA GEL DYNAMITE. *See* DYNAMITE.

AMMONIUM NITRATE (AN). Colorless to white crystalline solid used as an **oxidizer** in **improvised explosives**, such as **ammonium nitrate-fuel oil mixture**. Ammonium nitrate has the chemical formula NH_4NO_3 and is odorless. In addition to use in explosives, ammonium nitrate is used extensively for legitimate commercial purposes, particularly in agriculture as a fertilizer. Accidental explosion of bulk storage containers of ammonium nitrate has been wholly or largely responsible for several mass casualty events that included extensive structural damage. These events include ones in **Oppau, Germany** (1921; more than 500 known to have been killed), and **Texas City, Texas** (1947; 581 known dead). Such events have continued to happen during the twenty-first century, with accidental explosions in Toulouse, France (2001, 31 killed); West, Texas (2013; 15 killed); at the Ports of Tianjin, **China** (2015; 173 known dead); and Beirut, Lebanon (2020; 218 known dead). In 1867, Swedish patent 59–1867 was issued to Carl Johan Ohlsson (ca. 1802–1876) and Johan Henrik Norrbin (1835–1907) for the invention of an explosive mixture of ammonium nitrate and a carbonaceous material, such as **charcoal**, **nitrobenzene**, or **sawdust**. This invention failed to experience widespread use as a manufactured explosive, but the concept came into use as a **homemade explosive** owing to the ease of acquiring the ingredients.

AMMONIUM NITRATE-FUEL OIL MIXTURE (ANFO). Ammonium Nitrate-Fuel Oil Mixture. Explosive comprised of porous, pelletized **ammonium nitrate** and fuel oil. This mixture, often referred to by the acronym ANFO, is used in mining as a **blasting agent**, where the typical proportions are 94 percent ammonium nitrate and 6 percent fuel oil, such as diesel fuel. Because of the relative ease of acquiring ammonium nitrate and fuel oil, ANFO has been used for criminal purposes in **improvised explosive devices**, most notably in the 1995 **Oklahoma City Bombing**.

AMMONIUM PERCHLORATE. Strong **oxidizing agent** used as an ingredient in many **explosives** and **propellants**. It is a white or colorless crystalline powder and has the chemical formula NH_4ClO_4. Swedish electrochemist and mining engineer Oscar Carlson (1844–1916) recognized its usefulness as an explosive in the 1890s. In 1892 and 1897, he was granted Swedish patents for its production and use while working for Stockholms Superfosfat Fabrik AB, Månsbo. This firm marketed it as an explosive under the trade name "Carlsonit." Ammonium perchlorate has seen substantial use in solid propellant rocket boosters for both civilian and military applications. On 4 May 1988, it was involved in a fire and explosion at the Henderson, Nevada, facility of PEPCON (Pacific Engineering and Production Company, Nevada), resulting in two deaths, nearly 400 injuries, and more than $100 million in damages. *See also* POTASSIUM PERCHLORATE.

AMMONIUM PICRATE. Ammonium salt of 2,4,6-trinitrophenol ($C_6H_6N_4O_7$). Ammonium picrate is a bright yellow crystalline solid that explodes readily from heat or shock. It has been used directly as an **explosive** and in combination with other explosives. On 1 November 1911, Italian

Army Lieutenant Giulio Gavotti dropped three ammonium picrate-filled bombs on Turks encamped at the Ain Zara oasis, Libya, marking the first delivery of explosives from an aircraft. *See also* BINARY EXPLOSIVE; EXPLOSIVE D.

AMMONPULVER. Type of **black powder** notable for its lack of **sulfur**, lack of flash on **detonation**, and modest production of smoke. Invented in Germany about 1885 and originally called "amidpulver," ammonpulver (German for "ammonia powder") contains 40–45 percent **potassium nitrate** and 35–38 percent **ammonium nitrate**, with the residual being **charcoal**. Prior to the development of ammonpulver, most black powder formulations contained sulfur as a **fuel**. Substitution of ammonium nitrate for sulfur was novel, and the resulting product proved useful to the military because its lack of flash and its relatively modest amount of smoke produced on detonation improved the ability of marksmen to see their targets while lessening the visual clues of the position of those doing the shooting. Ammonpulver had the added advantages of being both cheap to produce and insensitive to shock and friction. Austria used ammonpulver extensively during the First World War.

ANFO. *See* AMMONIUM NITRATE-FUEL OIL MIXTURE (ANFO).

ANILITE. French designation for a **Sprengel explosive** comprised of **nitrogen tetraoxide**, as the **oxidizing agent**, and butane (C_4H_{10}), as the **fuel**. The French used anilite during the First World War in aviation bombs. The two components are flammable but were considered safe for handling, storage, and transport until mixed. The components were placed into separate compartments in the bomb. When the bomb was released, a propeller on its tip would rotate. This rotation would drive a mechanism that opened a valve allowing the nitrogen tetraoxide and butane to mix. The resultant mixture was highly sensitive to shock. The bomb lacked a fuze; rather, the shock of impact with the ground was expected to be sufficient to cause **detonation**. The use of butane as a fuel is surprising in that it is a gas that can be liquefied under pressure or by reducing the temperature to around 30°F (-1°C). This requirement for liquefying the butane is likely to have proven awkward during the First World War. In any event, use of anilite appears to have been discontinued after the war. *See also* PANCLASTITE.

ARMSTRONG, WILLIAM GEORGE (1810–1900). British inventor, industrialist, philanthropist, and scientist. His 1855 invention of a breech-loading gun designed to fire a shell rather than a ball helped to revolutionize artillery. He surrendered his patent for this gun to the British government, for which he was knighted in 1859. After serving as Engineer of Rifled Ordnance in the British War Department, Sir William, as he had become, moved to the private sector, eventually heading the Sir W. G. Armstrong & Company, which manufactured assorted armaments, **explosives**, ships, and locomotives. In 1887, he was created a hereditary peer, taking the name Baron Armstrong of Craig. *See also* NOBLE, ANDREW (1831–1915).

ARMSTRONG'S POWDER. Explosive made of 67 percent **potassium chlorate**, 27 percent red phosphorus, 3 percent **sulfur**, and 3 percent calcium carbonate. It was developed by **William George Armstrong** prior to 1872 for use in explosive shells. However, its extreme sensitivity to **shock** and **friction** limited its usefulness in military munitions. It did find use as the ammunition in toy cap guns, in which the mixture is suspended with a binder and dropped onto paper, and the resulting dots explode with smoke and sound when struck. Even this use carries risks, however, as shown by the events of the evening of Tuesday, 14 May 1878, when a fire at 22 Rue Béranger, Paris, caused an inventory estimated between six million and eight million paper caps to explode, killing 14 and injuring more than 16. The paper caps were stored at a toy factory belonging to Monsieur Blanchon. An **explosion** of these same toy caps took place on 3 August 1888, at a factory near Wandsworth, England. In this case, the explosion occurred in a shed where four girls were working, most likely separating the paper strips of toy caps using scissors. This explosion destroyed the shed and claimed the lives of three of the girls, with the fourth suffering severe injuries.

ARZAMAS-16. Code name given by the Soviet Union to a town that became the primary Soviet facility for designing **nuclear weapons**. The town is located in the Nizhny Novgorod region, northeast of Moscow. In 1946, the Soviet Union selected this site for constructing the All-Union Scientific Research Institute of Experimental Physics, with **Yuri Khariton** as its inaugural director. Arzamas shortly became the primary Soviet facility for designing nuclear weapons, while the town itself disappeared from unclassified maps of the Soviet Union. In August 1995, Boris Yeltsin decreed that Arzamas-16 would return to its historic name of Sarov, Russia. Despite the name change, the town remains "closed," and visitors require special permission for entry. It is the current home of the Russian Federal Nuclear Center, which is the primary Russian research institute for work involving nuclear weapons. *See also* KURCHATOV, IGOR VASILIEVICH (1903–1960); SAKHAROV, ANDREI DMITRIYEVICH (1921–1989).

ATLAS POWDER COMPANY. *See* POWDER TRUST.

ATOMIC EXPLOSIVE. *See* NUCLEAR EXPLOSIVE.

ATTEMPTED ASSASSINATION OF NAPOLÉON. Bombing notable as an early terrorist use of a **vehicle-borne improvised explosive device (VBIED)**. On 24 December 1800, an **explosion** rocked the rue Saint-Nicaise, Paris. It happened shortly after a carriage carrying Napoléon Bonaparte, First Consul of France, and his wife, Josephine, passed by. The attempt failed to assassinate Napoléon. Investigation revealed that a wine cask filled with **gunpowder** and fitted with a **fuse** had **detonated**. The cask was in a horse-drawn cart, held in place by a 14-year-old girl. The blast killed her and an unknown number of bystanders. Napoléon and Josephine were shaken but otherwise unharmed. The attempt on his life became known as the "Plot of the rue Saint-Nicaise" and was tied to Royalists and Catholics opposed to his rule. A number of suspected plotters were arrested, and four were executed for their role.

ATTEMPTED ASSASSINATION OF NAPOLÉON III. Bombing notable as the earliest known terrorist use of a **high explosive**, **mercury fulminate**. On 14 January 1858, Napoléon III, the Emperor of France, and his wife, Empress Eugénie, left for a night out at the opera. Arriving with their entourage outside the Paris Opera around 8:30 p.m., their carriage was knocked on its side by a bursting bomb, one of three **detonating** in rapid succession. They emerged unhurt, saved perhaps by iron plates in the floor of their carriage, and headed into the opera house so that they might reassure the public that they were safe. They left a scene of devastation. Their carriage had been preceded by another carriage carrying officers of the household, followed by mounted lancers. The bombs targeted the Emperor's carriage, though, landing in its front, beneath it, and to its left. The explosion left more than 150 people injured, eight of whom would die eventually of their injuries. The explosion also claimed the lives of two horses.

Police soon had four men in custody and charged them with the attempt on the life of Napoléon III. Felice Orsini confessed to leading the bombing plot, assisted by Giovanni Pieri and Camillo di Rudio as well as Orsini's servant Antonio Gomez. The four men were united in their dedication to reunifying Italy, blaming Napoléon III for betraying their cause. They hatched their plot while living in England, aided by a loose network of assorted radicals and revolutionaries. The unfilled iron bombs were made in Birmingham, while the filling of mercury fulminate was prepared in London by **Simon Bernard**. Orsini and Pieri were executed on 13 March 1858, while Gomez and di Rudio were sentenced to life at hard labor. Di Rudio, though, eventually escaped from prison and made his way to the United States, where he took the name

Charles De Rudio. Under his new name, he joined the US Army, serving during the **American Civil War** and in the American West, where he was one of the Union Army survivors of the Battle of the Little Bighorn, 25–26 June 1876, that took the life of Lieutenant Colonel George Armstrong Custer (1839–1876).

AUGUSTA POWDER WORKS. *See* AMERICAN CIVIL WAR (1861–1865).

B

B41. Designation for a **thermonuclear** weapon deployed by the US Air Force from 1961 until July 1976. Originally designated the Mk-41, the B41 was the most powerful **nuclear weapon** ever deployed by the United States, with a theoretical maximum yield of 25 megatons **TNT-equivalent**. It was the only three-stage thermonuclear weapon fielded by the United States, using a **boosted fission weapon design**. Development of the B41 began in 1955, and its manufacture began in September 1960, entering service the following year. Approximately 500 were made, with each one weighing a little over 5 tons, having a cylindrical shape slightly, and being over 12 feet long and 4 feet in diameter. Although components of the B41 were tested, no complete B41 was ever **detonated**.

BABY TOOTH SURVEY. A scientific study of deciduous teeth ("baby teeth") to assess whether children were being exposed to radiation from the fallout of atmospheric testing of **nuclear weapons**. Launched in December 1958 by the Greater St. Louis Citizens' Committee for Nuclear Information in conjunction with St. Louis University, the study collected baby teeth, which were analyzed for radioisotopes. The analyses focused on strontium-90, a known carcinogen. Among its findings was that children born in 1963 had 50 times more strontium-90 than did children born in 1950, prior to the advent of large-scale atmospheric testing. The study would continue until 1970, by which time more than 3,000 individuals had donated baby teeth that were studied. Results are believed to have informed US president John F. Kennedy in his support for a Limited Test Ban Treaty.

BACHMANN PROCESS. Method for producing **RDX** and **HMX**. The primary ingredients are acetic anhydride, **ammonium nitrate**, and **hexamethylene tetramine**. The process was perfected at the University of Michigan in early 1941 by John C. Sheehan (1915–1992) and Werner E. Bachmann (1901–1951), inspired in part by similar work done at McGill University, Montreal, Canada, in 1940 by James Hamilton Ross (1895–1976) and Ross's graduate student, Robert Walter Schiessler (1918–2008). The process saw widespread use during the Second World War; for example, by the end of the war, a plant operated by Tennessee Eastman, Kingsport, Tennessee, was producing 360 tons of RDX per day. It also became an important means of producing HMX, especially after the war.

When Sheehan began his research into RDX production, he had recently completed his doctoral thesis in chemistry, working under Bachman, and was pursuing postdoctoral studies. Although the United States was nearly a year away from entering the Second World War, the US government had already begun research into problems of military significance. One such problem was developing a more efficient process for making RDX. The basic method then in use involved the direct nitration of hexamethylene tetramine with nitric acid. The maximum yield for this method was one molecule of RDX for each molecule of hexamethylene tetramine. But the method had certain drawbacks, such as requiring large excesses of nitric acid. The method conceived by Sheehan and Bachmann, if successful, would produce two molecules of RDX for every molecule of hexamethylene tetramine; that is, it would double the theoretical yield over the prevailing method. Moreover, it would eliminate the need for nitric acid.

In late 1940, Bachmann had learned of the work by Ross and Schiessler, who used acetic anhydride, ammonium nitrate, and formaldehyde. However, Bachmann lacked details on the process used by Ross and Schiessler. In March 1941, Bachmann and Sheehan succeeded in producing a yield of greater than one molecule of RDX for every molecule of hexamethylene tetramine. The results were shared with others researching this same problem in the United States, the UK, and Canada, including Ross's group at McGill University and a group at the University of Toronto under George F. Wright (1904–1976). Both Wright's group and the team of Bachmann and Sheehan noted the presence of impurities in what eventually became known as the Bachmann Process, with Wright designating one such impurity "HMX."

The wartime collaboration between Canada and the United States on RDX, HMX, and other **explosives** production was both extensive and remarkable for the willingness each nation expressed to give the other the use of intellectual property arising from this collaboration. This willingness was entered into formally in a bilateral treaty titled "Exchange of Notes constituting an agreement relating to patent rights in connexion [*sic*] with RDX and other explosives." The exchange took place on 3 and 27 September 1946 and was registered with the United Nations on 10 December 1948. In this exchange of notes, each nation gave to the other a non-exclusive, royalty-free license to inventions covered by nearly 20 patent filings made during the war.

At the close of the Second World War, Canada, the UK, and the United States investigated the military programs of Nazi Germany. One such investigation covered German production of munitions and explosives. This investigation uncovered evidence that German scientists had independently developed a process identical to the Bachmann Process. That process was dubbed the "KA Process," after the two chemists responsible for its development, C. Knöffler and A. Apel. This development apparently took place in 1943, although results were never published in the scientific literature.

Although the development of the Bachmann Process was an important event in the scientific careers of Bachmann and Sheehan, both chemists are better known for research done in fields outside explosives. Bachmann, who spent his career at the University of Michigan, is remembered for the first successful total synthesis of the steroid hormones equilenin and estrone. Sheehan moved to the Massachusetts Institute of Technology (MIT) shortly after the war, where he developed and patented a process for making synthetic penicillin. This process brought more than $30 million in licensing income to MIT. Sheehan remained at MIT for the rest of his career.

BACON, ROGER (ca. 1214–1292). English Franciscan friar, philosopher, scientist, and educational reformer. Bacon conducted research related to **alchemistry** and is believed to have learned of the **explosive** and **pyrotechnic** nature of **black powder** through this research. The treatise *De nullitate magiae* (*The nullity of magic*), attributed to Bacon, contains a cipher that, when decoded, provides a recipe for black powder in terms of charcoal (referred to as "the bones of Adam"), sulfur ("the vapor of pearl"), and potassium nitrate ("the stone of Tagus"). This recipe is often cited as the earliest description of black powder in a European text, although this claim is open to dispute. For instance, some scholars cite *Liber ignien ad conburendos hostes* (*The book of fires for burning the enemy*), by an author writing under the pseudonym **Marcus Graecus**, which gives a similar recipe. In any event, the treatise by Bacon does note the explosive nature of black powder, observing that it may be useful in war. It fails to speculate further, however, and is silent on the possible use of black powder as a **blasting agent** or **propellant**. Bacon's teaching and writings were influential and may have inspired others to experiment with black powder. *See also* SCHWARZ, BERTHOLD.

BALI BOMBINGS. Around 11:00 p.m. on 12 October 2002, a suitcase bomb was **detonated** outside the US consulate, Denpasar, the capital region of Bali, Indonesia. One person was slightly injured. The **explosive** within the suitcase was surrounded by excrement, which splattered the consulate exterior and sidewalk. At 11:05 p.m. on 12 October 2002, a suicide bomber **detonated** an **explosive**-laden vest in Paddy's Pub, Kuta district. Many patrons and workers were able to flee and were standing outside when a **vehicle-borne**

improvised explosive device (VBIED) parked opposite the pub at the Sari Club was detonated 20 seconds later, claiming the life of its driver as well as the lives of many of those from Paddy's Pub. The death toll reached 202, plus two suicide bombers. The bombings are attributed to Jemaah Islamiyah, a terrorist group dedicated to establishing an Islamic state in Southeast Asia. The investigation will tie together all three bombings.

The VBIED was both the largest explosive and the most sophisticated, having an estimated mass of 1,000 kilograms (2,250 pounds). The main explosive was initially thought to be **Composition C4**, but **forensic analysis** will determine that it was a **homemade explosive**, comprised of **potassium chlorate**, **aluminum powder**, and **sulfur**. Other components, though, pointed to diverted commercial and military explosives as well as bomb-making skill. For instance, **TNT** was added as a booster, the **detonator** cord contained **PETN**, plus an array of 94 electrically operated **RDX** detonators was present to ensure simultaneous functioning of the main explosive. Further investigation will identify Ali Amrozi bin Haij Nurhasyim (1962–2008), an avowed Islamic terrorist, as the primary bomb-maker and his workshop. He is arrested, found guilty of charges related to these bombings as well as the Christmas Eve 2000 bombings that claimed 18 lives and injured more than 100, and executed by the government. In the coming months, other Jemaah Islamiyah terrorists will be identified, and many of these will be taken into custody, arrested, tried, and convicted. However, three more suicide bombers will strike south Bali on 1 October 2005, killing 20 and injuring more than 100. This later event became known as the Second Bali Bombings.

BALLISTITE. Propellant comprised of **nitroglycerin** and either **collodion** or some other form of soluble **nitrocellulose**, with or without a **stabilizer**. The proportions of nitroglycerin and collodion varied over time as experience with the manufacturing process and the properties of the product increased. Initially, Ballistite had two parts nitroglycerin and one part collodion, with camphor added as a **plasticizer** for the collodion and benzol added as a solvent. Various other chemicals were used as solvents and stabilizers, including amyl acetate and aniline. Eventually, a formulation having one part nitroglycerin and two parts collodion and using diphenylamine as a stabilizer came into common use.

Ballistite was invented by **Alfred Nobel** around 1886 and patented shortly thereafter in France (1887) and Britain (1888). He apparently was inspired to perform research with collodion by the publicity surrounding celluloid, as suggested by the title of the US patent he received on 21 July 1891 for Ballistite: "Celluloidal Explosive and Process of Making the Same" (US patent

456,508). His interest in celluloid may account for his initial use of camphor as a plasticizer for collodion.

At the time of his discovery of Ballistite, Nobel was living in Paris, where he had a laboratory. He offered his invention to the French government for use by its military. The French declined his offer, perhaps because they had recently adopted **Poudre B** for use as a **smokeless powder** and viewed Ballistite as a competing product. Thereafter, Nobel offered Ballistite to the Italian government and on 1 August 1889 signed a contract licensing Ballistite to Italy, which promptly began using it as a replacement for **black powder** in rifle cartridges. The French government regarded this development as unwelcome, and Nobel was publicly accused of treason and of having spied on **Paul Vieille**, the developer of Poudre B. Eventually, the French government banned Nobel from conducting any further research in France. This move, plus the general ill will that he faced from the public in France, prompted Nobel in 1891 to depart France, moving to San Remo, Italy.

In the 1890s, Nobel brought suit in Britain, claiming that the patent granted to Sir **Frederick Abel** and Sir **James Dewar** for **Cordite** infringed on his 1888 British patent for Ballistite. In 1895, the British government ruled against him, noting that his patent referred to the use of "soluble nitrocellulose," thereby failing to anticipate the use of **insoluble nitrocellulose**, which the government deemed to be a key ingredient to Cordite.

BANGALORE TORPEDO. Explosive device used to clear barricades, barbed wire, booby traps, and mines. Development of the Bangalore torpedo is credited to Robert Lyle McClintock (1874–1943) in 1912. At the time, McClintock was a Brevet Major, Royal Engineers, serving in Bangalore, India, with the Queen's Very Own Sappers and Miners, Indian Army. Aware of difficulties experienced during the **Russo-Japanese War** with clearing barricades and booby traps, he experimented with using connectable, explosive-filled metal tubes that might be extended across suspected mine fields and under barricades. He published his results in the March 1913 issue of the *Royal Engineers Journal* in an article titled "The Destruction of Wire Entanglements—A Suggested Method." His article used the term "Bangalore Torpedo" for a 6-foot, 6-inch-long, and 2.5-inch-diameter metal tube filled with **dynamite**. Variants on this design saw extensive use during the First World War and the Second World War, with the United States fielding the M1A1, Bangalore Torpedo, using **Amatol** as the explosive charge. This fill was later changed to **Composition C4**.

BARATOL. Explosive comprised of barium nitrate, **TNT**, and wax. The amount of TNT typically is in the range of 25–33 percent, while wax is present at one percent, and the residual 66–74 percent is barium nitrate. The wax acts as a **phlegmatizing agent**. Baratol was used by the United States as the **low velocity explosive** in the **explosive lenses** used in some of its **implosion-type nuclear explosives**, such as **Gadget** and **Fat Man**. The Soviet Union and India are believed to have used baratol as well as **Composition B** for their first implosion-type nuclear explosive, which the United States dubbed **Joe-1** and **Smiling Buddha**, respectively.

BARONAL. Ternary explosive comprised of 50 percent barium nitrate, 35 percent **TNT**, and 15 percent **aluminum**. Baronal was developed by the United States during the Second World War as a castable filler for underwater explosives, and was intended to be a replacement for **torpex**. Testing proved it failed with regard to being less sensitive than torpex, however, and it was rejected for use as a replacement.

BATH SCHOOL BOMBING. Around 9:45 a.m. on 18 May 1927, an explosion rocked the Bath, Michigan, Consolidated School. One wing of the school building was destroyed, and 39 students and teachers were killed. Around 10:10 a.m., a pickup truck driven by Andrew Kehoe arrived at the site of the blast. He gestured for the superintendent of the local school board to come over, which he did, along with two others. When they walked up, Kehoe took a shotgun and shot a batch of **explosives** concealed nearby. The ensuing **detonation** killed Kehoe and four others—the superintendent, a student, and two townspeople. The bombings would mark the worst attack on a school in US history.

Subsequent investigation would establish that Kehoe acted alone. A 55-year-old farmer, he had held various positions with the local school, serving as the school board treasurer, school bus driver, and maintenance man for the school itself. He was embittered for various reasons: higher local taxes, which he attributed to unwise spending by the school board on construction; his April 1926 defeat in the campaign for town clerk; and the impending foreclosure on the mortgage on his farm. That same morning, he murdered his wife, blew up his farmhouse and barn, and then set out on his normal school bus route. Survivors of the bombings who had been on that bus recalled that Kehoe told them, "Well kids, this is your last day." The meaning of this otherwise cryptic comment became clear after the bombings. His work as the school's maintenance man afforded him time and access such that he could emplace explosives throughout the building. Most detonated that morning, but another 504 pounds of unexploded materials were recovered later during rescue operations. Kehoe used a simple clock-and-battery setup to detonate the explosives, but some of the batteries proved too weak to set off their load.

Kehoe's choice of explosives was one that enabled him to avoid arousing suspicion. The US government, through the Department of Agriculture, had made available 1,000 pounds of **Pyrotol** to any farmer wishing to have this **blasting agent** for use on his property to make ditches, remove tree stumps, or otherwise clear added acreage for cultivation. The only costs to the farmer were for shipping and packaging. The notion behind Pyrotol was to enable the government to rid itself of excess **smokeless powder**, a residual of the First World War. The loss of these students and teachers was felt nationally, inspiring a campaign by Michigan children, in particular, to donate pennies to pay for a memorial. That memorial, a bronze statue titled "Girl with a Cat," was dedicated on 18 August 1928 and is displayed currently in the Bath School Museum.

BATTLE OF CRÉCY. On 26 August 1347, English forces under King Edward III defeated a much larger French force under King Philip VI at the site of present-day Crécy, France. The English forces included some guns, marking one of the earliest uses of **black powder** by the English. This battle is occasionally cited as the earliest military use of black powder, but that claim is dubious, especially in light of Chinese accounts from three centuries earlier of weapons using black powder. In addition, German forces are thought to have used guns and black powder during the siege of Cividale, Italy, in 1331, and Edward's forces may have used guns during their earlier victory at the Battle of Sluys on 24 June 1340. The use of guns at the Battle of Crécy, in any event, is unlikely to have contributed significantly to the English victory, which was due largely to English archers.

"B" BLASTING POWDER. Black powder containing **sodium nitrate** rather than **potassium nitrate**. **Lammot du Pont** developed this formulation of black powder in 1857, and it was initially dubbed "Soda blasting powder." Later, this designation was changed to "B" blasting powder, while the original formulation of black powder, made with potassium nitrate, was designated **"A" blasting powder**. By contrast with "A" blasting powder, "B" blasting powder is less expensive, making it attractive for applications such as mining, debris removal, demolition, and quarrying of rocks. "A" blasting powder continued to be preferred for conventional **explosives**, including in firearms and as a **propellant** for military use.

BÉCHAMP, PIERRE JACQUES ANTOINE (1816–1908). French scientist credited with distinguishing **nitrocellulose** from **nitrostarch**. Béchamp began his career as a chemist and was granted a doctorate from the University of Strasbourg in 1853 on the basis of his research into **pyroxylin**. He pursued this research for the next decade. In 1855, he published a scientific article presenting the first proof that nitrocellulose and nitrostarch are separate materials. Until this time, scientists had imagined that the nitrated products obtained from cellulosic materials, such as cotton and sawdust, were the same as those obtained from corn starch and potato starch. In 1862, Béchamp published results in which he isolated nitrostarch, such that it could be characterized and studied separately from nitrocellulose. Sometime thereafter, Béchamp's research interests moved on to the study of silk and, especially, the silkworm. He identified the parasites responsible for certain silkworm diseases then plaguing the silk industry in France and is best remembered for this work. *See also* BRACONNOT, HENRI (1780–1855).

BERNADOU, JOHN BAPTISTE (1858–1908). American naval officer credited with aiding US development of **smokeless powder**. Bernadou, an 1880 graduate of the US Naval Academy, was assigned to the **Torpedo** Station, Newport, Rhode Island, from September 1894 to December 1897. While there, he used his fluency in French, German, Russian, Swedish, Spanish, and other foreign languages to inform the military on the novel **explosives** being tested and produced in Europe and Russia. In particular, he translated articles by Russian chemist **Dmitri Mendeleyev** on **pyrocollodion**. These translations provided a technical basis for United States attempts at formulating a smokeless powder, culminating in the 1896–1897 construction and operation of a manufacturing facility at the Torpedo Station. Bernadou served with distinction during the **Spanish–American War**, during which he suffered serious wounds. From late 1898 until sometime in 1899, he resumed work on smokeless powders at the Bureau of Ordnance, Washington, D.C. Thereafter, he transferred to other naval duties.

BERNARD, SIMON FRANÇOIS (1817–1862). French-born bomb-maker. Bernard was forced to flee France in January 1851 because of his opposition to Louis Napoléon, soon to become Emperor Napoléon III. Bernard arrived in England that January, describing himself as a professor of orthophonics (speech). He settled in London and joined a loose association of like-minded radicals and revolutionaries, including Felice Orsini, Giovanni Pieri, and Camillo di Rudio. In 1857, he worked with these three men on a plot to **assassinate Emperor Napoléon III**. In particular, Bernard manufactured **mercury fulminate** in sufficient quantity for it to be loaded into five bombs, each having 4 pounds of this **high explosive**. He made the mercury fulminate from ethanol, concentrated nitric acid, and mercury, each of which was available commercially and could be purchased in large quantities without arousing suspicion. Mercury, for example, was used to treat syphilis, and

Bernard introduced himself as "Dr. Bernard," suggesting that he would use the mercury on his patients. Once the bombs were ready for use, he gave them to Orsini in Brussels when they met at the Café Suisse. Three bombs were used in the assassination attempt on 14 January 1858, but Napoléon survived. Bernard's role soon became known. He was arrested and tried in April 1858 as an accessory before the fact in the murder of Nicholas Battie, a Paris policeman who died from injuries received in the blast. He was acquitted on 17 April. In early 1862, he was committed to an insane asylum in Surrey, England, and then moved to one in London. He died there on 25 November 1862. His career as a bomb-maker is noteworthy because of the ease with which he could acquire the ingredients for what amounted to a **homemade explosive** and because this episode marked the earliest known use of a high explosive by terrorists.

BERTHELOT, PIERRE-EUGÈNE MARCELLIN (1827–1907). French chemist, government official, and historian of science. Berthelot made fundamental contributions to organic and physical chemistry, including pioneering studies of thermodynamics. These studies proved especially important in advancing the field of **explosives**, placing the development of explosives on a firm scientific basis in lieu of the more trial-and-error approach that had dominated the advances in explosives. A prime example is his research, conducted jointly with **Paul-Marie-Eugène Vieille**, on the velocity of **detonation**. Berthelot's contributions to the science of explosives were recognized with his appointment as President of the French Powder and Explosives Commission. In addition to his pursuit of science, he became active in government affairs. In 1876, he was Inspector of Public Education for France. In 1881, he was elected a Senator-for-life and served in government as Minister of Public Instruction (1886–1887) and Minister of Foreign Affairs (1895–1896). Later in life, Berthelot developed an interest in the history of science, translating various Arabic, Greek, and Syriac treatises on **alchemy** and chemistry. He was a prolific contributor to scientific journals, publishing approximately 1,600 articles and many books.

BERTHOLLET, CLAUDE LOUIS (1749–1822). French chemist credited with many fundamental contributions to science. These contributions include the discovery of cyanogen chloride and **potassium chlorate**; determining the composition of ammonia (NH_3), hydrogen cyanide (HCN), and hydrogen sulfide (H_2S); and introducing the use of aqueous chlorine as a bleaching agent.

Berthollet received his doctorate in medicine in Turin, Italy, in 1770. On returning to France, he became associated with **Antoine Lavoisier**, and they collaborated on various activities, most notably the reformation of chemical nomenclature. Through his association with Lavoisier, Berthollet became aware of the critical importance of **gunpowder** to France. At the time, **black powder** was used for gunpowder, and the manufacture of black powder required **charcoal**, **potassium nitrate**, and **sulfur**. While supplies of charcoal and sulfur were freely available in France, supplies of potassium nitrate were more precious. In particular, the settlement of the Seven Years' War (1754–1763) limited access by France to **India Saltpeter** and harvesting potassium nitrate from domestic sources through ***droit de fouille*** was problematic.

Berthollet recognized that his recent discovery, potassium chlorate, might be a suitable substitute as an **oxidizer** in **explosives**. With the support of Lavoisier, Berthollet set about preparing experimental batches of an explosive containing potassium chlorate, charcoal, and sulfur. Results proved tragic. On the morning of 26 October 1788, a group gathered at the gunpowder mill in Essonne, near Paris, for a demonstration of this novel explosive. The group included Berthollet; Lavoisier and his wife; one of the commissioners of the Gunpowder and Saltpeter Administration, Pierre de Chevrand (1742–1794), and his sister; and Pierre Edme Le Tors de Chessimont (1748–1788), director of the gunpowder mill. A hand-cranked mortar-and-pestle was used to mix the ingredients. Lavoisier warned the group of the hazards inherent in the mixing process and urged them to remain behind a protective barrier whenever mixing was underway. While most members of the group lingered over breakfast, Le Tors and Mademoiselle de Chevrand returned to the site of the mixing. As they did so, a 20-pound batch of powder exploded. Mlle. de Chevrand was thrown 30 feet against a wall and killed outright. Le Tors suffered massive injuries, including the traumatic amputation of a leg, and died shortly thereafter. As a result of this tragedy, Berthollet and Lavoisier ceased developing potassium chlorate-based explosives. *See also* FULMINATE.

BICKFORD, WILLIAM (1774–1834). English merchant credited as the inventor of the safety **fuse**. Bickford was a tradesman lacking direct involvement in any enterprise that made or used **explosives**. But he was distressed by the numerous accidents that occurred in the tin mines in the Cornwall region of England, where he was living, and was interested in improving the safety of these mining operations.

At the time, **black powder** was used to blast away rock. The powder was **detonated** by means of some strands of straw or a bundle of goose quills, set alight with a match or flame. These measures proved unreliable. Bickford tried various schemes to improve the process of igniting the black powder, but these failed. In 1831, he observed rope being made by a friend and hit upon the scheme of twisting jute around a central core of **gunpowder**. The mass of jute and gunpowder was then varnished to make it waterproof. The finished product had the advantage of burning at a comparatively steady and predictable rate. Therefore, someone needs only cut the desired length to give sufficient time to clear before the burning fuse causes the black powder to detonate.

Bickford received a patent on his device on 6 September 1831. Although he initially dubbed his fuse a "safety rod," he would change that name to "safety fuse." A safety fuse is sometimes called a Bickford fuse in his honor. A factory to produce Bickford's invention opened in 1834 in Cornwall, shortly after his death. At its peak, the factory produced over 100,000 miles of safety fuse in a single year. As one example of its popularity, construction of the 36-mile-long Manchester Shipping Canal linking Manchester, England, to the Irish Sea, from 1887 to 1894, used around 340 miles (548 kilometers) of the safety fuse, all of it produced by this factory. Such demand spawned many imitators as well as challenges to Bickford's patent. None of these challenges succeeded. Although safety fuses continue to be manufactured and used, Bickford's Cornwall operation is closed.

BINARY EXPLOSIVE. Term having two possible and quite different meanings. (1) **Composition explosive** comprised of two explosive compounds. Examples of modern binary explosives include: the **amatols**; the various formulations of **Composition B**; the ednatols (mix of **EDNA** and **TNT**), the octols (mix of **HMX** and TNT); the pentolites (mix of **PETN** and TNT); picratol (52 percent **ammonium picrate** and 48 percent TNT); the **tetrytols**; and tritonal (80 percent TNT and 20 percent **aluminum**). (2) Products consisting of two components that separately fail to meet the definition of explosives used by the United States for regulatory purposes, most notably that given in 27 Code of Federal Regulations (CFR) 555.11. As such, these products fall outside regulations governing their sale and distribution. Such products typically contain an **oxidizer**, such as **ammonium nitrate**, and a **fuel**, such as **aluminum**. Certain binary explosives of this type are available commercially in pre-packaged, two-component systems, as is the case for **Tannerite**. Another example of this type of binary explosive is the **ammonium nitrate-fuel oil mixture (ANFO)**.

BINDER. Any substance used to hold together dry solids. A familiar example of a binder is cement, which binds the aggregate, sand, and other materials used in making concrete. In the context of **explosives**, binders are used most often in **propellants** and **pyrotechnics**. Binders prevent the segregation of the explosive components in the blended final product, thereby enhancing its uniformity both as a material and in performance. In many instances, binders also permit the final product to achieve maximum packing density and maximum burning efficacy. Historically, paraffin, shellac, and various waxes saw extensive use as binders, especially in pyrotechnics. These substances are readily available from natural sources, but their properties vary because they are themselves mixtures rather than singular materials. Currently, binders are likely to be synthetic polymers, whose properties are reproducible and controllable. Examples include styrene-butadiene elastomeric copolymer, epoxy resins, and various polyvinyl acetates.

BLACK POWDER. Explosive typically comprised of **charcoal**, **potassium nitrate**, and **sulfur**. Charcoal and sulfur serve as **fuels**, while potassium nitrate serves as an **oxidizer**. Most likely, the combustible nature of a mix of charcoal, potassium nitrate, and sulfur was recognized by **alchemists** in China, perhaps as early as 142 and certainly by the third century.

European interest in black powder lagged developments in the Orient by several centuries. Such interest dates at least as far back as the thirteenth century, and this interest likely stemmed from knowledge communicated from the Orient or the Middle East, perhaps from trade on the Silk Road or through Mongol invasions. Many scholars credit **Roger Bacon**, an English scholar and alchemist, with stimulating interest in black powder among Europeans. His texts contain a description written in a cipher that can be decoded as describing a formula for black powder. Moreover, he accurately notes the explosive properties of black powder, although he failed to note its possible use as a **propellant** for military purposes. A text by **Marcus Graecus** describes a pyrotechnic material comprised of two-thirds potassium nitrate, two-ninths charcoal, and one-ninth sulfur. The authenticity and dating of this text, *Liber ignien ad conburendos hostes* (*The book of fires for burning the enemy*), are widely questioned. Some traditions hold that the honor of discovering the use of black powder as such a propellant—that is, as gunpowder—belongs to **Berthold Schwarz**, a fourteenth-century monk. Many scholars, however, believe Schwarz to be apocryphal and dismiss any claims attached to his work. Furthermore, such claims fail to account for Chinese texts from 1040 to 1044 describing the use of gunpowder in weapons.

The proportions of charcoal, potassium nitrate, and charcoal used in formulating black powder have varied with time and intended use. Early texts mentioning black powder or what is thought to be black powder specified using about 50 percent potassium nitrate and 50 percent a mix of charcoal and sulfur. For example, deciphering the thirteenth-century text of Roger Bacon yields a mix having approximately 40 percent potassium nitrate, 30 percent charcoal, and 30 percent sulfur. Over the next several centuries, the proportion of potassium nitrate in the total mix steadily increased, with equivalent drops in the proportions of charcoal and sulfur. By the late eighteenth century, the amount of potassium nitrate had nearly doubled while the amounts of charcoal and sulfur dropped by half or more. European texts of that time specified a black powder containing 75 percent potassium nitrate, 15 percent charcoal, and 10 percent sulfur. These proportions are often taken as representative of those used in black powder, although variations in these proportions continue as a function of intended use and local practices. For instance, the presence of sulfur in black powder used as a **primer** for certain **smokeless powders** was discovered to contribute to corrosion. This discovery spurred the development of sulfur-free black powders, such as **sulfur-free mealed powder**.

When black powder burns, its combustion includes many particulates. A significant fraction of the combustion products are solids such as carbon, potassium carbonate, potassium sulfate, and potassium sulfide. In addition, there are bits of charcoal, potassium nitrate, and sulfur that failed to be combusted. These solids help account for the dense gray or black smoke associated with the use of black powder and constitute a significant drawback to the use of black powder, especially by the military.

In its classic forms, black powder does, indeed, appear black, accounting for its name. However, other mixtures of these same ingredients have had different hues, with the differences most likely attributable to the nature of the charcoal used. For instance, chocolate powder and **brown powder** were made using charcoal that had been prepared at lower temperatures. Despite the differences in color, such powders are types of black powder.

BLAST EFFECTS. Damage or destruction suffered by people and structures from the force of an explosion. *See also* PROPELLANT.

BLASTING AGENT. Criminal law in the United States defines a blasting agent as a material or mixture consisting of a **fuel** and an **oxidizer**, intended for use in blasting and not otherwise classified as an **explosive**. A widely used blasting agent is the **ammonium nitrate-fuel oil (ANFO)** mixture. *See also* BLASTING CAP.

BLASTING CAP. An **explosive** device, typically small, used to **detonate** a larger, more powerful explosive. An early blasting cap was a simple electric spark. This design was demonstrated successfully in 1745 by British physician and scientist Sir William Watson (1715–1787), who used an electric spark provided by means of a **friction** machine to detonate **black powder**. His friend and professional colleague, American inventor and political figure Benjamin Franklin (1706–1790), had a similar idea. Franklin marketed one of the earliest commercial blasting caps, which consisted of black powder in a paper tube. A wire protruded from each end. The black powder detonated when an electric spark was applied between the two wires. Nowadays, substances used in blasting caps are **primary explosives** such as **lead azide** and **mercury fulminate**, while the more powerful, **secondary explosives** are substances such as **TNT** and **dynamite**. *See also* NOBEL, ALFRED BERNHARD (1833–1896).

BLASTING GELATIN. An **explosive** made of **collodion cotton** dissolved in either **ethylene glycol dinitrate** or **nitroglycerin**. It was invented in 1875 by **Alfred Nobel** and is more powerful and more stable than **dynamite**. Other names used for blasting gelatin include gelignite and jelly. *See also* PLASTIC EXPLOSIVE; SIMS-DUDLEY DYNAMITE GUN.

BLUE DANUBE. The first **nuclear weapon** deployed by the UK. The initial plans for Blue Danube called for producing 800, each having a nominal yield of 10–12 kilotons **TNT-equivalent**. Other names used for this weapon were "Smallboy" (perhaps an allusion to **Little Boy**), "Mk.1 Atomic Bomb," and "OR.1001" (a reference to Operational Requirement 1001, which established the need for the weapon). The initial design for Blue Danube featured a **plutonium** core, later changed to a combined plutonium and **uranium**-235 core. The weapon was tested successfully in October 1956 in Australia. Fifty-eight Blue Danubes were produced before the British government shifted to the **Red Beard**, deemed more capable and having an improved service life. By 1962, all remaining Blue Danubes had been retired. Blue Danube was also the basis for a truck-mounted land mine that was conceived but apparently never produced. That land mine went through at least three names while being considered for development and fielding: "Big Bertha," then "Brown Bunny," and finally "Blue Peacock." *See also* HURRICANE.

BN POWDER. *See* POUDRE B.

BOER WAR. *See* SECOND BOER WAR.

BOLIVA. *See* WAR OF THE PACIFIC (1879–1884).

BOOSTED FISSION WEAPON DESIGN. Nuclear weapon design concept using a small amount of **fusion** as fuel to increase the rate and yield of a **fission** reaction. The design was developed in the United States from late 1947 to late 1949. A primary benefit of the design is that it enables a weapon to be made smaller yet to provide equivalent yield to that of a non-boosted weapon. This concept was used successfully with the **Ivy Mike** device, in which the first stage involved fission of **plutonium**-239 that in turn caused a liquid deuterium secondary stage to undergo nuclear fusion. High-energy neutrons produced by this fusion caused fission of **uranium**-235 in the third and final stage. *See also* B41.

BOOSTER. A **secondary explosive** that reinforces the **detonation** wave from the **primary explosive**. The result of incorporating a booster into the overall explosive mixture is that a more powerful detonation wave is delivered to the main explosive charge, which typically is a high-energy but low-sensitivity explosive. Therefore, the proper functioning of the booster ensures that the main charge detonates.

BOSCH, CARL (1874–1940). German chemist, industrialist, and inventor who perfected the **Haber-Bosch process** for **nitrogen fixation**. While working as an industrial chemist for the German chemical firm Badische Anilin- und Soda-Fabrik AG (BASF), Bosch adapted the laboratory-scale nitrogen fixation process developed by **Fritz Haber** into an industrial-scale process for producing ammonia. His work had important consequences for the ability of Germany during the First World War to eliminate its dependence on **Chilean nitrates** and other overseas sources as feedstocks for its production of fertilizers and **explosives**. The adaptation of Haber's laboratory-scale method to an industrial scale required the use of high pressures previously unknown on an industrial scale. As a result, Bosch became a specialist in high-pressure methods and made many fundamental and practical contributions to working at high pressures, for which he shared the 1931 Nobel Prize in Chemistry. The success of his high-pressure work helped propel Bosch into senior management ranks, and he eventually rose to the presidency of the I.G. Farbenindustrie AG, the successor to BASF.

BÖTTGER, RUDOLF CHRISTIAN (1806–1881). German chemist who recognized the military potential of **guncotton**. Böttger originally trained for the ministry and had taken an appointment with a church before realizing that his true interest was in science rather than religion. In 1835, he was appointed lecturer in physics and chemistry at the Physical Society in Frankfurt-am-Main, where he would spend the rest of his career. His work was primarily in applied chemistry. In the course of this work, he made several notable contributions, including his work on guncotton. This work coincided with that of **Christian Schönbein** at the University of Basel, and in 1846, these two collaborated on a series of experiments in England to demonstrate the use of guncotton for the military. While their demonstrations succeeded in awakening interest in guncotton as a **propellant** or **blasting agent**, the processes for the manufacture of guncotton were imperfect, resulting in an unstable product that often **detonated** accidentally. *See also* ABEL, FREDERICK AUGUSTUS (1827–1902); COTTON POWDER COMPANY, LTD.; JOHN HALL & SON; LENK VON WOLFSBERG, WILHELM FREIHERR (1809–1894).

B POWDER. *See* POUDRE B.

BRACONNOT, HENRI (1780–1855). French chemist; perhaps the first scientist to publish results on the preparation and characteristics of **nitrocellulose** and **nitrostarch**. As was typical of many in France in the eighteenth and nineteenth centuries, Braconnot learned the basics of chemistry through an apprenticeship to a pharmacist. After serving briefly in the military and attending lectures in Paris, Braconnot moved to Nancy, taking up a position with the Botanical Gardens. He became director of the gardens in 1807 and remained there for the rest of his professional career, where he made many important discoveries in the chemistry of substances derived from plants.

Results of his studies into the nitration of starch were published in *Annales de Chimie et de Physique* starting in 1833. According to these publications, Braconnot dissolved potato starch in concentrated nitric acid. When he added water, he obtained a curd-like precipitate. On washing the precipitate with water and drying, he obtained a white mass that he described as crumbly and tasteless. He reported that similar results were obtained using cotton, **sawdust**, and linen in place of potato starch. These latter materials contain cellulose rather than starch. Therefore, Braconnot was also the earliest scientist to report the preparation of nitrocellulose, although the distinction between nitrocellulose and nitrostarch would be recognized some decades later, most notably by **Antoine Béchamp**.

When Braconnot boiled the product obtained by adding nitric acid to either starch or a cellulosic material such as sawdust, he obtained a sticky mass. He named this sticky mass *xyloidine*, from the Greek ξυλον (wood). Braconnot, in his publications, noted that what he referred to as xyloidine (most likely, nitrocellulose in this instance rather than nitrostarch) dissolved in acetic acid, giving a transparent film when the acid was evaporated. He also noted that the mix of acetic acid and xyloidine could be applied to paper and cloth, giving a brilliant varnish-like coating impervious to water.

His research into the nitration of starches and cellulosic materials constitutes one of the earliest preparations of a polymer. His publications received little attention at the time, and Braconnot went on to pursue other topics. Most notably, he failed to note the **explosive** nature of his nitrated starch. He does deserve credit, though, as one of the first to prepare and characterize nitrocellulose and nitrostarch. *See also* PELOUZE, THÉOPHILE-JULES

(1807–1867); PYROXYLIN; SCHÖNBEIN, CHRISTIAN FRIEDRICH (1799–1868).

BRANDYWINE POWDER. *See* DUPONT (E. I. du PONT de NEMOURS & COMPANY).

BRAZILIAN NAVAL REVOLT OF 1893–1894. *See* PNEUMATIC GUN.

BRIMSTONE. Archaic term for **sulfur**.

BRISANCE. Shattering effect from the sudden release of energy, as with an explosion. Brisance of an **explosive** is measured by comparing the weight of graded sand shattered when a charge of test explosive is **detonated** in a standard manner to that shattered by **TNT** detonated in the same manner.

BROWN CHARCOAL. Charcoal made from slightly charred straw, sugar, or a similar material and having a dark-red or brown color. Brown charcoal was used in making **brown powder**. As the name suggests, brown charcoal is typically brownish in color rather than the gray-black typical of conventional charcoal. In contrast to conventional charcoal, brown charcoal has lower carbon content, ignites at a lower temperature, and burns more slowly. This latter property was exploited in making brown powder.

BROWN POWDER. A type of **black powder** made from reddish or **brown charcoal** and containing a lesser quantity of **sulfur** than was present in conventional black powder mixtures. During the nineteenth century, weapon makers attempted to field larger naval guns and artillery pieces capable of firing more massive projectiles. These efforts were stymied by the conventional black powders then in use, which burned at a very fast rate, causing strain on the gun barrels. Brown powder was developed in a successful effort to reduce the rate of burning. Typical black powders in use at the time contained nearly 30 percent sulfur. By contrast, brown powder had 3 percent. This reduced sulfur content, coupled with the residual moisture in brown charcoal, reduced the burn rate, permitting an increase in both the caliber of guns and the mass of projectiles. Brown powder became the most successful form of black powder for use in rifled guns of long range. It was made in hexagonal or octagonal prisms resembling pieces of milk chocolate, and this resemblance accounts for its being referred to as "chocolate powder" and "cocoa powder."

Brown powder was used by the United States during the **Spanish–American War**. This use in combat surfaced a fatal flaw in brown powder, as it proved to be more sensitive to **friction** than ordinary black powder. Stocks were reported to have ignited from the simple act of shaking or handling the canvas bag in which it was stored. This experience caused the United States to destroy its inventory of brown powder following the war. *See also* CHARCOAL; MAMMOTH POWDER; PRISMATIC POWDER; RODMAN, THOMAS JACKSON (1816–1871).

BROWN'S ISLAND EXPLOSION. On 13 March 1863, an explosion destroyed an ordnance laboratory on Brown's Island, located in the James River off the southern shore of Richmond, Virginia. The explosion claimed at least 50 lives, making it the largest noncombat disaster to happen anywhere in the Confederacy during the **American Civil War**. The victims were mostly women and girls who worked in the ordnance laboratory assembling cartridges containing **primary explosives** for use by the Confederacy as **primers**.

A formal investigation into the explosion revealed a lax attitude toward personal safety, inadequacies in training, and an absence of proper supervision. The proximate cause of the explosion was striking a block holding small tubes of explosives—**friction** primers—to loosen one that had become stuck. A survivor, Mary Cunningham, reported she witnessed Mary Ryan, a teenage Irish immigrant, doing this striking. Ryan had lingered for two days before dying from the blast. The investigation revealed that the laboratory was occupied routinely by both workers and visitors, who would socialize while loading cartridges took place. It also revealed that it was common practice to sweep the floor with an ordinary broom—a practice that was at odds with the friction-sensitive nature of the explosives that were present. The lessons learned from this event appear to have had the desired effect because no further explosions were reported in any of the ordnance laboratories of the Confederacy.

BSX. Term for 1,7-diacetoxytetramethylene-2,4,6-trinitramine, an **explosive**. It has the chemical formula $C_6H_8N_6O_{10}$. BSX was first isolated as an impurity in RDX produced by the **Bachmann Process**. The precise meaning of "BSX" is uncertain. Contemporary records note that "BS" is an abbreviation for "Bachmann-Sheehan," the developers of the Bachmann Process. It is uncertain from these same records, however, whether "X" was used for "explosive" or "impurity." *See also* HMX.

BUNKER BUSTER BOMB. *See* MASSIVE ORDNANCE PENETRATOR (MOP).

C

C-86. German designation for a **smokeless powder** comprised of 38 percent **ammonium nitrate**, 48 percent **potassium nitrate**, and 14 percent **charcoal** and made at the Rhenish-Westphalian Explosives Works. The mixture was also made at the Chilworth Powder Company, a branch of the firm near the town of Chilworth, Surrey, England, where it became known as "Chilworth Powder."

CALICHE. Crude **sodium nitrate**. Caliche is a naturally occurring deposit rich in sodium nitrate. Historically, it was found primarily in the northern deserts of Chile, the southern deserts of Peru, and arid areas in the southwestern US Caliche builds up on the surface of the earth owing to complex chemical processes, and deposits can accumulate to significant depths. Caliche was used directly as a fertilizer or treated with water to remove most impurities. After evaporating the water, sodium nitrate appears as rhomboid-shaped crystals, and this shape caused the product to be referred to as "cubical saltpeter." Additionally, because Chile was a major source for sodium nitrate produced from caliche, sodium nitrate came to be known in the **explosives** and fertilizer industries as **Chile Saltpeter**.

South American caliche deposits became important in the later half of the nineteenth century, owing in part to the depletion of **guano** deposits on various Pacific islands. Competition among Bolivia, Chile, and Peru for the revenue from these deposits sparked the **War of the Pacific**. The 1857 invention by **Lammot du Pont** of a process for the direct use of sodium nitrate in manufacturing **black powder** added to the competition for caliche. By 1895, more than one million tons of sodium nitrate were being exported from Chile. Although the caliche deposits remain in South America and elsewhere, they were greatly depleted by the extensive harvesting during the late nineteenth and early twentieth centuries. The importance of caliche as a nitrate source diminished greatly with the development of an economical industrial process for **nitrogen fixation**. *See also* INDIA SALTPETER.

CANARY GIRL. Female worker in a munitions factory suffering from chronic exposure to **TNT**. The British used the term beginning in the First World War to describe the appearance of munitions workers, whose skin turned bright yellow-orange. Chronic exposure to TNT causes this change, as well as causing ginger-colored hair to turn green. During the war, most workers in shell-filling factories were female owing to the need for men to serve in the military or in other wartime industries. For instance, the British employed women exclusively at the National Shell Factory, Chilwell, and these women became known as "Chilwell canaries." Women working at the Royal Shell Filling Factory, part of the Royal Arsenal, Woolwich, were similarly affected and were known as canary girls or, more simply, as canaries. Although the acute toxicity of TNT was known when the First World War began, the effects of long-term, low-level exposure were not appreciated. The changes in skin and hair color were dramatic, however, and the British government responded in 1916 by surveying the health of the female workers in the munitions factories. Results pointed to various ill health effects, such as loss of appetite and changes in menstrual cycle, but the more insidious effects of anemia and liver damage were recognized much later. *See also* SHELL CRISIS OF 1915.

CARBAMITE. British term for **Centralite**.

CARBON DISULFIDE. Clear colorless or faintly yellow liquid used as a **fuel** in **Sprengel explosives**, such as **anilite**. Carbon disulfide (chemical formula CS_2) has a strong, disagreeable odor and is highly flammable.

CARBONITES. Class of **permissibles** having **nitroglycerin**, carbonaceous material such as flour, wood pulp, or **sawdust**, and **potassium nitrate** or **sodium nitrate**. The class was developed in Germany in 1887 in response to mining disasters attributable to **firedamp**. The incorporation of the woody material and nitrate had the desired effect of reducing the temperature of the explosion, thereby reducing the prospect of unplanned, secondary explosions following an intentional explosion in a mine. But this reduction in temperature arose from the incomplete burning of the ingredients. Therefore, the use of Carbonites was accompanied by the release of carbon monoxide (chemical formula CO), an asphyxiating gas whose presence posed a danger in all but the best ventilated areas. Carbonites and **Monobels** were the first explosives to be certified by the US Bureau of Mines as permissibles, meaning their use was judged safe for mining operations.

CARDOX. Carbon dioxide-based **mechanical explosive**. Cardox is a trademark for a system developed in the UK around 1920. It uses liquid carbon dioxide (chemical formula CO_2), which is heated to cause rapid vaporization. The gaseous carbon dioxide exerts tremendous internal pressure that can be used as a heaving force (e.g., in mining to dislodge rock). The pressure can also be used to rupture a disc, allowing a jet of high-pressure gaseous carbon dioxide to be released and directed against debris to unclog blockages in vessels. Cardox is also suitable for use underwater for demolition, drilling, and tunneling. *See also* HYDROX.

CELLULOID. *See* NITROCELLULOSE.

CELLULOSE NITRATE. *See* NITROCELLULOSE.

CENTRALITE. German term most often associated with diethyl diphenylurea, which was used as a non-volatile solvent, **stabilizer**, and **gelatinizing agent**. Research and development of various dialkyl diphenylureas took place prior to the First World War at the Central War Laboratory, Spandau, Germany, and the term "Centralite" honors this facility. The original diakyl diphenylurea to be used as a gelatinizing agent appears to have been dimethyl diphenylurea. Later research found that diethyl diphenylurea was superior, and this substance came to be called "Ethyl Centralite" or "Centralite No. 1," while the dimethyl analogue was known as "Centralite No. 2."

CHADWICK, JAMES (1891–1974). British physicist who played a prominent role during the Second World War efforts of the UK to develop **nuclear weapons**. Chadwick had been awarded the 1935 Nobel Prize in Physics for his discovery of the neutron. In October 1943, he replaced **Wallace Alan Akers** as the head of the UK team collaborating on the effort by the Allies to develop a nuclear weapon. He worked directly with his American counterparts and others, relocating along with his family to New Mexico, and was present on 16 July 1945 for the detonation of **Gadget** at the **Trinity Test Site**. Earlier that same year, he had been knighted in recognition of his services.

CHAMBER LYE. *See* NITER AND MINING BUREAU.

CHARCOAL. Dark residue of impure carbon resulting from the destructive distillation of organic matter and used as a **fuel** in **explosives**. The heating involved in distillation removes water, various oils, and many other substances contained in the organic matter. In terms of elemental analysis, the resultant charcoal is predominantly carbon, with the primary minor elements being hydrogen, nitrogen, and oxygen. Charcoal, **potassium nitrate**, and **sulfur** are the three principal ingredients in conventional **black powder**. While the potassium nitrate and sulfur used in formulating black powder have nearly uniform properties, charcoal can vary considerably in its properties. The process by which charcoal is prepared determines these properties and, hence, influences the properties of the black powder into which the charcoal is incorporated.

The process of creating charcoal is carbonization, and the temperature at which carbonization occurs heavily influences the nature of the resultant charcoal. As the temperature of carbonization increases, the percent carbon in the resultant charcoal increases, and the percent of hydrogen, nitrogen, and oxygen decreases. Similarly, as the heat of carbonization increases, the heat of ignition (i.e., the temperature at which charcoal will burn) increases, as does that of the final black powder made from this charcoal. The differences in temperature of carbonization often produce differences in the hue of the charcoal, perhaps attributable to the amounts and types of residual substances, especially those containing nitrogen and oxygen. Charcoal formed by heating organic matter to 140–175°C has a reddish hue, resembling cocoa or chocolate. Heating to 280–350°C produces a brownish charcoal, while heating to 350–450°C produces the darker, gray-black shade associated with the word "charcoal." Black powder made with reddish or brownish charcoal ignites at a lower temperature than does black powder made with black charcoal. Terms used for black powder made from reddish and brownish charcoal include "chocolate powder," "cocoa powder," and "**brown powder**."

Charcoal is often produced from wood, with non-resinous woods being superior to resinous woods such as pine. Hardwoods, such as beech, elm, and oak, tend to support a higher temperature of carbonization than do softwoods, such as alder, dogwood, hazel, poplar, and willow. Therefore, charcoal made from hardwoods will likely have a higher carbon content and will require a higher ignition temperature than charcoal made from softwoods. Similarly, other organic materials, such as meal, **sawdust**, straw, and **sugar**, can be used to make charcoal. For example, certain brown powders were made using charcoal produced from wet straw or sugar, resulting in a **propellant** that ignited at a relatively low temperature and burned at a lower rate than did conventional black powder. *See also* MAMMOTH POWDER; PRISMATIC POWDER; RODMAN, THOMAS JACKSON (1816–1871).

CHEDDITES. Family of explosives made from **potassium chlorate** or sodium chlorate and a nitroaromatic compound. The original preparation of Cheddite is credited to E.A.G. Street, a British-born French chemist working for Bergés, Corbin et Cie., in Chedde, Haute Savoie, France, using 80 percent potassium chlorate and 20 percent **1-nitronaphthalene** and coating the potassium chlorate in a lubricant such as paraffin or castor oil to reduce its sensitivity to **friction** and **shock**. In 1897, Street obtained French patents 9,970 and 13,724, as well as German patent 100,522 for his discovery. This discovery took on the name "Cheddite" to honor the village where Street conducted his research. Further work expanded the formulations for Cheddites to include sodium chlorate and other nitroaromatics, such as **nitrobenzene**. Cheddites are most often yellow granules. Although the earliest uses of Cheddite were for industrial purposes, such as mining, they came to be used during the Second World War by the Polish resistance in occupied Poland as a fill for **improvised explosive devices**, especially hand grenades, owing to the relative ease with which the ingredients could be obtained by certain resistance members without arousing suspicion.

CHEMICAL EXPLOSIVE. An **explosive** in which chemical reactions cause a sudden increase in pressure and volume. The chemical composition of the substance or substances undergoing the reaction is altered, but the elemental composition remains unchanged. An example of a chemical reaction in which there is a change in chemical composition but not in elemental composition can be found with liquid **nitroglycerin**, which can decompose rapidly to form gaseous carbon monoxide (CO), molecular nitrogen (N_2), molecular oxygen (O_2), and water vapor (H_2O). In this case, the decomposition is accompanied by the release of a large amount of energy in the form of heat as well as by a rapid rise in pressure and volume. *See also* DEFLAGRATION; DETONATION.

CHILE. *See* CHILEAN NITRATES; CHILE SALTPETER; WAR OF THE PACIFIC (1879–1884).

CHILE SALTPETER. Synonym for **sodium nitrate**. *See also* CALICHE; DU PONT, LAMMOT (1831–1884); INDIA SALTPETER; SALTPETER; WAR OF THE PACIFIC (1879–1884).

CHILEAN NITRATES. Naturally occurring inorganic compounds confined principally to the Atacama Desert of northern Chile. These compounds include **sodium nitrate**, which was desired for use in fertilizers and **explosives**. In 1857, **Lammot du Pont** invented a process for the direct use of sodium nitrate in making **black powder**. Prior to his invention, sodium nitrate had to be converted into **potassium nitrate** prior to its use in making black powder. This invention heightened the economic importance of Chilean nitrates and contributed to the **War of the Pacific**, which was fought by Bolivia, Chile, and Peru for control of the Atacama Desert region and its mineral deposits. The invention and industrial development of **nitrogen fixation** reduced the international demand for Chilean nitrates. *See also* CONVERTED SALTPETER.

CHILWELL CANARIES. *See* CANARY GIRL.

CHILWORTH POWDER. *See* C-86.

CHINA. Nuclear weapon state (NWS). China began a **nuclear weapon** program in the mid-1950s. It might have been prompted by the threat issued by US president Dwight Eisenhower to use nuclear weapons during the 1954–1955 Taiwan Strait Crisis. In issuing his threat, Eisenhower was responding to a recommendation given on 12 September 1954 by the US Joint Chiefs of Staff to use nuclear weapons against military targets on the Asian mainland. China began its nuclear weapon program with assistance from the Soviet Union, although that assistance waned as the relationship between Moscow and Beijing cooled. On 16 October 1964, China conducted its first successful **detonation** of a nuclear weapon. Given the code name 596 by China, it was an **implosion fission device** using **uranium**-235 as the **fuel** and had an estimated yield of 22 kilotons **TNT-equivalent**. On 28 December 1966, it detonated its first **thermonuclear** device. Both detonations took place at the Lop Nur test site. China conducted its 45th and, to date, final nuclear weapon test on 29 July 1996.

CHOCOLATE POWDER. *See* BROWN POWDER.

CHRISTY GADGET; CHRISTY PIT. The initial design for the **Fat Man implosion device** specified use of a hollow sphere of **plutonium**-239 that would form a **supercritical** mass when imploded by **conventional explosives**. Fabricating such a sphere and ensuring its uniform implosion proved troublesome for the **Manhattan Project** weapon designers. An alternative approach was proposed by Robert Frederick Christy (1916–2012). He proposed using a solid sphere rather than a hollow one, thus eliminating some of the problems the designers anticipated regarding uniformity of implosion. This proposal was accepted, despite its known drawbacks of requiring a special initiator, consuming more plutonium, and being less efficient. The resultant sphere became known as the Christy gadget or, more commonly, the Christy pit. The **detonation** of **Gadget** on 16 July 1945 proved the design was a success. *See also* EXPLODING BRIDGEWIRE DETONATOR (EBW).

CHROME DOME. An airborne alert program in which US Air Force B-52 bombers carrying **nuclear weapons** were flown to and along the borders of the Soviet Union. When the program was begun in 1960, it ensured 12 bombers carrying nuclear weapons were always aloft, providing the United States a nuclear deterrent meant to counter a Soviet attack with nuclear weapons. The number of bombers was reduced to four in 1966, and the program was terminated in 1968 following the **Greenland incident** in which one of the B-52 bombers caught fire and crashed, causing its nuclear payload to be scattered in the sea ice in North Star Bay, Greenland.

COCOA POWDER. *See* BROWN POWDER.

COLEMAN, VINCENT P. *See* HALIFAX EXPLOSION.

COLLODIN. *See* SCHULTZE POWDER.

COLLODION. A solution of **pyroxylin** in an ether-alcohol mix. Collodion was discovered in 1846 by Louis-Nicolas Ménard (1822–1901), a French writer and chemist. When Ménard learned of the work of **Henri Braconnot** and **Théophile-Jules Pelouze** on the newly discovered **xyloidine**, he conducted his own experiments on this material. He dissolved xyloidine in a mixture of ether and alcohol and then allowed the solvent to evaporate. In the process, he was left with a sticky but transparent and highly flammable film. He named this substance "collodion," from the Greek κολλα (glue). Eventually, collodion came into wide use, especially in wet-plate photography. It continues in use as a liquid-skin bandage, and a specification for this application is given in *The United States Pharmacopeia*: 40 grams of pyroxylin, 750 milliliters (mL) of ether, and 250 mL of alcohol. Collodion was used extensively in **smokeless powders**, such as **Cordite RDB**.

COMMITTEE'S MODIFICATION OF BALLISTITE. *See* CORDITE.

COMPOSITION A. Explosive comprised of **RDX** and a **phlegmatizing agent**, such as a natural or synthetic wax. Several formulations of Composition A were developed and used. Compositions A, A2, A3, and A3 type II were similar in that each contained 91 percent RDX and 9 percent wax but differed with regard to the wax that was used. Composition A4 contained 97 percent RDX and 3 percent wax; Composition A5 type I, 98.5 percent RDX and 1.5 percent stearic acid; Composition A5 type II, 98 percent RDX, 1.6 percent stearic acid, and 0.4 percent lubricant; and Composition A6, 86 percent RDX and 14 percent wax. The Composition A group of explosives was originally developed by the British and came into use during the Second World War.

COMPOSITION B. Binary explosive comprised of **RDX** and **TNT**. In some cases, D2 composition wax (a mix of 84 percent polyethylene, 14 percent **nitrocellulose**, and 2 percent lecithin) is present as a **phlegmatizing agent**. Several formulations of Composition B explosives have been developed and used, including Composition B, Composition B2, Composition B3, and the **cyclotols**. However, the term "cyclotol" is more commonly used to refer to those formulations containing at least 70 percent RDX, while Composition B is used for those formulations containing nominally 60 percent RDX and 40 percent TNT.

Composition B was used by the United States as the **high velocity explosive** in the **explosive lenses** used in some of its implosion-type **nuclear explosives**, such as **Gadget** and **Fat Man**. The Soviet Union and India are believed to have used some form of Composition B as well as **baratol** for their first **implosion-type nuclear explosive**, which the United States dubbed **Joe-1** and **Smiling Buddha**, respectively. Composition B became a standard explosive for use in munitions early in the Second World War and saw extensive use by the military of the United States and many of the NATO nations. It remained in use until the early 1990s, when it was replaced by less sensitive explosives.

COMPOSITION C. Composition explosive comprised of **RDX** and either an inert **plasticizer** or an explosive plasticizer. Various formulations and classes of Composition C have been developed. The British developed the original formulation of Composition C during the Second World War as part of a search for a **plastic explosive** that could be shaped by hand and had high **brisance**. The original formulation contained 88.3 percent RDX and 11.7 percent inert, oily plasticizer. Composition C had a useful range of 0°C to 40°C, turning brittle at temperatures below freezing and becoming gummy and exuding oil at temperatures above 40°C.

Composition C2 replaced Composition C and contained 80 percent RDX and 20 percent explosive plasticizer. The explosive plasticizer was itself a mixture of mononitrotoluene, various dinitrotoluenes, **TNT**, **nitrocellulose**, and dimethylformamide. This composition had a useful range of -30°C to 52°C, superior to that of Composition C. But Composition C2 lost plasticity if stored at high temperatures owing to the evaporation of some of the chemicals used in the explosive plasticizer. This deficiency prompted the development of Composition C3, containing approximately 77 percent RDX and 23 percent explosive plasticizer. In this case, the plasticizer was similar to that used for Composition C2, but the dimethylformamide was replaced by **tetryl**. In addition, two classes of Composition C3 were identified (class A and class B), with the classes being identical in chemical composition but with class B having a lower residual acidity.

While Composition C3 was an improvement over Composition C2, it had a few undesirable characteristics. It hardened at temperatures below -29°C, had an undesirable volatility, and picked up water from the environment. As a result, Composition C4 was developed, and three classes of Composition C4 were fielded. Classes I and II contained 92 percent RDX and 9 percent polyisobutylene, an inert plasticizer. However, these two classes differed in the properties of the RDX used. The third class, Composition C4 Class IV, contains about 90 percent RDX, 10 percent polyisobutylene, and a dye. All three classes of Composition C4 have a superior useful range to that of the three earlier Composition C formulations, remaining plastic over the range -57°C to +77°C. Additionally, Composition C4 is essentially non-hydroscopic, making it unlikely to pick up water from the environment. The superior attributes of Composition C4 have made it a favored choice for use as a demolition explosive, and use of Composition C4 eventually supplanted use of Compositions C, C2, and C3.

COMPOSITION EXPLODING (CE). Synonym for **tetryl**, used especially by the British military.

COMPOSITION EXPLOSIVE. Explosive in which two or more explosive compounds are mixed to produce a material having more desirable characteristics for a particular use than is the case for the individual explosive compounds. Composition explosives can be categorized according to the number of explosive compounds present in the final product. For example, **binary explosive** refers to a composition explosive having two explosive compounds; **ternary explosive**, three; and **quaternary explosive**, four. A **plastic-bonded explosive** (PBX) is also a composition explosive, although the number of explosive compounds used in preparing a PBX varies.

COMPRESSED AIR GUN. *See* PNEUMATIC GUN.

CONCENTRATED HYDROGEN PEROXIDE (CHP). Hydrogen peroxide that has been concentrated to 40 percent or greater. Hydrogen peroxide concentrated above 68 percent is often referred to as **high-test peroxide (HTP)** to distinguish it from concentrated hydrogen peroxide. CHP is available commercially and finds use in particular for bleaching pulp and paper. Aqueous solutions of hydrogen peroxide for household use typically have

3–6 percent hydrogen peroxide and might be used for bleaching hair and as an antiseptic. Such solutions, however, can be converted to CHP by distilling off the water, as has been done for their use in **homemade explosives**, where the CHP serves as an **oxidizer** while a common kitchen material, such as coffee, flour, spices, or tea, is used as the **fuel**. CHP made in this manner was implicated in the **7/7 Terrorist Bombings** on the London transit system.

CONFEDERATE POWDER WORKS. *See* AMERICAN CIVIL WAR (1861–1865).

CONFEDERATE STATES OF AMERICA. *See* AMERICAN CIVIL WAR (1861–1865); NITER AND MINING BUREAU.

CONGREVE, WILLIAM. Father and son who made notable contributions to the field of **explosives**. The elder William Congreve (1742–1814) was a career officer in the British Army, specializing in artillery and eventually reaching the rank of Lieutenant General. He devoted much of his career to postings at the Woolwich Royal Laboratory, overseeing assorted improvements in the manufacture of **gunpowder**. In recognition of his contributions, he was created a hereditary peer on 7 December 1812, becoming the first baronet of Walton.

On his death on 30 April 1814, his son succeeded to the title as the second baronet. By this time, the younger William Congreve (1772–1828) had distinguished himself as an inventor, including a rocket that became known as the Congreve rocket. This invention was inspired by rockets he had witnessed in use by the Kingdom of Mysore. Using his own funds and working at Woolwich, Congreve developed rockets having gunpowder as the **propellant** with a sheet iron projectile, launched from a wooden stick. Although the Congreve rockets could not compete with smoothbore guns in range or accuracy, their simplicity enabled a greater sustained rate of fire. By 1813, he had developed three classes of rockets—light, having a 6- to 18-pound warhead; medium, having a 24- to 42-pound warhead; and heavy, with a 100- to 300-pound warhead. Congreve rockets saw extensive use in the nineteenth century and were responsible for the "Rockets' red glare" seen by Francis Scott Key during the bombardment of Fort McHenry outside Baltimore, Maryland, on 14 September 1814 and immortalized in the "Star-Spangled Banner." Variants on this rocket for which Congreve received patents included a means of launching a line to the crew of a ship in distress and an explosive harpoon for killing whales.

CONVENTION ON THE MARKING OF PLASTIC EXPLOSIVES FOR THE PURPOSE OF DETECTION. International agreement to control and limit the use of undetectable **plastic explosives**. This convention was opened for signature in Montreal, Canada, on 1 March 1991 and entered into force on 21 June 1998. It is atypical of measures adopted to combat global terrorism in that it identifies no specific offenses to be criminalized and is silent on obligations to prosecute or extradite a suspected offender. Presently, more than 100 nations are party to this convention, including the United States and the UK.

The convention obligates each party to the convention to:

- Take the necessary and effective measures to prohibit and prevent the manufacture of unmarked plastic explosives.
- Exercise strict and effective control over possession and transfer of unmarked explosives made or imported prior to entry into force (EIF) of the convention.
- Prevent movement of unmarked explosives into and out of its territory.
- Ensure all stocks of unmarked explosives not held by the military or law enforcement are destroyed, consumed, marked, or rendered ineffective within three years of EIF for that State.
- Take necessary measures to ensure all stocks of unmarked explosives held by the military or law enforcement and that are not incorporated as an integral part of duly authorized military devices are destroyed, consumed, marked, or rendered permanently ineffective within 15 years of EIF for that State.
- Ensure the destruction as soon as possible of any unmarked explosives manufactured after EIF for that State.

The convention includes a Technical Annex that both defines "plastic explosive" and designates **detection agents** whose use will satisfy requirements for marking an explosive. According to the Technical Annex, a plastic explosive is any substance that (a) is formulated with one or more high explosives which, in their pure form, have a vapor pressure less than 0.0001 pascals (Pa) at 25°C; (b) is formulated with a **binder**; and (c) is, as a mixture, malleable or flexible at normal room temperature. The destruction of **Pan Am Flight 103** on 21 December 1988 directly led to the development and adoption of this convention. The event prompted the United Nations to adopt UN Security Council Resolution 635 on 14 June 1989, followed by UN General Assembly Resolution 44/29 on 4 December 1989. Both resolutions urged the International Civil Aviation Organization (ICAO) to elaborate a regime for marking plastic explosives. The ICAO responded promptly, culminating in the convening of the International Conference on Air Law from 12 February to 1 March 1991. At the end of that conference, the Convention on the Marking of Plastic Explosives for the Purpose of Detection was opened for signature at the Montreal headquarters of the ICAO. The ICAO also agreed to serve as the repository for all instruments of ratification, acceptance, approval, or accession to the convention.

CONVERTED SALTPETER. Potassium nitrate obtained by the chemical reaction of **sodium nitrate** and potassium chloride (chemical formula KCl). Sodium nitrate is hygroscopic, meaning it will absorb water, such as the moisture in the air. This affinity for water makes sodium nitrate unsuitable for direct use in formulating **black powder** because the resultant mixture would absorb water and be unreliable for use in **explosives**. Potassium nitrate, however, is not hydroscopic.

The chemical reaction between sodium nitrate and potassium chloride involves dissolving them in water heated to 100°C. On mixing, these two substances produce potassium nitrate and sodium chloride (NaCl). At this temperature, sodium chloride is insoluble in water. As a consequence, the sodium chloride precipitates, and the reaction continues producing potassium nitrate until all the sodium nitrate and potassium chloride have been consumed. The process of producing "converted saltpeter" became important in the nineteenth century as an alternative to importing potassium nitrate from sources that were largely controlled by the British. At that time, plentiful and relatively cheap supplies of sodium nitrate had been found, especially in South America and on various Pacific Ocean islands. In 1859, **Lammot du Pont** invented a process permitting direct use of sodium nitrate in manufacturing black powder, eliminating the need for a separate chemical reaction to convert sodium nitrate into potassium nitrate. *See also* CHILEAN NITRATES; INDIA SALTPETER.

COPPER FULMINATE. Green crystalline salt of fulminic acid, having the chemical formula $C_2H_2CuN_2O_2$. In August 1867, Jabez Biggs Dowse (1829–1878) received US patent 68,055, covering the use of copper fulminate as a **detonator**. In his patent, he referred to this substance as "copper amalgam." At the time, Dowse, a native of County Wiltshire, England, was working as a grocer and miller in Lockport, Illinois. The source of his interest and expertise in chemistry is unknown; however, he grew up and lived in a mining district of England prior to immigrating to the United States and might have learned of detonators, especially ones using **mercury fulminate**, while there. The use of copper fulminate detonators soon became popular in the United States, especially during the construction of the **Hoosac tunnel**, when they were referred to as "Dowse fuses." Copper fulminate is less impact sensitive than mercury fulminate but is more **friction** sensitive. It is stable when dry but is hygroscopic and will decompose when moist. Experience with the Hoosac tunnel suggested that mercury fulminate detonators were superior, with **George Mowbray** criticizing copper fulminate detonators as "oversensitive." Consequently, the use of copper fulminate detonators had virtually disappeared by the twentieth century.

CORDITE. Any of various formulations of **smokeless powder**. The initial formulation adopted for military use was composed of 58 percent **nitroglycerin**, 37 percent **nitrocellulose**, and 5 percent mineral jelly, with this last ingredient serving to lubricate the gun barrel following **detonation**. Acetone was used as a solvent to soften the nitroglycerin and nitrocellulose, allowing them to be kneaded together in a paste. This paste was passed through a die, forming strings that could be produced to a desired length and, depending on the size of the hole in the die, to a desired diameter. Narrow strings were relatively fast-burning and useful as **propellants** for cartridges fired in small arms. Thicker strings burned more slowly and proved useful as propellants for longer barrels, such as artillery and naval guns.

The original name selected for these strings was "the Committee's Modification of **Ballistite**," with the committee being a special group set up to find a smokeless powder suitable for use by the British military. But the resemblance of the strings to cords caused them to get the name "cord powder." The strings soon came to be called "Cordite," a name that stuck for both the original formulation and future formulations. By the end of the nineteenth century, Cordite had become the standard British smokeless powder.

The invention of Cordite is credited to the British scientists Sir **Frederick Abel** and Sir **James Dewar**. It was patented in Britain and the United States in 1889. The British patent became a matter of dispute, as both **Hiram Stevens Maxim** and **Alfred Nobel** sued the British government for patent infringement. In 1895, the British House of Lords decided the dispute in favor of Abel and Dewar. The courts ruled that Maxim's patent for a similar smokeless powder had lower percentages of nitroglycerin and used castor oil as a lubricant rather than mineral jelly. They also ruled that Nobel's patent had referred to the use of "soluble nitrocellulose," thus failing to anticipate the use of **insoluble nitrocellulose**, as was the case for the **guncotton** in Cordite.

The original formulation for Cordite was superseded by other formulations. Consequently, the original formulation was designated "Cordite Mark 1" (Cordite Mk 1) to differentiate it from the others. These other formulations include:

- Cordite MD, also known as "modified Cordite" (hence, "MD"), comprised of 30 percent nitroglycerin, 65 percent nitrocellulose, and 5 percent mineral jelly. Cordite Mk 1 was modified in an effort to reduce corrosion of gun barrels. Cordite MD became the standard British smokeless powder early in the twentieth century.
- Cordite RDB, comprised of 42 percent nitroglycerin, 52 percent **collodion**, and 6 percent mineral jelly, with an ether-alcohol mixture used as the solvent to aid in softening the ingredients so that they might be kneaded together. This formulation was developed by Britain during the First World War when shortages of acetone forced a search for alternative solvents to aid in mixing. The Research Department, Royal Arsenal, Woolwich, investigated various formulations and eventually settled on formula "B." This formulation became known as Cordite RDB (Research Department formula B) and was adopted in May 1915 as a substitute for Cordite MD for use in land guns. Cordite MD, which required the use of relatively scarce acetone as a solvent and had superior stability and uniformity of effect, was reserved for use in naval guns. The use of an ether-alcohol mixture as a solvent had several unintended consequences. First, to maintain ballistic performance, the nitroglycerin content was raised from 30 percent in Cordite MD to 42 percent, thereby increasing demand for glycerin. Second, the use of the ether-alcohol mixture as a solvent led to price increases for ordinary alcohol (such as those present in spirits and beer) and for the grains from which the alcohol was distilled. These price increases posed hardships for ordinary citizens, as well as putting a strain on the wartime budget of the British government. Third, in order to conserve supplies of all types of Cordite, the government converted to **single-base propellants** for use in rifle cartridges. Eventually, an improved process for producing acetone was developed, allowing the use of Cordite RDB to be discontinued.
- Cordite SC, also known as "**solventless** Cordite" (hence, "SC"), comprised 50 percent nitrocellulose, 41 percent nitroglycerin, and 9 percent diethyl diphenylurea.
- Cordite W, introduced about 1935 and comprised 65 percent nitrocellulose, 29 percent nitroglycerin, 3 percent mineral jelly, and 2 percent carbamite. This formulation was developed at the Royal Arsenal, Waltham Abbey (hence, "W").

CORNING. The process of making **black powder** into grains. Raw black powder, made by simply mixing certain proportions of **charcoal**, **sulfur**, and **potassium nitrate**, is most often a soft and loose powder. This product might be satisfactory for some applications, but more finishing is required to make a sturdier material. This finishing can be accomplished by mixing the raw black powder with a suitable liquid (alcohol, urine, or water was used for this purpose). The product of this wet mixing is a pasty granular material rather than a fine powder and is referred to as "corned" owing to its grainy appearance. The process is ancient, although not as old as black powder itself. Corning is described in *Fire Book*, written by Conrad von Schongau and published in 1429, and it may have been known well before then. *See also* CONGREVE, WILLIAM.

COSTON FLARE. A **pyrotechnic** device used as a signaling flare. US patent 23,536 was granted in 1859 for a system of pyrotechnic night signals. It was issued to Benjamin Franklin Coston (1821–1848), who is its eponym. However, successful development and large-scale manufacturing of the flare should be credited to his widow, Martha Jane Coston (née Hunt) (1826–1904), who completed the design, working from his notes. The flares were designed to be handheld, producing a red, green, or blue light sufficiently bright to be seen 15–20 miles away and facilitating ship-to-ship and ship-to-shore signaling. Coston flares were used extensively by the US Navy during the **Civil War**, demonstrating their utility. In 1871, Martha Coston obtained US patent 115,953 in her own right for improvements in the design. These flares came into routine use by the US Life-Saving Service (predecessor to the US Coast Guard) from its founding in 1878 in communicating with ships in distress or to warn ships of danger. Coston-type flares continue to be used internationally. In 2006, she was inducted into the National Inventors Hall of Fame, Alexandria, Virginia, an honor for which her husband has yet to receive. *See also* VERY LIGHT.

COTTON POWDER COMPANY, LTD. Early commercial producer of **guncotton** and other **nitrocellulose** products. The Cotton Powder Company, Ltd., was established in 1873 on the Uplees Marshes near Faversham, England. It came to dominate the British market for explosives used for civil blasting, such as mining and the construction of dams, roads, and tunnels. The major products were **Tonite** and **Potentite**, with the latter being produced at a plant in Melling, near Liverpool. Eventually, the company came to be controlled by the **Nobel Dynamite Trust Company**.

CRITICAL MASS. The least mass of **fissionable** material that will allow a self-sustaining nuclear chain reaction. Any **nuclear weapon** must have a critical mass for the nuclear fuel to act as an **explosive**. The critical mass of a fissionable material depends upon various factors, including density, enrichment, shape, temperature, and its nuclear properties.

CUBICAL SALTPETER. *See* CALICHE.

CYCLOTOL. Designation for certain formulations of **Composition B**. At least three formulations were developed and used, all containing **RDX** and **TNT**: cyclotol type I (75 percent RDX and 25 percent TNT); cyclotol type IIA (70 percent RDX and 30 percent TNT); and cyclotol type IIB (nearly the same as cyclotol type IIB but with the addition of 1 percent wax and 0.5 percent calcium silicate).

CYCLOTRIMETHYLENETRINITRAMINE. Chemical name commonly used for **RDX**.

CYCLONITE. Synonym for **RDX**. American chemist Clarence J. Bain (1892–1974) gave RDX the name "cyclonite" based on the cyclic structure of this substance.

D

DADP. Abbreviation for **diacetone diperoxide**.

DEFLAGRATION. A type of **explosion** involving rapid, self-propagating, and subsonic burning that propagates through thermal conductivity (i.e., the burning material heats the next layer, causing it to ignite). Deflagration differs from **detonation**, which involves supersonic burning and propagates through **shock**. In the general context, most fires are deflagration, including the burning of wood in a fireplace or that of a fuel-air mixture in an internal combustion engine. In the context of **explosives**, deflagration is the effect seen in **low explosives**, such as **gunpowder**, and in **pyrotechnics**, such as fireworks. However, some low explosives can detonate when confined and exposed to **friction** or impact. Deflagration is more easily controlled than detonation and, consequently, is more desired for applications such as **propelling** a bullet from a firearm or causing a firework to sparkle.

DEPTH BOMB EXPLOSIVE (DBX). Quaternary explosive comprised of 40 percent **TNT**, 21 percent **RDX**, 21 percent **ammonium nitrate**, and 18 percent **aluminum**. DBX was developed to be less sensitive to **shock** and **friction** than is **torpex**, which it replaced in many applications.

DETASHEET. Trade name for various formulations of a **sheet explosive**. One example of a formulation is Detasheet C, comprised of 63 percent **PETN**, 29 percent **plasticizer**, and 8 percent **nitrocellulose**. The trade name originally belonged to **DuPont**, which first used it for a sheet explosive developed around 1950. DuPont exited the explosives business in the 1990s and sold the trade name.

DETECTION AGENT. Substance added to an **explosive** to render it detectable. The concept of a detection agent is fundamental to the **Convention on the Marking of Plastic Explosives for the Purpose of Detection**. The Technical Annex to this convention calls out four substances that might be incorporated into an explosive to enhance detectability by vapor means.

These four are 2,3-dimethyl-2,3-dinitrobutane (DMDB); **ethylene glycol dinitrate**; *o*-mononitrotoluene (o-MNT); and *p*-mononitrotoluene (p-MNT). The annex requires that the detection agent be added in a manner such that it is distributed homogeneously in the final product. It also specifies the minimum concentration of each detection agent that shall be present in the final product. Another term often used for a detection agent is "taggant."

DETONATION. A type of explosion involving rapid, self-propagating, and supersonic burning that propagates through **shock**. Detonation differs from **deflagration**, which involves subsonic burning and propagates through thermal conductivity. Both **low explosives** and **high explosives** can detonate, although detonation is more common with high explosives.

DETONATOR. Device containing a **detonating** charge that is used to initiate an **explosion** reliably, at a specified time, and in a prescribed sequence. The three types of detonators are non-electric, electric, and electronic. An example of a non-electric detonator is a detonating cord, such as a plastic tube filled with **PETN**. An electric detonator uses electrical currents to initiate detonation. For example, in 1745, Sir William Watson used a **friction** machine to generate an electric spark that detonated **black powder**. Electric detonators were in use by 1839 for underwater demolitions, as illustrated by **Charles Pasley**'s use of a battery and platinum wire to ignite **gunpowder**. By the late 1860s, electric detonators had become more sophisticated and more portable, with **Henry Julius Smith**'s design of a plunger detonator being widely adopted. An electronic detonator uses electronic components to send a firing signal. Unlike an electric detonator, an electronic detonator has no pyrotechnic delay element, such as a safety fuse. *See also* EXPLODING BRIDGEWIRE DETONATOR (EBW).

DETONATOR LOCK. *See* FORSYTH, ALEXANDER JOHN (1769–1843).

DEWAR, JAMES (1842–1923). Scottish chemist, physicist, and inventor. On graduating from the University of Edinburgh, Dewar stayed on as an assistant in chemistry. In 1875, he was appointed to the faculty of the University of Cambridge as a professor of natural philosophy. Two years later, he moved to the Royal Institution, London, where he was appointed Fullerian professor of chemistry and where he spent the rest of his professional career. Dewar had a wide range of research interests in chemistry and physics. He served the British government in various capacities, including on the Committee on **Explosives** (1888–1891) and is credited, along with Sir **Frederick Abel**, with inventing **Cordite**. Dewar received many awards and honors for his contributions to science and was knighted in 1904. Dewar's research into the liquefaction of gases and the properties of substances at extremely low temperatures and pressures inspired his development of what came to be known as the Dewar flask. This vessel actually consists of two flasks, one inside the other, but separated by a vacuum. The vacuum minimizes the transfer of heat, as do the materials of construction. The Dewar flask was later adapted by German glassblowers and sold for commercial and home use as the thermos bottle.

DIACETONE DIPEROXIDE (DADP). An **acetone peroxide explosive**. Diacetone diperoxide (DADP) is a white crystalline powder. The preferred, systematic IUPAC name for DADP is 3,3,6,6-tetramethyl-1,2,4,5-tetraoxane, and its chemical formula is $C_6H_{12}O_4$. DADP is highly sensitive to **friction**, heat, **shock**, and vibration. It and the similar substance, **triacetone triperoxide** (TATP), are of concern owing to their use or feared illicit use in **homemade explosives** because of the relative ease with which they might be produced from commercially available chemicals.

DICKINSON, CHARLES S. (1809–1874). American inventor credited with designing a **steam gun** that was built at the time of the **American Civil War**. The design used steam to rotate spherical shot within a drum that was some 4 or 5 feet in diameter. The concept was that rotation would impart centrifugal force to the shot, which would then exit the drum when an aperture was opened. This concept was similar to one patented earlier by New York inventor Robert McCarty. On 31 December 1838, McCarty was issued US patent 1,046 for "a machine for throwing balls, shot, &c." The McCarty gun was hand-operated, requiring two men to turn the cranks. Its operation imparted scant centrifugal force. When demonstrated in New York City on 20 June 1861, the shot bounced harmlessly off the nearby target.

By contrast, Dickinson's design used steam in place of manual labor, potentially creating much greater forces. Dickinson may have developed the concept for a steam gun in partnership with William Joslin in Ohio. After these two had a falling out, Dickinson continued to pursue the concept and succeeded in building at least one. Most likely the gun was built in Boston to Dickinson's design, although Maryland industrialist, inventor, and Southern sympathizer Ross Winans (1796–1877) claimed it was built near Baltimore at his iron works. The completed gun, which was mounted on an armored carriage, was in Baltimore, Maryland, at the outbreak of the American Civil War. Forces sympathetic to the Southern cause attempted to transport the gun to Harper's Ferry, Virginia (now West Virginia), for use by the Confederate military. But a force of Massachusetts volunteers under the command of

Colonel Edward F. Jones seized the gun, and it remained in Union hands for the duration of the war.

Dickinson's steam gun attracted considerable notice early in the war, and it came to be known by various names, including "Baltimore Steam Battery," "Centrifugal Steam Gun," "Dickinson's Patent Centrifugal Gun," and "Winans Steam Gun." Writing about the gun in the 25 May 1861 issue of *Harper's Weekly*, Dickinson observed that it "would mow down opposing troops as the scythe mows standing grain." He predicted that "its very destructiveness will prove the means and medium of peace." His prediction appears to have been faulty, and the merits of his design for a steam gun remain a matter of controversy.

DINITROGEN TETRAOXIDE. *See* NITROGEN TETRAOXIDE.

DINITROGLYCERIN (DNG). Colorless or light-yellow oily liquid whose properties and behavior are similar to those of **nitroglycerin**. Two isomers of dinitroglycerin (DNG) exist: (a) 1,2-*bis*-nitrooxy-propan-1-ol, which is commonly called 1,2-dinitroglycerin (1,2-DNG), and (b) 1,3-*bis*-nitrooxy-propan-2-ol, which is commonly called 1,3-dinitroglycerin (1,3-DNG). The isomers have the empirical formula $C_3H_6N_2O_7$. Technical-grade DNG is typically a mix of these isomers. A common term used for DNG is glycerol dinitrate. DNG was important in the development of **low-freezing dynamite**. Specifically, DNG was used successfully to depress the freezing point of nitroglycerin. DNG is very soluble in water, unlike nitroglycerin, which is only sparingly soluble in water. This solubility in water spurred a search for DNG in the manufacture and use of **explosives**. *See also* MAXIM, HUDSON (1853–1927).

DINITROPHENOL (DNP). Yellow crystalline solid used in **explosives**. Three isomeric forms of dinitrophenol (DNP) exist in the commercial grade product: 2,4-dinitrophenol; 3,4-dinitrophenol; and 3,5-dinitrophenol. The isomers have the empirical formula $C_6H_4N_2O_5$. In contrast to most **picric acid**-based explosives, DNP is both less shock-sensitive and less powerful. It is also highly toxic, adding to the hazard of working with DNP in loading or handling munitions. The British used DNP in **shellite** to overcome the tendency of munitions containing pure **Lyddite** to **detonate** on contact with armor rather than to penetrate the armor.

DIPHENYLAMINE. Colorless to gray solid, used extensively as a **stabilizer** for **explosives**, especially in **smokeless powders**. The chemical formula for diphenylamine is $C_{12}H_{11}N$, and the preferred, systematic IUPAC name for it is N-phenylaniline. Diphenylamine and other polyamines are especially effective as stabilizers for **nitrocellulose**-based explosives. These amines readily take up nitrous acid (HNO_2), which is a major degradation product associated with nitrocellulose and similarly nitrated substances. By taking up the nitrous acid, the polyamines are themselves nitrated. Additionally, they reduce the accumulation of acidic products that speed up the further degradation of nitrocellulose and other smokeless powders. This use was discovered by **Alfred Nobel**, who received German patent 51,471 in 1889 for his invention.

DIRTY BOMB. *See* RADIOLOGICAL DISPERSAL DEVICE (RDD).

DIVINE, SILAS R. *See* RACK-A-ROCK.

DOPE. Term for inert material added to **nitroglycerin** as an absorbent. Kieselguhr is a prominent example of such a material, seeing use in **dynamite**. Other examples of inert materials used or proposed for use include ground glass and mica.

DOUBLE-BASE POWDER. Smokeless powder comprised of some form of **nitrocellulose** plus as much as 50 percent **nitroglycerin**. Other substances might be present, such as **binders**, **stabilizers**, and residual solvents from the manufacturing process. An example of a double-base powder is **Ballistite**. *See also* SINGLE-BASE POWDER; TRIPLE-BASE POWDER.

DOWSE, JABEZ BIGGS (1829–1878). *See* COPPER FULMINATE.

DREYSE NEEDLE-GUN. Prussian .607 caliber, single-shot bolt-loading rifle credited as the first widely used military firearm to fire a cartridge containing a type of **smokeless powder**. The rifle was designed by Johann Nikolaus von Dreyse (1787–1841) in 1836 and was adopted by the Prussian army shortly thereafter. Known in German as the *Zündnadelgewehr* (*firing-pin rifle*), it fired an acorn-shaped lead bullet encased in a paper cartridge containing a **propellant**. The innovation of the Dreyse needle-gun was that it featured a long needle on its bolt. On pulling the trigger, the needle was released, driven through the paper cartridge, and into a **primer** attached to the base of the cartridge. This contact **detonated** the primer, which in turn detonated the propellant. The original propellant used in the cartridge was ordinary **black powder**. However, during the late 1860s, the Prussian army replaced black powder as the propellant in some or all of these cartridges with a type of **Schultze powder**. Unlike the detonation of black powder, the detonation of this Schultze powder produced no smoke and no residue. The military usefulness of this smokeless feature was demonstrated during the **Franco-Prussian War**. During that conflict, Prussian forces firing the Dreyse needle-gun failed to give away their positions and kept their lines of sight unclouded, in contrast to the experience of the French forces, which were using conventional black powder in their rifle cartridges. *See also* LEBEL MODEL 1886 RIFLE.

DROIT DE FOUILLE. French term for "the right of excavation." The term *droit de fouille* applied specifically to the right of one person to enter another person's property for the purpose of harvesting the **nitrate**-rich efflorescences from places such as the surfaces of caves as well as the spaces beneath latrines, chicken coops, and dovecotes. Nitrates, other salts, and assorted other substances would rise to the surfaces on the excrement accumulating in such places. Persons exercising this right of excavation were called "**saltpeter** farmers." The harvested efflorescences were used to support the production of **black powder** as part of the Gunpowder Farms that operated in seventeenth- and eighteenth-century France. The process of exercising the right of excavation was intrusive and led to abuses by the saltpeter farmers, who would solicit bribes to exempt a property from a harvest or would demand accommodations while on the premises during a harvest.

Additionally, the assorted non-nitrate impurities in such harvested materials had to be removed. The extent to which saltpeter farmers removed these impurities varied, resulting in a final product of varied quality. When this material was used to produce black powder, the resultant mixture had unpredictable performance as an **explosive** and was often inferior to black powder made with imported potassium nitrate. The combination of poor quality of final product and public resentment over exercising *droit de fouille* helped spur the creation in France of the **Gunpowder and Saltpeter Administration** in 1775. *See also* NITER AND MINING BUREAU.

DUALIN; DUALINE. Explosive consisting of **nitroglycerin**, **potassium nitrate** or **sodium nitrate**, and nitrated **sawdust** or wood pulp, both of which are forms of **nitrocellulose**. This explosive is important in that it became the subject of litigation concerning the basic patents granted to **Alfred Nobel** for **dynamite**. British patent 2,542 covering the basic formulation of dualin was granted on 14 August 1868 to **Eduard Schultz**, although Carl F. W. E. Dittmar (1837–1883) would later claim credit for the idea for this formulation. Dittmar's claim may have been accurate, as he and Schultz had been in business together around 1865 and again in 1867. Dittmar was granted US patents in 1870 for formulations nearly identical to those of dualin. However, these patents were invalidated in 1880 when the US courts ruled that Nobel's basic patent had priority.

DUNN, BEVERLY WYLY (1860–1936). American ordnance officer and government executive responsible for the development of **Explosive D**. While serving as laboratory director, Frankford Arsenal, Philadelphia, Dunn proposed the use of **ammonium picrate**, the ammonium salt of **picric acid**, as an **explosive** filling for armor-piercing shells. At the same time, he and his staff perfected a **fuze** that was sufficiently insensitive that it would withstand the **shock** of impact on armor but sufficiently powerful to **detonate** the otherwise insensitive ammonium picrate. In 1901, this combination of fuze and explosive was demonstrated successfully, leading to its adoption by the military. From 1907 on, he devoted most of his career to the work of the **Bureau of Explosives** of the American Railway Association as its chief inspector. In 1948, Dunn was honored along with John Pitman (1842–1933), a long-serving US Army officer, by the redesignation of the Laidley Laboratories, Frankford Arsenal as the Pitman-Dunn Laboratories (later Pitman-Dunn Research Institute).

DUNNITE. Common or unofficial name for **Explosive D**, after **Beverly Wyly Dunn,** who proposed its use.

DUPONT (E. I. du PONT de NEMOURS & COMPANY). International chemical firm; formerly, a major producer of **explosives**, especially **black powder**, **smokeless powder**, and **dynamite**. In 1801, Eleuthère Irénée du Pont (1771–1834) met with Marquis de Lafayette, Thomas Jefferson, and others. At their meeting, Jefferson urged E. I. du Pont to establish a gunpowder mill in America. Jefferson knew that du Pont had learned the art of making gunpowder at Laboratoire Centrale des Poudres et Salpêtres, Paris, where he was tutored by the famed French chemist **Antoine Lavoisier**. Du Pont had visited the Lane-Decatur powder works outside Philadelphia in November 1800 and knew that American methods for making powder were inferior to those used in France.

In following up on his conversation with Jefferson, E. I. du Pont, members of his family, and others founded a partnership on 19 July 1802. Over time, it would become known as the E. I. du Pont de Nemours & Company, now commonly referred to as DuPont. The firm sited its first manufacturing facilities in the Brandywine Valley of Delaware, along the Brandywine River. The black powder produced here was branded "Brandywine Powder" and became prized for its quality. Sales were brisk, and by 1805, DuPont was selling products to Spain, its first export market. By 1808, a competitor in Connecticut was using this same name for its product, causing confusion in the marketplace. This situation prompted DuPont to retitle its product as "Du Pont Powder," using a capital "D." This spelling became permanent for the company, but most family members retained "du Pont" as the spelling for their surname.

Over the nineteenth century, DuPont became the dominant manufacturer of black powder, smokeless powder, and dynamite. During the **American Civil War**, DuPont manufactured four million pounds of explosives, accounting for 40 percent of all supplies purchased by the Union. Shortly after the war, DuPont entered into an arrangement with other manufacturers to control prices, agree on sales territories, and otherwise manage trade. Called the **Gunpowder Trade Association**, the arrangement proved enormously successful but set the stage for an antitrust suit brought by the United States in the early twentieth century. This suit against what was popularly known as the **Powder Trust** forced DuPont to divest much of its black powder and **dynamite** assets in 1912.

Despite the breakup of the Powder Trust, DuPont retained its near monopoly on manufacturing smokeless powder, giving it a strong position when the First World War erupted in 1914. By the end of that war, DuPont had manufactured and sold nearly 1.5 trillion pounds of explosives, accounting for more than 40 percent of the standard explosives used by the Allies during the conflict. The dominance of the explosives market would rebound on DuPont, however, as some in the public, as well as certain politicians, portrayed it as a "merchant of death." The situation was highlighted by the **Nye Committee**, 1934–1936. Gerald Nye, Committee Chairman and a US senator, used this term in describing DuPont and referred to an insidious "DuPontocracy" having undue influence over American foreign affairs.

By the outbreak of the Second World War in 1939, DuPont had broadened its manufacturing beyond explosives. This broadening began with products exploiting chemistry related to chemicals used in explosives. For instance, DuPont's expertise in **nitrocellulose** aided its development of Pyralmin, a **pyroxylin** derivative used in celluloid shirt collars and cuffs. While it remained an important manufacturer and supplier of conventional explosives to the Allies, it also turned its engineering expertise to help the United States with the **Manhattan Project**. Known internally as the "TNX Project," DuPont designed, built, and operated the Hanford Engineer Works, which became the world's first facility for the mass production of plutonium, the fuel used in **nuclear explosives** such as the **Fat Man** bomb that was detonated successfully over Nagasaki, Japan, on 9 August 1945.

DuPont continued to be a manufacturer of explosives and to support the US nuclear weapons program in the decades immediately following the end of the Second World War. However, explosives became progressively less important in terms of corporate revenues, as DuPont shifted its research and manufacturing to other areas. In 1977, DuPont ceased making dynamite. In 1988, it exited the explosives business entirely, ending 186 years as a manufacturer of explosives. *See also* DU PONT, LAMMOT (1831–1884); MAMMOTH POWDER; PYROTOL.

DU PONT, LAMMOT (1831–1884). American industrialist and inventor of **explosives**. Lammot du Pont was the grandson of E. I. du Pont, founder of **DuPont**, the manufacturer of explosives, where Lammot spent most of his professional career. Educated in chemistry at the University of Pennsylvania, Lammot du Pont brought a rigorous scientific approach to the search for new formulations and types of explosives. That approach was exemplified by US patent 17,321, which Lammot du Pont received on 19 May 1857. Titled "Improvement in Gunpowder," the invention permitted the use of **sodium nitrate** in lieu of **potassium nitrate** in formulating black powder. Sodium nitrate was plentiful in South America, where it was called "**Chile Saltpeter**," as the result of centuries of bird droppings that had accumulated in arid regions, creating nitrate-rich deposits. Potassium nitrate, by contrast, was imported from Asia, chiefly the Indian subcontinent, where it was called "**India Saltpeter**." At the time of his invention, trade in India Saltpeter was dominated by the British. The invention eventually shifted international trade in nitrates away from Asia and toward South America.

While the effects of this invention would be delayed until after the **American Civil War**, the effects of another invention were more immediate. In 1858, he and **Thomas Rodman** collaborated on the development of **Mammoth powder**, which burned at a slower rate. The practical result of this slower rate was that a naval gun or field artillery piece could fire a larger projectile, thus allowing the caliber of such guns to be increased. Consequently, the Civil War witnessed the fielding and successful use of larger naval guns and artillery pieces than had been possible previously.

Lammot du Pont was also engaged in an unusual business and diplomatic mission shortly after the start of the Civil War, aimed at cornering the market in India Saltpeter and other commercial sources of potassium nitrate. Shortly after the Civil War ended, he recognized that the explosives industry was changing. In particular, he recognized the importance of the 1867 invention of **dynamite**. He attempted to interest DuPont in manufacturing dynamite, which was already outselling black powder in the civilian market by a margin of three-to-one. At this time, DuPont was run by his uncle, Henry du Pont, who was steadfastly opposed to entering the dynamite business, citing the unsafe nature of manufacturing and handling **nitroglycerin**, a primary ingredient in dynamite.

In 1878, Lammot du Pont resigned from the firm out of frustration with his uncle's management practices. In 1880, he formed Repauno Chemical Company for the express purpose of manufacturing dynamite. Initial investors in this enterprise included DuPont and Laflin & Rand, although he would eventually exchange his interest in DuPont for its shares in Repauno. On 29 March 1884, du Pont was working in the Repauno laboratory in Thompson's Point, New Jersey, when he learned of a runaway reaction in a cooling tank containing 2,000 pounds of nitroglycerin. He raced to the building housing the tank and attempted to dump its contents. He realized, though, it was too late and raced out of the building. He was too late. The contents exploded, killing him and four other workers. Eventually, DuPont would take control of Repauno, merging that business into its own and becoming a major manufacturer of dynamite. *See also* GUNPOWER TRADE ASSOCIATION; POWDER TRUST; WAR OF THE PACIFIC (1879–1884).

DUST EXPLOSION. Deflagration or **detonation** of tiny particles of a combustible material, most often occurring accidentally. In November 2006, the US government's Chemical Safety & Hazard Investigation Board (CSB) released a report documenting results of its investigation of 281 dust explosions in the United States between 1980 and 2005. These explosions killed 119, injured 718, and caused extensive structural damage and loss of property. The investigation was prompted by dust explosions at three separate facilities in 2003 that killed 14, injured 77, and destroyed all three facilities. The CSB report noted that combustible materials having a particle size of 420 micrometers (μm) or smaller can react rapidly when exposed to oxygen and a source of ignition, including static electricity. The reaction often produces an explosion or a series of explosions occurring in waves. Dust that has settled on floors, within suspended ceilings, and atop beams, ductwork, and pipes, is disturbed by the initial explosion, thrown into the air, mixes with oxygen, and is ignited by the first explosion, thereby causing the explosion to propagate within a facility. This CSB report helps to explain the phenomenon of explosions of dust, where the dust is some combustible material such as coal, grain, plastics, and **sawdust**. The handling of materials routinely viewed as benign, such as scrap metal and sugar, can be hazardous, as proven by the 7 February 2008 dust explosion at the Imperial Sugar Company, Port Wentworth, Georgia. This explosion was attributed to sugar and resulted in at least 11 deaths and more than 40 injured.

Similar explosions have occurred at sawmills. For example, on 20 January 2012, the Babine Sawmill exploded in Burns Lake, British Columbia,

Canada, killing two men and injuring 20 others. Three months later, on 23 April, a sawmill at the Lakeland Mills, Ltd., plant in Prince George, British Columbia, exploded, killing two men and injuring 22. In both cases, the explosion was attributed to dust from milling dry, beetle-killed pine trees.

In response to such explosions, the National Fire Protection Association (NFPA) in 2020 issued NFPA 664, "Standard for Prevention of Fire and Dust Explosions in Wood Processing and Woodworking Facilities." This document provides guidance for sawmills and similar facilities where combustible dust might be present as a result of routine operations. Such explosions also point to the prospect of using readily available materials in making **homemade explosives**. *See also* ICING SUGAR.

DWARF DUD. A **nuclear explosive** that **detonates** but fails to provide the estimated or specified yield when used under normal operating parameters. *See also* ABSOLUTE DUD; PREINITIATION.

DYNAMITE. Solid **explosive** prepared by absorbing a liquid explosive onto a suitable material. The primary liquid explosive used in preparing dynamite has been **nitroglycerin**. In a narrow sense, dynamite is a **plastic explosive** because the final product is malleable and capable of being molded. Various types of dynamite have been developed, with some of the most prominent types being:

- Ammonia dynamite: Dynamite in which some of the nitroglycerin has been replaced by **ammonium nitrate**. The development of ammonia dynamite was pioneered by **Russell Sylvanus Penniman**, whose patented process of coating ammonium nitrate proved essential to excluding water from the final product. Ammonia dynamite came on the market in the 1880s. Its use expanded during the First World War owing to the need to find a substitute for nitroglycerin, which was in short supply.
- Ammonia gelatin dynamite: Gelatin dynamite in which some of the nitroglycerin has been replaced by ammonium nitrate.
- **Blasting gelatin**: Nitroglycerin that has been made plastic by mixing with **nitrocellulose**. Blasting gelatin is the base for gelatin dynamite, although it is not typically considered a type of dynamite.
- Extra dynamite: Ammonia dynamite. The name was originally a brand name and may have been selected because the presence of ammonium nitrate was featured as boosting the overall explosive power; that is, this form of dynamite was advertised as having something extra. Eventually, extra dynamite became a synonym for ammonia dynamite, losing its brand name status.

- Gelatin dynamite: Blasting gelatin that has been diluted by the incorporation of **sodium nitrate**, **sulfur**, and some combustible carbonaceous material, such as **sawdust** or **charcoal**. The final product has the consistency of jelly.
- Granular dynamite: Dynamite in which the final product is a solid. The name distinguishes it from gelatin dynamite.
- Guhr dynamite or Kieselguhr dynamite: Dynamite in which the sorbent is Kieselguhr, and the liquid explosive is nitroglycerin. Kieselguhr, also referred to as "diatomaceous earth," is a soft, bulky, nontoxic solid material comprised of the skeletons of tiny prehistoric aquatic plants related to algae. Its highly absorbent nature proved important in the successful development and manufacturing of dynamite.
- **Low-freezing dynamite**: Dynamite to which a suitable liquid has been added to depress the freezing point of nitroglycerin.
- Straight dynamite: Dynamite having nitroglycerin and, possibly, an additional **fuel**, an **oxidizer**, and an alkaline substance as a neutralizer but lacking ammonium nitrate and nitrocellulose. The name arose to distinguish the product from extra dynamite and gelatin dynamite.

Discovery and development. **Alfred Nobel** is credited with inventing dynamite and with coining the term "dynamite." The development of dynamite was spurred by an interest in converting nitroglycerin into a form that would enhance its safety in handling, transport, storage, and use. Nobel had learned of the explosive properties of nitroglycerin in 1848, shortly after its 1847 discovery by **Ascanio Sobrero**. He had learned, too, of the difficulties associated with its production and storage. Several explosions attended his production of nitroglycerin, including one in Sweden in 1864 that killed Emil, his brother. By 1866, Nobel had relocated to Krümmel, Germany, where his newly formed firm **Alfred Nobel & Co.** was producing nitroglycerin for sale on the European continent. This location would prove auspicious. The area was on the river Elbe and was bordered by high sand dunes, which would help shield in the event of an explosion. The sands were largely comprised of diatomaceous earth, known as Kieselguhr. During 1866–1867, Nobel experimented with various additives in an effort to find a means of **stabilizing** nitroglycerin. At some point in 1866 or 1867, he used some of the Kieselguhr-rich sands and found that this material combined with nitroglycerin to form a moldable mass that was relatively insensitive to shock. He found, too, that this mass could be formed into a size and shape suitable for inserting drill holes for blasting. In 1867, Nobel traveled to England, where he filed for patent protection on his invention. His patent application described his invention as nitroglycerin "absorbed in porous unexplosive [*sic*] substances such as charcoal, silica, paper, or similar material, whereby it is converted into a powder, which I call dynamite, or Nobel's safety powder." This application was granted on 7 May 1867 as letter patent 1,345 and became the basis for subsequent developments in dynamite. On 26 May 1868, he received US patent 78,317 covering the same invention.

Eventually, Nobel and others learned that the liquid explosive could be absorbed onto a fuel (such as charcoal, sawdust, or **sulfur**), oxidizer (such as ammonium nitrate or **sodium nitrate**), or both. The use of a fuel or oxidizer in lieu of an inert material as the sorbent boosted the explosive power of the final product. Additionally, developers learned to add an alkaline substance, such as calcium carbonate, to neutralize the acidic products associated with impurities in nitroglycerin as well as its decomposition. Finally, substances such as **ethylene glycol dinitrate** (EGDN) were added to depress the freezing point of nitroglycerin, thus extending the useful range of the final product and eliminating hazards accompanying the thawing of any nitroglycerin that had become frozen. The invention of dynamite prompted Nobel to invent a suitable **detonator**, which he did in the form of a **blasting cap** ignited by a lighted **fuse**.

Competing claims. The commercial success of dynamite prompted various challenges to the various patents obtained by Alfred Nobel, with various parties alleging priority of discovery. Most challenges lacked credibility, and none succeeded in overturning Nobel's basic patents, especially his British and US patents. The challenge mounted by Carl Dittmar and his business associates, however, merits examination. The narrative presented by Dittmar in pleadings before a federal court in the United States describes a situation where it is conceivable that experiments conducted by Dittmar influenced Nobel in the search for a suitable additive. Dittmar's narrative is contained in the case of The Atlantic Giant Powder Co. *v*. The Dittmar Powder Manufacturing Co. This case involved competing claims of patent infringements. The Atlantic Giant Powder Co. claimed **dualin**, a product of the Dittmar Powder Manufacturing Co., infringed on Alfred Nobel's US patent for dynamite. The Dittmar Powder Manufacturing Co. countered that Nobel's patent was invalid owing to prior art on the part of Dittmar.

To support its case, the Dittmar Powder Manufacturing Co. submitted a narrative prepared by Dittmar himself. In it, he claimed that he had been hired by Theodor Winkler, one of the four founders of Alfred Nobel & Co., in July 1866 to ascertain the causes of accidental explosions of nitroglycerin packed in sawdust. He further claimed that he was hired thereafter to supervise the construction of a nitroglycerin factory for Alfred Nobel & Co., remaining in its employ until fall 1867. These claims are reasonable. Dittmar was an experienced engineer, having served as such in the Prussian Army. While in the army, he served in 1861–1862 as technical director of the Prussian government gunpowder factory at Spandau. After resigning from the army in 1862, he continued to work in the explosives industry, including a brief association with **Eduard Schultz**, during which they manufactured **Schultze powder**. Therefore, his professional experience made him a logical person to be hired by Alfred Nobel & Co. Further, the nitroglycerin factory in Krümmel suffered a devastating explosion in July 1866, and a new factory was needed. Dittmar went on to claim that he informed Winkler of his own idea of changing liquid nitroglycerin into a powder and that he began experiments in this direction in August 1866. According to Dittmar, he experimented with lampblack, ground bricks, charcoal, cement, and something he called "infusorial earth" as possible additives with nitroglycerin. This last material is an archaic term for diatomaceous earth, such as Kieselguhr. Dittmar stated that his inspiration for using diatomaceous earth was its common use by Alfred Nobel & Co. to pack around cans of nitroglycerin. Dittmar went on to claim that Nobel ridiculed the suggestion of using infusorial earth, labeling it "nonsense" and instructing Dittmar to refrain from conducting any further experiments with additives. Finally, Dittmar claimed that Nobel pretended not to notice Dittmar's experiments, but that he did witness public tests conducted in October 1866 of nitroglycerin absorbed into diatomaceous earth. According to Dittmar, the tests were successful. These claims relating to his search for an additive that would turn liquid nitroglycerin into a powder are the strongest ones advanced by Dittmar. If accurate, then Dittmar either anticipated the work of Nobel in discovering dynamite or independently discovered it.

Dittmar left Alfred Nobel & Co. in the fall of 1867, several months after Nobel had been granted the British letter patent 1,345, and rejoined Eduard Schultz at his Berlin factory. Thereafter, Dittmar continued to experiment with additives that might turn nitroglycerin into a powder, and it was during this period that he claimed to have invented dualin, which contained no diatomaceous earth. This point was important to the court, which interpreted it as a sign that Dittmar considered the use of diatomaceous earth to be a failure as an additive. The court also noted that Dittmar failed to show that Alfred Nobel was not conducting his own experiments with various additives independently of those of Dittmar during the time they were both present in Krümmel. In light of these two facts (Dittmar's continuing experimentation with additives and Dittmar's failure to demonstrate that Nobel was not pursuing similar research independently), the court on 13 March 1880 ruled against Dittmar and his associates with the Dittmar Powder Manufacturing Co., invalidating its patents on dualin, upholding Nobel's US patent on dynamite, and ruling that dualin infringed on Nobel's patent.

Another individual who might have asserted priority to the invention of dynamite is Vasil Fomich Petrushevski (1829–1891). As a Russian artillery officer serving during the Crimean War, Petrushevski was aware of the nitroglycerin-filled grenades produced at the direction of Nikolai Zinin. These grenades had failed to be adopted by the Russian military, most likely owing to the handling and storage issues associated with nitroglycerin. Petrushevski, however, continued work on finding a means of enhancing the safety of nitroglycerin. At some point, he hit upon the scheme of mixing nitroglycerin with magnesium oxide, a white solid, later replacing it with diatomaceous earth. From 1860 to 1863, he oversaw the production of nearly 3 tons of this combination of liquid nitroglycerin and solid diatomaceous earth, thus anticipating Nobel's invention by several years. However, the Russian military failed to adopt Petrushevski's invention for general use, rewarding him instead with a grant of 3,000 rubles and a lifetime pension of 1,000 rubles annually. Additionally, Petrushevski made no formal filing for priority of invention, challenging Nobel's claims.

Uses. The invention of dynamite took place at the same time that diamond drill bits and the pneumatic drill came into use. This combination of a powerful but safe explosive, durable drill bits, and an automatic drill had important consequences for civil engineering projects such as clearing fields and building canals, roads, and tunnels. The combination was important, too, for industrial applications such as mining coal and precious metals. But direct use of dynamite by the military in combat munitions has proved largely unsuccessful. When used in a conventional artillery piece or naval gun, dynamite-containing munitions risk detonating from the heat generated by the **propellant**. Additionally, dynamite can **sweat** in storage, and this sweating can coat the dynamite in nitroglycerin. This nitroglycerin coating proved to be especially vulnerable to detonation from the heat generated in the gun barrel. Therefore, another form of propulsion was needed in lieu of conventional **chemical explosives**. A solution emerged with the development of the **pneumatic gun**, which employs a **mechanical explosive** in the form of compressed air as the propellant. The most successful military weapons using dynamite-containing munitions appear to have been the **Sims-Dudley dynamite gun** and the pneumatic gun installed on the ***Vesuvius***. In both cases, success was limited to a brief period of fighting in Cuba in 1897–1898.

DYNAMITE GUN. *See* PNEUMATIC GUN.

E

EDNA. *See* ETHYLENE DINITRAMINE (EDNA).

EDNATOL. *See* BINARY EXPLOSIVE.

EGDN. *See* ETHYLENE GLYCOL DINITRATE (EDGN).

EGG, JEAN JOSEPH (1775–1837). Early nineteenth-century gunsmith credited by some as the inventor of the **percussion cap**, circa 1815–1817. Egg was born on the European continent in Alsace but immigrated to England in his youth. The claim that he invented the percussion cap may be accurate, and the record certainly supports the idea that Egg helped to popularize the use of percussion caps as a replacement for the flintlock. However, a contemporary account by Henry Wilkinson states that "Mr. Egg purchased [the idea of the percussion cap] from Mr. Roantree, a gun-maker at Barnard Castle, Durham, who had it from a workman employed by Mr. **Joshua Shaw**." If Wilkinson's account is accurate, then Egg's claim to the right of priority for the invention of the percussion cap appears to be discredited.

ENHANCED RADIATION WEAPON. A specialized type of **thermonuclear** weapon that, by comparison with other **nuclear weapon** designs, produces enhanced amounts of potentially lethal radiation but reduced blast and heat. A prime example is the **neutron bomb**, where the design calls for allowing a burst of neutrons to escape the weapon rather than being absorbed by its other components. In most thermonuclear weapon designs, a **nuclear fission** reaction is used to initiate **nuclear fusion**, releasing energy, neutrons, and other forms of prompt radiation. Most of the radiation is absorbed in the casing, causing more fission events, making it a **boosted fission weapon design**. In the neutron bomb, the weapon casing is intentionally transparent to neutrons, allowing them to be released and becoming the primary weapon effect. The design has been portrayed as cleaner owing to lessened fallout and reduced damage from blast and heat effects, although it has been incorrectly overstated as lethal to humans while leaving structures intact.

The enhanced radiation weapon concept most likely arose with the United States, designed and produced at least three such weapons: the W66, used in the Sprint antiballistic missile; the W70, used in the Lance missile; and the W79, used in an 8-inch artillery shell. The W66 had been retired by the mid-1970s, possibly earlier, and the W70 and W79 were retired in 1992. On 21 October 2011, the final step in the dismantlement of the W70 was accomplished at facilities in Oak Ridge, Tennessee. The Soviet Union is reported to have tested an enhanced radiation weapon on 11 November 1978 and may have produced such weapons. **China** and France are speculated to have had designs for such weapons, but no public record exists to indicate that testing or production occurred. In 1998, **India** publicly claimed that it could produce an enhanced radiation weapon, but nothing further is known publicly about this claim. *See also* SALTED WEAPON.

ETHYLENE DINITRAMINE (EDNA). White or buff-colored crystalline substance used as an **explosive**, either singly or in **composition explosives**, such as the ednatols. Ethylene dinitramine (EDNA) has the chemical formula $C_2H_6N_4O_4$, and its preferred, systematic IUPAC name is 1,2-dinitrodiaminoethane.

George C. Hale (1891–1948) is credited with the original synthesis and discovery of EDNA. His work took place in the early 1930s, while Hale was serving as chief of the chemistry department at Picatinny Arsenal, New Jersey. On 20 August 1935, Hale was granted US patent 2,011,578 for his discovery. EDNA was dubbed "Haleite" in honor of Hale's discovery, which is frequently cited as the first explosive to have been discovered by an American. Bulk production of EDNA lagged, however, owing to problems in the manufacturing of ethylene urea, a key intermediate in the production of EDNA. By 1943, these problems had been overcome. Bulk production was beginning when the Second World War ended in September 1945. EDNA continues to be an important explosive, owing to various characteristics, such as high **brisance**, low sensitivity to **shock**, ease of exploding by heat, and low explosion temperature.

ETHYLENE GLYCOL DINITRATE (EGDN). Transparent, oily liquid **explosive**. Ethylene glycol dinitrate (EGDN) has the chemical formula $C_2H_4N_2O_6$. It is often referred to as nitroglycol and is a nitrate ester of ethylene glycol, from which it can be prepared. Its chemistry and properties are similar to those of **nitroglycerin**, although it is more volatile, which can be undesirable for certain uses. The earliest preparation of EGDN is attributed to Belgian chemist Louis Henry (1834–1913) in 1870. Several other chemists studied EGDN over the next decades. In 1904, German chemist Conrad

Clässen obtained German patent 1789,789 for its use as an explosive. EGDN has a freezing point significantly lower than that of nitroglycerin; for this reason, it was mixed with nitroglycerin in some formulations of **low-freezing dynamite**. EGDN is one of the four **detection agents** named in the **Convention on the Marking of Plastic Explosives for the Purpose of Detection**, the other three named detection agents being 2,3-dimethyl-2,3-dinitrobutane, *o*-mononitrotoluene, and *p*-mononitrotoluene.

EXPANDING BULLET. *See* EXPLOSIVE BULLET.

EXPLODER. Obsolete term for **detonator**.

EXPLODING BRIDGEWIRE DETONATOR (EBW). A type of electric **detonator**. An exploding bridgewire detonator (EBW) uses a rapid jolt of voltage, vaporizing the wire, creating an arc that causes an **explosive** to detonate. It functions 10,000 times faster than an ordinary electric detonator, typically taking 0.1 microsecond or less to go from initiation to detonation. The concept of EBW is credited to Luis W. Alvarez (1911–1988) and Lawrence H. Johnston (1918–2011), nuclear physicists who were working in the South Mesa Detonator Group at **Site Y** during the **Manhattan Project**. In planning for the detonation of an **implosion-type nuclear weapon**, a need arose for a detonator that was extremely fast, very precise, and unlikely to fail. Alvarez and Johnston conceived of the EBW to meet this need. In particular, 32 Model 1773 EBWs were used successfully with both **Gadget** and **Fat Man**, creating a shockwave that detonated the **explosive lenses** used to compress the sphere of **plutonium**-239, thus forming a supercritical mass. Johnston was granted US patent 3,040,660 for the invention of the EBW, although he shared credit with Alvarez, his mentor. EBWs continue in use, although the United States regulates its sale owing to the possibility of its use in nuclear weapons.

EXPLOSIF NX. *See* TNT.

EXPLOSIF S, EXPLOSIF SC. *See* SCHNEIDERITE.

EXPLOSIFS ANTIGRISOUTEUX. French term for **permissibles**; literally, "anti-**firedamp** explosives." *See also* GRISOUTITES.

EXPLOSIFS S.G.P. Belgian term for **permissibles**. The initials *S.G.P.* stand for *sècuritè, grisou, poussierè* (*safety [from]* ***firedamp*** *[and]* ***dust***).

EXPLOSIVE. A discrete substance or a mixture of substances producing a dramatic increase in pressure and volume owing to the evolution of gas that expands rapidly and in an extremely brief period. Explosives might be divided into three fundamental types, according to the processes responsible for producing the increase in pressure and volume:

- **Chemical explosive**, in which chemical reactions cause the rise in pressure and volume and in which the chemical composition of the substance or substances undergoing the reaction is altered, but without any alteration to the elemental composition of this substance or these substances;
- **Mechanical explosive**, in which physical processes cause the rise in pressure and volume but in which there is no change in the chemical or elemental composition of the substance or substances responsible for these processes; and
- **Nuclear explosive**, in which the atomic-level processes of nuclear fission and nuclear fusion cause the rise in pressure and volume and in which the elemental composition is changed.

An example of a chemical reaction in which there is a change in chemical composition but not in elemental composition can be found with liquid **nitroglycerin**, which can decompose rapidly to form gaseous carbon monoxide (CO), molecular nitrogen (N_2), molecular oxygen (O_2), and water vapor (H_2O). An example of a physical process in which there is a change in physical state but where there is no change in chemical composition can be found with carbon dioxide (CO_2), which can expand rapidly from the liquid form to the gaseous form, as is done in **Cardox**. An example of an atomic-level process in which there is a change in the elemental composition can be found with the isotope uranium-235, a radioactive form of the element **uranium**. This isotope can undergo nuclear **fission** when it absorbs a neutron, splitting into the elements barium and krypton.

Any explosive can be a combination of types of explosives. For example, **hybrid explosives** involve both chemical reactions and physical processes to achieve the intended or desired effect. Additionally, most nuclear devices incorporate one or more chemical explosives to initiate the nuclear explosion. *See also* DEFLAGRATION; DETONATION.

EXPLOSIVE BULLET. Projectile containing an **explosive** charge intended to function on contact with the target and sized to be shot from a handgun or a rifle. Designs for explosive bullets date from at least 1856. Some reporting suggests explosive bullets were in use prior to 1856 for hunting game, although such reports might relate to expanding bullets, which have been called exploding bullets, but which lack an explosive charge. Rather, expanding bullets are incised, normally in an X-pattern on the nose. These incisions cause the bullet to expand and fragment once it hits a target. Designs for expanding bullets date from the early nineteenth century—perhaps earlier. Among the best-known designs from that period are the work of Claude-Étienne Minié (1804–1879), eponym of the Minié ball that was used to devastating effect during the **American Civil War**; French Army Captain Henri-Gustave Delvige (1799–1876); and British Army captain John Norton (c. 1788–1867).

Explosive bullets emerged later. A verified example is the work of John Jacobs (1812–1858), then serving as an officer with the British East India Company. He proposed a .22 caliber bullet containing an explosive charge that could be used with a long-range rifle. The intended target was artillery caissons, the goal being that the explosive bullet would cause munitions within the caisson to explode. In Jacobs's words, the explosive bullet "would revolutionize the art of war. Two good riflemen so armed could annihilate the best battery of field artillery in ten minutes."

Jacobs proposed that **mercury fulminate** be used as the explosive charge in the bullet. He designed a copper tube that was embedded in the tip of the bullet and that held the mercury fulminate. Satisfied with his design, Jacobs had bullets fabricated to his specifications. His next step was to arrange a demonstration of his invention. He did this on 5 September 1856, inviting fellow officers in the British Army serving in India and others to observe. On Jacobs's order, a rifleman fired a musket loaded with one of the bullets. It hit a wooden box containing 500 pounds of gunpowder at a distance of 1,800 yards. The box, intended to simulate an ammunition case such as one used with artillery pieces, was obliterated. An observer pronounced Jacobs's bullet "the most formidable missile ever invented by man." Jacobs, however, died 15 months later without seeing his invention adopted widely.

During the **American Civil War**, Samuel Gardiner, Jr. (1816–1880), received US patent 40,468 for an explosive bullet having a chamber at its base into which an explosive was loaded and to which a time fuze was attached. The identities of the explosive charge and the fuze remain uncertain. The Union Army ordered 100,000 Gardiner explosive bullets, 75,000 .58 caliber for use in infantry rifled muskets, and 25,000 .54 caliber for cavalry service. Unlike the Jacobs bullets, the intended targets for the Gardiner bullets were both ammunition caches, such as caissons, and personnel. Accounts from survivors indicate that their use caused great suffering. Given the scale of the conflict, however, their use was limited and failed to prove decisive in any single engagement. Reports surfaced, too, that Jacobs's bullets and the specially designed rifles to go with these bullets were used. If such use did occur, it was sporadic and failed to prove decisive in any battle or campaign.

The **Saint Petersburg Declaration of 1868** arose from Russian concerns over explosive bullets developed initially for use against hard targets, such as ammunition wagons, and later for softer targets such as personnel. Explosive bullets failed to see use to any notable extent during conflicts from 1868 onwards, notably the First World War and the Second World War. However, explosive bullets were used during the 30 March 1981 attempt by John Hinckley, Jr., to assassinate US president Ronald Reagan. Hinckley fired six shots of .22 caliber "Devastator" brand ammunition, each containing a **lead azide** explosive charge, at Reagan as he walked to his limousine following a speaking engagement at the Washington, D.C., Hilton. None of the six shots struck Reagan directly, but one ricochet off the limousine did strike him. Three others were hit as well—a Washington, D.C., police officer, a Secret Service agent, and Press Secretary James Brady. The explosive charge functioned solely with the bullet that struck Brady, whose injuries were the most severe and included permanent brain damage. The use of explosive bullets by Hinckley emerged some days after the shootings, during the forensic examination of the evidence from the scene, and prompted a risky procedure to remove an otherwise intact bullet from the D.C. police officer. *See also* MEAD, SAMUEL HOLMES, JR. (1848–1875).

EXPLOSIVE D. Name adopted by the US military for **ammonium picrate** to protect its identity. The letter "D" honored **Beverly Wyly Dunn**, the military officer who had proposed its use. The United States adopted Explosive D in 1906 as an **explosive** filling for armor-piercing artillery shells and projectiles, replacing **maximite**, which had proved unreliable. Explosive D or **Dunnite**, as it was commonly known, had the advantage of being less **shock**-sensitive than other explosives then in use, especially **Lyddite**, which tended to **detonate** when the shell contacted the target. By contrast, with Explosive D, the shell could penetrate the target, and then a **fuze** would trigger the Explosive D to detonate. By the time the United States entered the First World War, Explosive D had become the standard charge for armor-piercing shells.

EXPLOSIVE LENS. A highly specialized **explosive** precisely formed so that it will change the shape of a **detonation** wave passing through it, with the intent being to amplify or to direct the overall effects of the explosion. The name "lens" was applied to this type of explosive charge because it focuses the detonation in a way that is similar in concept to how an optical lens bends and focuses light. While the concept of an explosive lens is straightforward and physics is well known, the engineering and machining required to manufacture successful explosive lenses present significant technical challenges.

In an explosive lens, two or more explosive charges are used, in which each has a distinctly different velocity of detonation. The charge that has the higher velocity of detonation is often referred to as a **high velocity explosive** or merely the "fast explosive," while the charge having the lower velocity of detonation is referred to as the **low velocity explosive** or the "slow explosive." In this context, the terms "fast" and "slow" are relative, as the velocities are typically supersonic. Explosive lenses are used in **implosion-type nuclear explosives**, where the lenses are used to focus the detonation waves such that the mass of nuclear fuel is compressed to achieve a supercritical state. For instance, the United States used **baratol** as the low velocity explosive and **Composition B** as the fast velocity explosive for its first implosion-type nuclear explosives, **Gadget** and **Fat Man**. *See also* KISTIAKOWSKY, GEORGE BOGDANOVICH (1900–1982).

EXPLOSIVE ORDNANCE DISPOSAL (EOD). The discipline of **rendering safe** an **explosive** device. Explosive Ordnance Disposal (EOD) personnel might be drawn from law enforcement, other public safety groups, or the military. EOD might be performed on a suspect **improvised explosive device (IED)**, a facility where an IED or other **homemade explosive** is made, or an item of military ordnance that failed to detonate and has been recovered. In some jurisdictions and written accounts, EOD is referred to as the bomb squad. EOD took on special significance during the Second World War in noncombat areas when unexploded ordnance (UXO) littered cities and urban areas. UXO from that conflict and from others continues to be found and to require the assistance of EOD.

EXPLOSIVE TRAIN. Sequence of igniting and **initiating** elements arranged to cause a main **explosive** charge to function. The elements are arranged in order of decreasing **Figure of Insensitivity**, such that the first element is most sensitive to **friction**, heat, or **shock**, while the final or main charge is least sensitive. This arrangement enhances safety in handling, storage, and transport. The first element is typically a **primary explosive**, while an intermediate element is a **booster**. The final, bursting charge is a **secondary** or, perhaps, **ternary explosive**.

EXTRA DYNAMITE. *See* DYNAMITE.

F

FACING THE FLAG. Work of fiction by French author Jules Verne (1828–1905) in which an inventor of **explosives**, Thomas Roch, and his inventions, the *Fulgurator* and the *Deflagrator*, play a central role. The book was originally published in French in 1896 under the title *Face au Drapeau* (*Facing the Flag*), and the English edition came out the following year. The fictional character Roch caused Verne to become embroiled in a libel suit brought by **Eugène Turpin**, a well-known and somewhat notorious French inventor of explosives. Verne was defended by Raymond Poincaré (1860–1934), later to become President and Prime Minister of France. Verne was acquitted, although he would write to his brother, Paul, that Turpin was his inspiration for Roch.

In the novel, Roch is portrayed as an insane scientist who promised that his Fulgurator would generate shock waves that would destroy everything in a large area associated with the explosion. Efforts by Roch to interest various European governments and of the United States in his invention failed, and he was confined to an insane asylum in the United States. He leaves confinement, though, to join the pirate Ker Karraje and his followers on an island, where he constructs and successfully detonates the Fulgurator. A tiny amount of explosive shatters a large target, with the promise that a larger amount would create destruction on a massive scale. According to the story, the Fulgurator is an inert powder that must be mixed with the liquid Deflagrator in order to be explosive. This description of a fictional explosive is similar to actual **Sprengel explosives**, such as **Rack-a-Rock**, that were in use when Verne wrote the book. Roch shares the secret of the Fulgurator with others but holds the formula for the Deflagrator secret. In the end, Roch chooses to commit suicide when he explodes the Fulgurator, taking the secret of the Deflagrator with him to his death.

During the First World War, news of a novel French explosive, **Turpinite**, would bear a remarkable resemblance to Verne's description of the Fulgurator.

FAT MAN. Initially, the nickname for the second **nuclear weapon** used in warfare; later, the nickname for bombs of a similar design. On 9 August 1945, the United States **detonated** the original Fat Man bomb over Nagasaki, Japan, with a yield estimated at 21 kilotons **TNT-equivalent**. The weapon used a solid sphere of an alloy of **plutonium** and gallium in an **implosion**-type **fission** design. This design called for a **shaped charge** to crush the solid sphere—known as the **Christy gadget** or the Christy pit and containing 14 pounds of plutonium-239—into a beryllium-polonium core, known as the urchin. Although the mass of the Christy gadget did not change because of the initial explosion, its volume decreased, thus creating conditions needed for a supercritical process, with the urchin providing the neutrons needed to initiate a fission reaction. Studies of the Nagasaki detonation revealed that the explosion consumed a fraction of the nuclear fuel—that is, the plutonium-239—with various estimates of its efficiency ranging from 13 to 21 percent.

The bomb used on Nagasaki weighed slightly over 5 tons and was 128 inches long and 60 inches in diameter at its widest, giving it a distinctive bulbous shape. Its official designation was the Mark-III (Mk-III). Although the design proved successful, it suffered from various defects in that the bomb was aerodynamically unstable at high speeds, causing it to wobble during its descent when it reached terminal velocity approaching the speed of sound. To minimize wobble, drag plates were installed in the tail box to slow down the bomb during its drop. These drag plates proved unreliable. Additionally, the design for the Mk-III Fat Man bomb required that each weapon be assembled by hand, taking an average of two days for a 39-person team. Once assembled, the weapon could remain combat-ready for only 48 hours, after which it had to be partially disassembled to recharge or replace batteries that powered the bomb's **fuzes**. In total, approximately 120 Mk-III Fat Man bombs were added to the US stockpile, and all had been retired by 1950.

To address the various deficiencies of the Mk-III, development of an improved bomb—the Mark-IV—began in 1947. When deployed in 1949, the Mk-IV Fat Man bomb had improved aerodynamic properties that eliminated the wobble problem. Moreover, the new design permitted the **nuclear explosive** core to be inserted during flight, enhancing in-flight safety and adding considerably to the service life of each bomb. The Mk-IV Fat Man bomb remained in the US nuclear weapon stockpile until 1953. By the time the last of the Fat Man series of bombs was removed from service, about 250 Fat Man series bombs had been produced. *See also* GADGET; LITTLE BOY.

FATHER OF ALL BOMBS (FOAB). Large-yield non-nuclear **explosive** device, believed to have been the largest non-nuclear weapon ever **detonated** when it was tested in 2007. "Father of all bombs" (FOAB) is the nickname given to a Russian-made weapon that was successfully tested on 11 September 2007. The Russian government claimed that the weapon contained roughly 7.8 tons of a novel explosive and detonated with a yield of 44 tons **TNT-equivalent**. This yield is four times that of the **Massive Ordnance Air Blast Bomb**, which carries nearly 9 tons of explosive and was the largest non-nuclear explosive device when it was detonated in 2003. It is widely reported to be a **fuel-air explosive**. However, verified information on the FOAB is scarce, and the Russian government has released no details on the nature of the explosive.

FIGURE OF INSENSITIVITY. Relative measure of the ability of an **explosive** to withstand natural forces that might cause **deflagration** or **detonation**. The Figure of Insensitivity (FofI) is calculated on a relative basis, with a lower value indicating heightened sensitivity to **shock** or **friction**. Originally, **TNT** was given an FofI value of 100, and other substances were calculated versus TNT. Later, **RDX** was taken as the baseline for purposes of comparison, and its value was set at 80. Therefore, a substance having an FofI less than that of RDX has less ability to withstand poor storage or rough handling conditions than does RDX. *See also* FRICTION SENSITIVITY; SHOCK-SENSITIVE.

FIREDAMP. A combustible gas, chiefly methane (CH_4), occurring naturally in coal mines and forming **explosive** mixtures with air and coal **dust**. The term grew out of the experience of miners that certain vapors (in German: *Dampf*) in the mines could catch on fire. Similar experiences gave rise to the related terms *afterdamp* (chiefly the asphyxiating gas carbon monoxide [CO]) and *blackdamp* or *chokedamp* (a mix of carbon dioxide (CO_2), nitrogen (N_2), and water vapor) for hazardous vapors present in the aftermath of mining operations and explosions. Numerous mine explosions caused by **black powder** and other, newer explosives prompted the appointment of special "Firedamp" commissions in France (1877), England (1879), and Germany (1880). These commissions identified the flame accompanying the use of powder explosives in mines as at least partly responsible for these mine disasters. An eventual outgrowth was the adoption by various European nations and the United States of a class of explosives known in the United States as **permissibles**, meaning that their use in mines was considered safe.

FIREWORKS. *See* PYROTECHNICS.

FIRST LIGHTNING. *See* JOE-1, JOE-4, JOE-111.

FISSION. The process of splitting the nucleus of an atom into two or more lighter nuclei. This process is accompanied by the release of substantial energy. This release of energy is fundamental to using fission in **nuclear explosives**. *See also* FUSION; IMPLOSION DEVICE.

FLARES. *See* PYROTECHNICS.

FLOOD ROCK. On 10 October 1885, the US Army **detonated** over 150 tons of **explosives** (both **Rack-a-Rock** and **dynamite**) to demolish Flood Rock, a nine-acre obstacle to ship traffic in New York's East River. The detonation was successful, and Flood Rock was eliminated as a hazard to navigation. Some sources credit this blast as the largest planned non-**nuclear explosion** experienced as of that date. The demolition was overseen by Major General John Newton (1822–1895), US Army Corps of Engineers. *See also* RIPPLE ROCK EXPLOSION.

FORENSIC ANALYSIS OF EXPLOSIVES. Scientific and law enforcement discipline involving the analysis, identification, and attribution of **explosives** in a manner that can withstand legal challenge. The discipline extends from the point at which evidence is collected through its examination by laboratory specialists deemed competent and qualified in their field. On-scene collection must be performed in a manner that avoids contamination of the evidence and that preserves its integrity such that any results can be attributed unambiguously to the scene of collection and to the substances found. Laboratory techniques that might be involved in examining evidence are likely to be drawn from various disciplines, especially chromatographic and spectrometric methods. **Sir Vivian Dering Majendie** is cited as an early practitioner of forensic analysis of explosives in his capacity as the Chief Inspector of Explosives for the UK from 1871 to 1898.

FORSYTH, ALEXANDER JOHN (1769–1843). Scottish minister and inventor. Forsyth was a Presbyterian minister for a congregation along the northeastern coast of Scotland, in Belhelvie near Aberdeen. He was an avid shooter of wild fowl, and he became frustrated that the flash of the spark from his flintlock firearm would alert his prey so that they might alter their flight path and avoid being hit by his shot. In an effort to minimize the visible flash, Forsyth devised a hood to fit over the flintlock. This arrangement helped, but it proved clumsy, especially for sighting the firearm. He then experimented with various **detonating** substances, hoping to find a means of igniting the main charge of **gunpowder** in the weapon with a minimum of spark or flash. By 1805, he had designed a bottle-shaped chamber that could be mounted near the flash hole of a firearm. The chamber held a pellet or "pill," as it came to be called, of a mixture of **charcoal**, **potassium chlorate**, and **sulfur**. When the chamber was revolved over the flash hole, it released the pill near the flash hole. The chamber was then rotated away from the flash hole. When the trigger was pulled, it released a hammer, which struck the pill. The pill would then detonate, and its detonation would touch off the main **explosive** charge in the firearm. Forsyth called this chamber arrangement a percussion lock, and others would refer to it as a detonator lock.

In 1807, Forsyth was granted a British patent that claimed "all forms of percussion locks with all suitable detonating materials." His broad patent claim proved a barrier to other inventors, such as **Joshua Shaw**, who sought to patent the **percussion cap**. While some sources credit Forsyth with inventing the percussion cap, his actual invention was the detonating or percussion principle for detonating gunpowder in firearms. There is no evidence that his invention extended to the percussion cap; rather, it was limited to the percussion lock as well as the use of a **shock-sensitive** mixture that could be harnessed to initiate the main charge. Forsyth also resisted the use of more shock-sensitive materials, especially the **fulminates**, for his pills, sticking with his chlorate-based mixture. He submitted his invention to the British master general of ordnance in 1806, but several decades would pass before the British government adopted the percussion principle to the firing of military weapons. In 1843, Forsyth was granted a pension for his services to the government, but the first installment of this pension arrived on the day of his death.

FOX-7. Name for a powerful **explosive** that is a bright yellow crystalline powder. FOX-7 is extremely insensitive to fire, impact, shock, or vibration and, thus, is difficult to **detonate** by accident. These characteristics make FOX-7 a prime candidate for use in an **insensitive high explosive (IHE)** where extreme safety is required, such as in **nuclear weapons**. In such cases, it is often used as a **plastic-bonded explosive (PBX)**. The name "FOX-7" is derived from the abbreviation of the name in Swedish for Sweden's Defense Research Agency, "FOI," but with the "I" replaced by an "X" for explosive. FOI claims that the original synthesis of FOX-7 took place there in 1998. A chemical name commonly used for FOX-7 is 1,1-diamino-2,2-dinitroethylene. The preferred, systematic IUPAC name for FOX-7 is 2,2-dinitroethene-1,1-diamine, and its chemical formula is $C_2H_4N_4O_4$.

FRACKING. *See* ROBERTS PETROLEUM TORPEDO COMPANY.

FRANCK, JAMES (1882–1964). German-born physicist. Franck focused his research on fundamental atomic properties, for which he shared the 1925 Nobel Prize in Physics with Gustav Hertz (1887–1975) for their discovery of the laws governing the impact of an electron upon an atom. Following the ascension of the Nazis to power in 1933, Franck departed Germany, eventually relocating to the United States and became a naturalized US citizen in July 1941. When the **Manhattan Project** was established in 1942, Franck was named to head the chemistry division of the Metallurgical Laboratory in Chicago, which had the mission of creating a workable nuclear pile that might create the **plutonium** needed for **nuclear weapons**. He also headed the Committee on Political and Social Problems set up by the staff of the Metallurgical Laboratory. That committee prepared and disseminated a report recommending against the use of nuclear weapons on **Japanese** cities without warning. The report, known as the Franck Report in recognition of Franck's leadership of the committee, was issued on 11 June 1945 and distributed to senior US decision-makers considering the fate of the newly developed nuclear weapons. The report advocated either no use of a nuclear weapon or use as a demonstration on an unidentified barren island or in the desert, to be conducted before representatives of many nations. The report also noted that the existence of nuclear weapons could not remain a secret in the exclusive disposal of the United States for more than a few years and predicted a nuclear arms race once their existence and destructive potential became known. As history shows, the report failed to alter the decision to use nuclear weapons on Japan but proved accurate regarding the nuclear arms race.

FRANCO-PRUSSIAN WAR. Conflict between France and Prussia, from 15 July 1870 to 10 May 1871. The conflict, sometimes referred to as the Franco-German War and, especially in France, as the 1870 War, began when France declared war against Prussia and ended officially with the signing of the Treaty of Frankfurt, although fighting had ceased some months previously. Prussia was backed by the North German Confederation and by the South German states of Baden, Bavaria, and Württemberg. The Franco-Prussian War witnessed extensive military use of several nineteenth-century technological developments, including breech-loading artillery, railways, and telegraph. It also witnessed the use of the mitrailleuse, a 1-ton gun having 35 barrels and capable of firing 370 rounds in one minute. The sole innovation relative to **explosives**, however, appears to have been the introduction of a type of **smokeless powder**. Its use does not appear to have been decisive relative to the outcome of the conflict.

The Prussian army was equipped with **Dreyse needle-guns**, which by 1870 had been updated to use a paper cartridge containing **Schultze powder** as the **propellant**. This powder emitted no smoke and left no residue when **detonated**. Use of this powder demonstrated the operational superiority of smokeless powder because personnel firing such cartridges could maintain a clear line-of-sight and avoid issuing puffs of smoke that might betray their positions. The French army was equipped with Chassepot rifles. Unlike the Dreyse needle-guns, the Chassepot rifles fired conventional, **black powder**-filled cartridges. Firing such a cartridge gave off a telltale puff of smoke, betraying the position of the shooter, and the accumulation of such puffs of smoke added to the difficulty of seeing the enemy. But the Chassepot rifle had far greater range and rate of fire, making it a superior weapon to the Dreyse needle-gun during this conflict. *See also* LEBEL MODEL 1886 RIFLE.

FRENCH AMMONAL. *See* AMMONAL.

FRICTION SENSITIVITY. The relative propensity of a substance to **deflagrate** or **detonate** as a result of rubbing. For example, **nitroglycerin** is highly sensitive to friction, while **TNT** has low-friction sensitivity. This property is exploited in the friction match, where the act of drawing the match-head against a surface can cause the match to ignite. *See also* FIGURE OF INSENSITIVITY; SHOCK-SENSITIVE.

FRISCH, OTTO ROBERT (1904–1979). Austrian-born scientist. Frisch made several fundamental contributions to nuclear physics, especially to the development of **nuclear weapons**. In 1939, Frisch and his aunt, Lise Meitner (1878–1968), provided the first mathematical explanation for the results of experiments performed by others. They correctly identified that these results confirmed the process of **nuclear fission**, a term chosen by Frisch himself. Later that same year, Frisch visited England, remaining there when the Second World War broke out. In 1943, he was granted British citizenship and became a member of the team of British scientists who traveled to the United States to join the **Manhattan Project**. Assigned to **Site Y**, Frisch determined experimentally the critical mass of **uranium**-235 and **plutonium**-239. Both results were essential to developing the nuclear weapons that were used successfully against Japan in August 1945. Following the war, Frisch returned to England and briefly headed the nuclear physics division of the Atomic Energy Research Establishment, Harwell. In 1947, he accepted a position as the Jacksonian Professor of Natural Philosophy, Cambridge University, where he remained until retiring in 1972.

FUEL. Any substance that releases energy in a controlled chemical or nuclear reaction. A familiar type of fuel is gasoline, which releases energy when burned in a controlled manner in an internal combustion engine. This energy can then be harnessed as a means of propulsion. In the context of a **chemical explosive**, a fuel is used with an **oxidizer** to evolve energy in the form of either **deflagration** or **detonation**. In the context of a **nuclear explosive**, a fuel is a radioisotope that can undergo either **fission** or **fusion**, with the resultant release of an enormous quantity of energy.

FUEL-AIR EXPLOSIVE (FAE). An **explosive** in which atmospheric oxygen serves as the predominant **oxidizer** necessary for a **fuel** to **detonate** or **deflagrate**. Fuels used in a fuel-air explosive (FAE) are often finely divided solids that are readily combustible or are combustible liquids that can be dispersed as droplets. One type of FAE is a weapon whose design deliberately exploits the use of atmospheric oxygen as the **oxidizer**. This design eliminates the need for any significant portion of the payload of the weapon to be an oxidizer. A typical FAE-based weapon includes a charge of a **conventional explosive** to rupture the container as well as a small quantity of oxidizer to aid in initiating the explosion. A FAE-based weapon has the advantage of producing a greater amount of explosive energy than would a conventional weapon of equivalent size. But it has the disadvantage of being dependent on the weather because high winds and precipitation can lessen its effects. Another, but most often accidental, type of FAE is a **dust explosion**.

Other terms used for a FAE include high-impulse thermobaric weapon, thermobaric weapon, and vacuum bomb. The latter term reflects an effect seen when the explosion of an FAE propagates. Because the FAE uses atmospheric oxygen as its oxidizer, it draws in the surrounding atmosphere as it burns. In some instances, such as enclosed areas, the draw on the surrounding atmosphere is sufficiently strong to create a dramatic drop in pressure, akin to a partial vacuum. *See also* FATHER OF ALL BOMBS (FOAB).

FÜLLPULVER. German term meaning "filling powder." A filling powder is an **explosive** or a mixture of explosives used as the filling within ammunition. Germany began using the term *füllpulver* in 1888 to designate explosives without disclosing the actual composition. The term was often abbreviated *Fp.* An elaborate system of designations evolved, using a basic system of *Fp. x/y/z*. In this system, *x* was the percent **TNT** in the mixture, *y* was the percent **ammonium nitrate**, and *z* was a one- or two-letter code for other materials added. These codes included *C* (for calcium nitrate), *N* (for **sodium nitrate**), *S* (for **potassium nitrate**), and *LV* (for either potassium nitrate or sodium nitrate). Therefore, under this system, *Fp. 60/40* is designated a mixture of 60 percent TNT and 40 percent ammonium nitrate. There are exceptions, however. *Füllpulver 1902* (*Fp. 02*) was the designation for 100 percent TNT, with *1902* denoting the year in which the German military adopted TNT for use in munitions. Similarly, *Fp. 5*, *Fp. 10*, *Fp. 15*, and *Fp. 20* designated TNT to which 5, 10, 15, and 20 percent, respectively, wax had been added to help coat the gun barrel and reduce corrosion. This system was in use through the First World War.

FULMINATE. A salt formed from fulminic acid. In general, such salts are unstable, tending to decompose or **detonate** from **friction**, heat, **shock**, or vibration, especially when dry. This sensitivity to friction, heat, shock, and vibration has been used to advantage with certain fulminates, especially **mercury fulminate**, which has seen application as an initiator in **blasting caps**, **percussion caps**, and **primers**.

The empirical formula for fulminic acid is CHNO. The structure of fulminic acid and, hence, of the fulminates was uncertain until the advent of modern spectroscopic techniques in the 1960s and 1970s. The discovery of fulminates is difficult to date precisely, but likely occurred in the seventeenth or eighteenth century. For example, an entry for 11 November 1663, in the diary of Samuel Pepys, mentions *Aurum Fulminans* (Latin for *gold fulminate*), and describes one grain as having blown a hole in a silver spoon when struck. Some of the difficulty in establishing the date for the discovery of fulminates arises from the inability of early experimenters to distinguish fulminates from substances that behave similarly but fail to contain the fulminic ion and, hence, are not fulminates. Therefore, Pepys may have been referring to another substance rather than to what chemists now know to be **fulminating gold** and which had been discovered by at least 1585.

Similar problems arise with references in other seventeenth-century manuscripts. *Basilica Chimica*, published in 1608 by the **alchemist** Oswald Croll (ca. 1561–1609), describes the preparation of a "fulminating gold ammonia." Croll refers to this substance as *Aurum Fulminans*, but its identity as a fulminate is in doubt. Another alchemist, Johann Thölde (ca. 1565–1625), writing under the pseudonym "Basil Valentine" in 1609, refers to a substance that some have taken to be mercury fulminate, although Thölde failed to use that term and, again, its identity as a fulminate is in doubt. The 1612 edition of *Tyrocinium Chymicium e Naturae Fonte et Manuali Experientia Depromptum* (*Chemical Essays. Acquired from Nature and Manual Experience*), written by Jean Beguin (1550–1620) and regarded by many as the earliest chemistry text, describes a substance referred to as *Aurum Fulminans*, although Beguin makes no claim to have produced this substance. Modern scholarship regards this substance, though, as either gold nitride (Au_3N) or gold imide (Au_2NH), either of which would have properties of an **explosive**. A 1621 manuscript and other writings by the Dutch alchemist and inventor Cornelis Drebbel

(1572–1633) show that he engaged in experiments that present-day scientists deem likely would have led to the production of fulminates. Drebbel made no explicit claim to this effect, however. Finally, a 1789 scientific report by French chemist Antoine François Fourcroy (1755–1809) describing mercury fulminate and another report that same year by French chemist **Claude Louis Berthollet** describing silver fulminate are now believed to deal with the corresponding nitride or imide rather than with the fulminate.

Swedish-German alchemist Johann von Löwenstern-Kunckel (1630–1703) reported results consistent with his having prepared mercury fulminate and silver fulminate, noting the spontaneous explosion of these materials the day following their synthesis. His report was published posthumously in 1716 and fails to refer to either of these substances as a fulminate. The 1800 scientific report by **Edward Charles Howard** is credited as the first report that is correct both in its stepwise directions for preparation of fulminates and in labeling the products silver fulminate and mercury fulminate, although Pierre Bayen (1725–1798) had discovered mercury fulminate a quarter century earlier but failed to note stepwise instructions for its preparation. The report by Howard and his public announcement of his discovery before the Royal Society on 13 March 1800 stimulated others to produce fulminates for use in explosives. *See also* COPPER FULMINATE.

FULMINATING GOLD. Light- and shock-sensitive **high explosive**. Fulminating gold has yellow to yellow-orange amorphous crystals. It is not a fulminate in the true chemical sense; rather, it is a combination of ammonia, chlorine, and gold, without a precise chemical formula. It is most likely the earliest high explosive discovered, with its discovery credited to German alchemist Sebald Schwärtzer (1552–1598) in 1585. He isolated it and, critically, left a written record of his discovery and of its explosive nature. Fulminating gold remained a curiosity for the next 200 years, injuring many who prepared it, including the famed Swedish chemist Jons Jacob Berzelius (1779–1848). Berzelius had a beaker of fulminating gold explode as he held it, injuring his hand and an eye. The cost of gold and the sensitive nature of fulminating gold limited its usefulness, but did encourage research into other compounds formed from a metal, ammonia, and chlorine. Such research led to the 1775 discovery of **mercury fulminate**, which proved sufficiently stable to be useful as a high explosive.

FUSE; FUZE. Device that functions to set other forces into motion, causing an **explosive** to **ignite**, **detonate**, or **deflagrate**. The spelling "fuse" is used to differentiate simple devices from more complex devices incorporating chemical, electronic, or mechanical components. An example of a fuse would be the length of cord used to ignite a firecracker or a stick of **dynamite**. An example of a fuze would be an altitude or proximity fuze for an artillery shell, which might use a barometric altimeter, an infrared rangefinder, or radar to determine height above sea level and then to function at the desired altitude. In practice, both spellings are encountered, the terms are interchangeable, and the distinction between "fuse" and "fuze" can be uncertain. NATO has standardized the use of "fuze" for its purposes, and that spelling is used here for most entries.

Fuzes might be classified according to the manner in which they function. Those fuzes that function at a pre-set time or delay after firing are time fuzes. For artillery-fired munitions, this delay might enable the munition to be calculated to allow it to reach a particular point in its trajectory. Developing time fuzes was credited to an 1846 invention by James Freeburn (1808–1876), a quartermaster serving in the British Army, Royal Artillery. However, such fuzes were likely in use elsewhere prior to 1846. Those fuzes that function on contact with a structure or the ground are contact fuzes. Developing contact fuzes was credited to an 1850 invention by Commander William Moorsom (1816–1860), British Royal Navy. He used the term "percussion fuzes" for his invention, and they came to be known as Moorsom percussion fuzes. Again, though, such fuzes were likely in use elsewhere prior to his 1850 invention. Those fuzes that function on the detection of a structure or other object are proximity fuzes. The concept of a proximity fuze is credited to scientists and engineers working at the Telecommunication Research Establishment in the UK during the Second World War. The concept proved successful, and proximity fuzes were used extensively during the war. *See also* BICKFORD, WILLIAM (1774–1834).

FUSION. In the context of **nuclear explosives**, fusion refers to the process of combining two light-nuclei atoms into a heavier nucleus. This process is accompanied by the release of substantial energy. *See also* FISSION; GEORGE; HYDROGEN BOMB.

G

GADGET. Name given to the first man-made **nuclear explosive** that successfully **detonated**. The detonation occurred at 5:29:45 a.m. Mountain War Time, 16 July 1945, at the **Trinity Test Site**, New Mexico. The spherically shaped device, about 6 feet in diameter, weighed 5 tons and exploded with a blast of about 20–22 kilotons of **TNT-equivalent**. Scientists working on the **Manhattan Project** named the device Gadget. It was an **implosion device**, with **plutonium** as the nuclear explosive. *See also* LITTLE BOY; FAT MAN.

GBU-43/B. *See* MASSIVE ORDNANCE AIR BLAST BOMB (MOAB).

GBU-57A/B. *See* MASSIVE ORDNANCE PENETRATOR (MOP).

GELATIN DYNAMITE. *See* DYNAMITE.

GELATINIZING AGENT. Substance or substances incorporated into an **explosive** to reduce the rate at which the final product burns. A gelatinizing agent typically forms a gel-like coating on the final product. This coating gives the product a more nearly uniform rate of burning. Reducing the burning rate also reduces the temperatures achieved during **detonation**, helping to lessen the flash that might otherwise accompany detonation. Finally, the gel-like coating reduces the uptake of moisture, which might otherwise decompose or harm the final product. *See also* CENTRALITE; PROGRESSIVE-BURNING POWDER.

GELIGNITE. *See* BLASTING GELATIN.

GEORGE. Code name for the **thermonuclear** device successfully **detonated** on 9 May 1951. The detonation was conducted by the United States and took place in the Pacific Ocean at Enewetok Atoll of the Marshall Islands. It marked the first successful thermonuclear detonation. George was estimated to have a yield of slightly over 200 kilotons **TNT-equivalent**. The detonation validated the Teller-Ulam configuration, a concept advanced by

Edward Teller and Stanislaw Ulam (1909–1984) in which the x-rays emitted from a primary **fission** event compress the nuclei in the secondary fusion core. *See also* IVY MIKE; JOE-1, JOE-4, JOE-111.

GERBOISE BLEUE. Code name for the first successful **detonation** of a **nuclear explosive** by France. The detonation occurred above ground at 6:04 a.m. Greenwich Mean Time (GMT), 13 February 1960, near Reggane, Algeria, in the Sahara Desert, and had an estimated yield of 70 kilotons **TNT-equivalent**. Gerboise Bleue (*blue jerboa* or *blue desert rat*) was followed by Gerboise Blanche (*white jerboa*) on 1 April 1960, Gerboise Rouge (*red jerboa*) on 27 December 1960, and Gerboise Verte (*green jerboa*) on 25 April 1961. All four explosions took place at the Centre Saharien d'Expérimentations Militaires (*Saharan Military Experiments Center*), located approximately 24 miles from the French military base at Hammoudia. At the time of the explosions, Algeria was an overseas territory of France.

GERMAN AMMONAL. *See* AMMONAL.

GLONOIN. Name given to **nitroglycerin** in the 1850s when it first came into use as a cardiac stimulant. Nitroglycerin was also referred to in that era as Oil of Glycerin. The use of nitroglycerin in medicine as a vasodilator predates its widespread use as an **explosive**. For example, "Spiritus Glonoini" is a term found in nineteenth-century pharmacy texts for a solution of one percent nitroglycerin in alcohol. Patent medicines marketed in the United States during the nineteenth century included chocolate-coated nitroglycerin tablets, distributed by The Ray Chemical Co., Detroit. Direct sale of nitroglycerin to consumers as a homeopathic medicine persists in some areas; see, for instance, Bakson's Homeopathic Glonoinum, which is marketed as a treatment for dizziness, headaches, heart palpitations, breathing difficulties, and vision-related disorders.

The earlier adoption of nitroglycerin for medical use rather than for use as an explosive might be tied to the consequences of an accidental **detonation** of the small quantities of nitroglycerin. The small quantities used in formulating medicine would have caused less damage than what would have occurred with the larger quantities needed for use in explosives. The word "glonoin" was constructed from the principal constituents then known to be present in nitroglycerin: Glycerin ("Gl") + Oxygen ("o") + Nitrogen ("n") + "oin" as a general ending indicating a fluid. *See also* MOWBRAY, GEORGE MORDEY (1814–1891).

GLYCEROL DINITRATE. *See* DINITROGLYCERIN (DNG).

GLYCERYL TRINITRATE. *See* NITROGLYCERIN (NG).

GO-DEVIL. Device for **detonating** an **explosive** used to improve flow in gas, oil, or water wells. A typical go-devil is made of iron or other dense material. It travels the same line that lowers the explosive device (known as a **torpedo**) into the well, trailing the torpedo. The most frequently used explosive is **nitroglycerin**, which fills the torpedo. A **percussion cap** is fitted atop the torpedo, and it will detonate when contacted by the go-devil. The go-devil usually survives the explosion and is thrown out of the well, enabling its reuse. *See also* ROBERTS, EDWARD A. L. (1829–1881).

GOLDSCHMIDT, BERTRAND (1912–2002). French chemist regarded as one of the founders of the French nuclear program, including its efforts to develop **nuclear weapons**. He studied at the Radium Institute, initially under Marie Sklodowska Curie (1867–1934), obtaining his doctorate in 1939. Thereafter, he joined the Institute faculty but lost this position in 1940 when the Vichy government forbade Jews from holding academic positions. In May 1941, he emigrated to the United States and joined the Free French Forces. In July 1942, Goldschmidt joined the **Manhattan Project**, working at the Metallurgical Laboratory in Chicago and becoming the sole Frenchman to work in the United States on the project. Later that same year, he joined the Free French Team at the Montreal Laboratory. After the Second World War, he returned to France and became one of the creators of the French Atomic Energy Commission (*Commissariat à l'énergie atomique*, CEA). He also played a critical role during 1956–1957 in assisting **Israel** to establish its Negev Nuclear Research Center near Dimona. *See also* GERBOISE BLEUE; HALBAN, HANS VON (1908–1964).

GOODE WRIGHT, EDWIN (1783–1859). English country gentleman credited with introducing the use of **mercury fulminate** as a **primer**. In 1822, Goode Wright attended chemistry lectures that included a demonstration of the **explosive** nature of mercury fulminate. While this property of mercury fulminate was well known, mercury fulminate itself was deemed too sensitive to **shock** and **friction** for use as a **propellant** in firearms or artillery. Goode Wright had the genius to see that this sensitivity to shock could be put to effective use in firearms. He was a devoted sportsman, and, like many sportsmen of the time, he used **percussion** primers based on **potassium chlorate** to ignite the **black powder** in his firearms. These primers are sensitive to moisture, impairing their reliability. Furthermore, when they do **detonate**, they liberate chlorine, which would deposit on the surface of the gun barrel and hasten its corrosion.

Inspired by the chemistry lectures, Goode Wright set about experimenting with mercury fulminate in an outbuilding of his Hereford, England, estate. Within a year, he had perfected the use of mercury fulminate as a primer for firearms. He published his results in 1823, stimulating an immediate industry in mercury fulminate percussion caps for firearms used by sportsmen. Adoption by the military was slower, but by the 1840s and 1850s, militaries throughout the world were converting to the use of mercury fulminate percussion caps for firearms and artillery. It remained the most commonly used primer until the 1920s, when more convenient alternatives, especially **lead azide**, became available. *See also* HOWARD, EDWARD CHARLES (1774–1816).

GRAECUS, MARCUS. Pseudonym of an otherwise unknown author whose text provides what some scholars believe is a recipe for **black powder**. The text, *Liber ignien ad conburendos hostes* (*The book of fires for burning the enemy*), gives a recipe for a **pyrotechnic** mixture. One possible interpretation of this recipe is that the mixture is made of 22 percent **charcoal**, 67 percent **potassium nitrate**, and 11 percent **sulfur**. If the document is authentic, then the text provides one of the earliest descriptions of black powder in Western literature. However, the authenticity and dating of this text as an eighth-century document are questionable, with most scholars deferring to the thirteenth-century texts of **Roger Bacon** as more deserving of the right to be designated as the earliest description of black powder in Western literature.

GRANULAR DYNAMITE. *See* DYNAMITE.

GRANULATED-WOOD GUNPOWDER. *See* SCHULTZE POWDER.

GREAT LAW WAR. Legal action taken by **Roberts Petroleum Torpedo Company** and its principal officers, the brothers **Edward A. L. Roberts** and Walter Brooke Roberts, in response to violations of patents on **torpedoes** used to **shoot** gas, oil, and water wells. The action was prompted by **moonlighting**, in which individuals and firms offered to shoot wells without fair compensation being paid to Roberts Petroleum Torpedo Company. More than 4,000 lawsuits were filed by or against that firm and, especially, by or against Edward A. L. Roberts during the 1860s and 1870s, leading some writers to observe that he was responsible for more lawsuits than any other man in the United States. This observation is difficult to prove, but the Roberts brothers and their firm certainly became notorious for their tenacity in pursuing legal action, most of which was pursued in the state courts of western Pennsylvania.

Violations of the patents granted to Edward A. L. Roberts were spurred in part by financial considerations by those wishing to avoid paying what they perceived as excessive fees demanded by Roberts Petroleum Torpedo Company. But some violators genuinely questioned the legality of these patents. Such violators typically believed that "prior art" could be proven and, thus, would invalidate the patents. The basis for this belief was their position that the use of explosives to stimulate increased flow in gas, oil, and water wells was well known prior to November 1864, when Roberts filed his initial patent claim. Many persons active in the explosives industry or in the oil production endorsed this position and denounced the Roberts brothers and their firm as exploiters. While the courts occasionally showed sympathy for the argument of prior art, they invariably ruled in favor of Roberts. As a result, some moonlighters were jailed for their illegal production of **nitroglycerin**, many others were forced out of business, and other individuals and firms settled to avoid the expenses of litigation.

GREENLAND INCIDENT. On 21 January 1968, a US Air Force B-52 bomber crashed following a fire that started in the navigator's compartment. At the time of the crash, the aircraft was 7 miles southwest of Thule Air Base, over North Star Bay, Greenland. Six crewmembers ejected safely, but a seventh was killed during his attempted escape. The bomber was carrying four **hydrogen bombs** as part of its **Chrome Dome** alert mission. The crash caused the conventional explosives in each **nuclear weapon** to **detonate**, rupturing the nuclear payload and dispersing radioactive material, including **plutonium**, the primary fuel in the bombs. Denmark, which governs Greenland, and the United States launched a recovery and cleanup operation. Over the following four months, more than 237,000 cubic feet of contaminated debris, ice, and water were collected and sent to the United States for secure storage pending disposal.

The incident had far-reaching consequences beyond the environmental damage done to Greenland. It prompted the United States to curtail and, eventually, to end the Chrome Dome airborne alert program. It also prompted the United States to undertake negotiations with the Soviet Union to avoid having any future, similar accident involving nuclear weapons, being interpreted incorrectly to mean that a preemptive nuclear strike was underway. These negotiations culminated in the 30 September 1971 signing of the bilateral Agreement on Measures to Reduce the Risk of Nuclear War.

The incident caused difficulties for the Danish government. The presence of nuclear weapons in Danish airspace violated that nation's 1957 policy declaring it to be a **nuclear-weapon-free zone (NWFZ)** during peacetime. When the news of the nuclear payload aboard the B-52 bomber became public, both the Danish and the US governments claimed that this particular bomber had entered Danish airspace owing to a one-time emergency, whereby the crew of the B-52 bomber attempted to land at Thule Air Base, Greenland. However, that explanation proved to be false. An investigation was commissioned by the Danish parliament and determined that the United States made such overflights routinely and had used Thule Air Base to store nuclear weapons until 1965, both of which violated the Danish NWFZ policy.

The detonation of the conventional explosives in the hydrogen bombs in this incident, as well as in an aviation accident over Palomares, Spain, in 1966, prompted a search by the United States for safer conventional explosives. This search for safer conventional explosives culminated in the development by 1979 of a replacement, called **insensitive high explosive**.

GRENADE. A small device, typically having an **explosive** fill and suitable for throwing by hand, shooting from a rifle, or delivering from a launcher designed expressively for that purpose. Explosive-filled grenades emerged in **China** as early as the eleventh century. They found use, in particular, where an adversary was in a confined space, such as when boarding a ship or when facing troops in trenches or dugouts. By the twentieth century, the design had evolved to a round or elongated body containing an explosive fill, a **detonator** that functioned by means of a striker, and a safety arming ring. The word *grenade* is thought to have come from the French word for pomegranate owing to a perceived similarity in size and shape. *See also* MILLS BOMB.

GRENITE. *See* NITROSTARCH.

GRISOUTITES. French term for a group of **permissibles** incorporating metal salts having waters of hydration. The term derives from *grisou*, the French word for **firedamp**, which prompted the need for such **explosives**. The search for explosives that would be safe to use in firedamp-laden atmospheres led to the development in the 1880s of **nitroglycerin**-based explosives having salts that naturally contain "waters of hydration," such as sodium carbonate decahydrate ($Na_2CO_3 \cdot 10H_2O$) and magnesium sulfate heptahydrate ($MgSO_4 \cdot 7H_2O$). The addition of the hydrated salts lowered the temperature of the explosion, reducing the possibility of an unplanned, secondary explosion. Later, the French government would adopt other permissibles, but without the incorporation of hydrated metal salts. Examples include Grisoudynamite, a blend of 70 percent **ammonium nitrate**, 29 percent nitroglycerin, and 1 percent **collodion**, and Grisoutetrylite, a blend of 88 percent nitroglycerin, 7 percent **tetryl**, and 5 percent **potassium nitrate**. *See also* WETTER DYNAMITE.

GROUGH. Impure **potassium nitrate**. *See also* INDIA SALTPETER.

GROUP SNPE. *See* GUNPOWDER SERVICE (*SERVICE DES POUDRES*).

GUANO. Excrement, primarily of sea birds and bats. Guano accumulates naturally in caves, on uninhabited islands, and in other areas inaccessible to humans. It is an important source of various elements and of ammonia and nitrates, especially **potassium nitrate** and **sodium nitrate**. Guano was harvested in many cultures as a source of potassium nitrate for making **black powder**, and such harvesting from caves was organized on a large scale by the **Niter and Mining Bureau** of the Confederacy during the **American Civil War**. Guano deposits in the Pacific Ocean, especially those close to the coast of Peru, were an important part of the South American economy, both for export and for direct use in agriculture as a fertilizer. Conflict over the revenue from these deposits contributed to the **War of the Pacific** fought among Bolivia, Chile, and Peru. The importance of guano as a nitrate and ammonia source diminished greatly with the development of an economical industrial process for **nitrogen fixation**.

GUNCOTTON. **Nitrocellulose** having 13 percent nitrogen content or higher. Guncotton is a **smokeless explosive**, most often used as a **propellant**. *See also* BRACONNOT, HENRI (1780–1855); SCHÖNBEIN, CHRISTIAN FRIEDRICH (1799–1868).

GUNPOWDER. An **explosive** used as a **propellant** in firearms; also, more expansively, one used with projectiles as either a propellant or the explosive charge and for non-military applications, such as mining, debris removal, demolition, and quarrying of rocks. Originally, gunpowder was some type of **black powder**, and the terms "gunpowder" and "black powder" have been used interchangeably, although their meanings can differ. Black powders have been used for applications beyond firearms, including blasting and in **pyrotechnics**. Although the original gunpowders were black powders, **smokeless explosives** and other explosives came to be used for the same applications.

GUNPOWDER FARMS. *See DROIT DE FOUILLE*; GUNPOWDER SERVICE (*SERVICE DES POUDRES*).

GUNPOWDER SERVICE (*SERVICE DES POUDRES*). French government agency responsible for manufacturing **explosives**. French government control over manufacturing explosives might be dated to 1336, when Philippe

VI of Valois created royal charters for gunpowder manufacturers and placed them under the authority of the Grand Maître des Arbalétriers ("Grand Master of Crossbowmen" or, less literally, "Grand Master of the Infantry"). In 1582, the government established the Corps de Commissaires et de Contrôleurs des Poudres ("Gunpowder Commissioners and Controllers Corps") as a further step to controlling the quality and quantity of explosives, especially **black powder**.

In 1665, Louis XIV took the additional step of creating the **gunpowder** and **saltpeter** farm (*ferme de poudres et salpêtres*), a privately owned company. This company received a six-year lease from the king, giving it the exclusive right to make and sell gunpowder and saltpeter. Individuals engaged in these pursuits were known as gunpowder and **saltpeter farmers**. The system was similar to that used in many countries for the collection of taxes and suffered similar defects: fraud, corruption on the part of the farmers, and excessive costs to the state for the quantity and quality of goods obtained. Among the most bothersome aspects of the gunpowder and saltpeter farming system was the right of excavation (***droit de fouille***) provided to the farmers. This right afforded a gunpowder and saltpeter farmer unimpeded access to private property for the purpose of collecting the nitrate-rich efflorescence to be found in caves, basements, wine cellars, and stables, as well as the spaces beneath latrines and chicken and pigeon coops. Exercising this right was a source of both resentment by the private property owners and bribe taking and bribe seeking by the farmers.

Louis XVI took a significant step toward improving the manufacture of explosives in France on 30 March 1775 when he established the Gunpowder and Saltpeter Administration (*Régie des Poudres et Salpêtres*) and, more importantly, named **Antoine Lavoisier** as one of its commissioners. The Gunpowder and Saltpeter Administration replaced the gunpowder and saltpeter farm. This decision to reform gunpowder manufacturing in France may have been influenced by the Seven Years' War (1754–1763). Poor performance by the French military during that conflict was blamed on a lack of quality gunpowder, and its outcome included a loss of ready access by the French to **India Saltpeter**. Lavoisier and his fellow commissioners embarked on various reforms.

By the 1780s, French gunpowder was being produced in five types, each of a different quality and each identified for a particular application: *poudre royale*, the most potent form of gunpowder, reserved for selected members of the royalty; military gunpowder, reserved for the arsenals; *poudre à giboyer*, hunter's gunpowder for wide use by the citizenry; blasting powder to be used in mines and quarries; and *poudre de traite*, trader's gunpowder. This last type was a deliberately inferior quality of gunpowder intended for sale exclusively to slave traders in French ports. The notion was that these traders would use the trader's gunpowder as a currency for buying slaves. The commissioners worried that gunpowder used in trade might wind up being used against the French military or other French interests internationally. Consequently, the trader's gunpowder was of an inferior quality to limit its effectiveness.

The Gunpowder and Saltpeter Administration succeeded in enhancing the quality and quantity of French gunpowder. For example, one means then in use to determine the quality of gunpowder was to measure the distance traveled by a projectile on firing a standard amount of gunpowder. By this measure, the distance increased by nearly double from the period of the Seven Years' War to 1778, from around 140–160 meters to 230–260 meters. Moreover, production increased to the point that France became an exporter of gunpowder, especially to Holland, Spain, and the American colonies during its rebellion against Britain. On this last point, Lavoisier wrote in 1789 that:

> The Kingdom [of France] was able to furnish gunpowder to all the King's friends and allies in the last war; it can truthfully be said that it is to these supplies that North America owes its freedom.

Lavoisier departed the Gunpowder and Saltpeter Administration in 1792. In February 1794, it was renamed the Revolutionary Gunpowder and Saltpeter Administration (*Administration Révolutionnaire des Poudres et Salpêtres*), in keeping with changes brought about by the French Revolution. The name was changed again in 1800 to the National Gunpowder and Saltpeter Agency (*Agence Nationale des Poudres et Salpêtres*) and, again, in 1819, to the National Gunpowder and Saltpeter Service (*Service National des Poudres et Salpêtres*). By 1873, it had become the Gunpowder and Saltpeter Service (*Service des Poudres et Salpêtres*). In May 1914, shortly before the First World War, the archaic term "saltpeter" was dropped, and it became simply the Gunpowder Service. In October 1971, the French government formed Group SNPE (*Société Nationale des Poudres et Explosifs*) and folded all industrial and commercial operations of the Gunpowder Service into this quasi-governmental entity. The government continues to hold a majority stake in this firm. *See also* BATTLE OF CRÉCY.

GUNPOWDER TRADE ASSOCIATION. Late nineteenth-century American business cartel formed to protect prices for and control over trade in **explosives**. On 29 April 1872, the Gunpowder Trade Association was formed in New York City. Seven firms were the original incorporators: **DuPont**, Laflin & Rand Powder Company, Hazard Powder Company, Oriental Powder

Mills, Austin Powder Company, American Powder Company, and Miami Powder Company. The association was formed in response to the unrestrained competition, rampant price-cutting, and the decline in military sales following the end of the **American Civil War** (1861–1865). The US government had entered the market in 1866, selling surplus explosives (mainly **black powder**) at steep discounts. These circumstances combined to reduce or eliminate profits for the private sector, and several firms went bankrupt.

Lammot du Pont was chosen as the first president of the Gunpowder Trade Association. He promptly gained agreement among the seven-member firms on a fixed schedule of prices for the New York market. Soon thereafter, penalties were enforced for any sales at prices below the schedule. In 1875, the association entered into an agreement with the California Powder Works to control trade. The association members banned themselves from trade in British territories west of the Rocky Mountains; in return, California Powder Works agreed to a ban of its trade for much of the territories to the east of the Rocky Mountains. In 1886, the association set quotas for the firms in an elaborate scheme that distributed sales quotas according to size, thus ensuring the dominance of the largest three firms (DuPont, Laflin & Rand, and Hazard), while promising survival for the rest.

Over the years, firms would depart, and others would join, owing to acquisitions and mergers within the explosives industry. The association was dissolved in June 1904. By then, DuPont dominated the US trade in black powder and **smokeless powder** (introduced in the 1880s), having acquired a controlling interest in most competitors. This dominance by DuPont meant the association no longer served any real business purpose, because agreements on prices, quotas, and sales territories were set by DuPont based on its dominance in the marketplace. *See also* POWDER TRUST.

GUN-SAWDUST. See NITROCELLULOSE; SAWDUST.

GUN-TYPE NUCLEAR DEVICE. Design for a **nuclear explosive** in which one sub-critical mass of **fissionable** material is propelled toward a second sub-critical mass. The concept is straightforward, although the engineering and physics needed for the design to succeed are complex. Once the two masses are in the proper configuration, **critical mass** is achieved, and a nuclear **detonation** results. In the early development of this type of nuclear weapon, an actual gun tube was used to both separate the masses and provide a path for one mass to be propelled toward a second. This design was used for the **Little Boy** bomb and was thought to be sufficiently straightforward that firing of a test device was not required. Such thinking proved correct, as witnessed by the successful detonation of Little Boy over Hiroshima. *See also* FAT MAN: GADGET; IMPLOSION DEVICE.

H

H-6. Ternary explosive comprised of approximately 45 percent **RDX**, 30 percent **TNT**, 20 percent powdered **aluminum**, and 5 percent D2 composition wax (a mix of 84 percent polyethylene, 14 percent **nitrocellulose**, and 2 percent lecithin). This composition is a derivative of **torpex**, which is now considered obsolete. Compared to torpex, H-6 is less **shock**-sensitive. Both attributes are considered advantageous in many applications. H-6 is the explosive charge used in the **Massive Ordnance Air Blast Bomb**.

HABER, FRITZ (1868–1934). German-born chemist, inventor, and Nobel Prize laureate. Haber made many practical and fundamental contributions to electrochemistry and physical chemistry. Among these contributions, he is given partial credit for inventing the **Haber-Bosch process**, which permitted ammonia to be synthesized directly from the nitrogen in air in a process known as **nitrogen fixation**. Haber's early twentieth-century invention is essential for producing abundant and relatively inexpensive synthetic fertilizers and earned Haber the 1918 Nobel Prize in Chemistry. But this process also provided a direct means of making the **nitrates** needed for manufacturing **explosives**. As such, it proved essential to Germany during the First World War by eliminating its dependence on imported **Chilean nitrates**. Haber also played a key role in Germany's development and use of chemical weapons and is seen by some as central to the introduction of chemical weapons onto the modern battlefield.

The award of the 1918 Nobel Prize in Chemistry was delayed owing to the First World War. When it was announced in late 1919, the selection of Haber was widely denounced, especially in England and France. In the view of some people, Haber's role in promoting chemical warfare and the role of the Haber-Bosch process in the manufacture of explosives were regarded as incompatible with the standards set for the Nobel Prize awards. Prior recipients of the Nobel Prize typically attend the award ceremonies, but several Nobel Prize laureates boycotted the ceremony for Haber in protest.

HABER-BOSCH PROCESS; HABER PROCESS. Industrial process for **nitrogen fixation**. In March 1909, **Fritz Haber** of the Technische Hochschule Karlsruhe (Karlsruhe Institute of Technology) synthesized ammonia from atmospheric nitrogen. Robert Le Rossignol (1884–1976), a British chemist, provided key assistance to Haber in this endeavor. Prior to joining Haber's staff, Le Rossignol had been working in the laboratory of Sir William Ramsay (1851–1939) on the chemistry of nitrogen while pursuing his undergraduate degree at University College London. His experience proved essential to the design and operation of the laboratory-scale apparatus for nitrogen fixation. The work of Haber and Le Rossignol was funded in large part by the German chemical firm Badische Anilin- und Soda-Fabrik AG (BASF). Following a successful demonstration of their nitrogen fixation apparatus to a representative from BASF in July 1909, Haber and Le Rossignol filed for patents in various countries. On 27 September 1910, they were granted US patent 971,501 for "Production of Ammonia," and received a German patent 135,875 on 3 October 1911. Haber and Le Rossignol later engaged in a bitter dispute over the sharing of royalties from their patents.

Shortly after the July 1909 demonstration, Carl Bosch (1874–1940), an industrial chemist working for BASF, undertook the task of successfully adapting Haber's laboratory-scale process to an industrial scale. Bosch experimented with various catalysts and found that relatively inexpensive iron could be substituted for the more expensive osmium used by Haber and Le Rossignol. Bosch also experimented with various steels before finding a composition that could withstand the needed high pressures. By the August 1914 outbreak of the First World War, BASF had set up a plant in **Oppau**, Germany, using the Haber-Bosch process to produce ammonia at a rate of 25 tons per day. This ammonia was used primarily to produce ammonium sulfate for use as a fertilizer. By late September 1914, the German government recognized its vulnerability to overseas sources for the **nitrates** essential in manufacturing **explosives**. This vulnerability was heightened by the stalemate that had developed on the western front, thereby ruining plans for a rapid conclusion to the conflict. Germany had insufficient domestic supplies of nitrates to sustain the production of explosives needed for a prolonged war. Consequently, Bosch and colleagues at BASF undertook the design and construction of Haber-Bosch process plants on a massive scale. By May 1915, BASF had completed a plant capable of producing ammonia at a rate of 150 tons per day. This ammonia was used to produce nitrates, especially **ammonium nitrate**, for use in making explosives. Several more Haber-Bosch process plants were constructed during the war to feed Germany's requirements for explosives and eliminate its dependency on overseas sources of nitrates.

HAGUE PEACE CONFERENCES OF 1899 AND 1907. *See* SAINT PETERSBURG DECLARATION OF 1868.

HALBAN, HANS VON (1908–1964). Austrian-born physicist. After studying in Germany, Switzerland, and Denmark, Halban joined the research group of Jean Frédéric Joliot-Curie at the Collège de France, working on **nuclear fission** chain reactions for energy production. Their work was interrupted by the Second World War. He escaped the German occupation of France in 1941, making his way to the UK, where he joined a Free French contingent working at Cavendish Laboratory, Cambridge University. In 1942, he was sent to Canada, where he became director of the Montreal Laboratory. Following the liberation of Paris in 1944, he made an unauthorized trip to Paris, meeting with Joliot-Curie, whose communist affiliation was now known widely. Although Halban denied sharing secrets with Joliot-Curie, he was removed as director of the Montreal Laboratory at the demand of the US government. After the war, he was not invited to return to the Collège de France for undisclosed reasons; instead, he returned to the UK, where he worked at the Clarendon Laboratory, Oxford University. In 1954, he was invited by the prime minister of France to return to direct the building of a laboratory that would be part of the French Atomic Energy Commission (*Commissariat à l'énergie atomique*, CEA). During Halban's tenure as director, this laboratory would oversee the development of French **nuclear weapons**. *See also* GERBOISE BLEUE; GOLDSCHMIDT, BERTRAND (1912–2002).

HALIFAX EXPLOSION. On the morning of 6 December 1917, a French cargo ship, *Mont-Blanc*, **detonated** in the harbor of Halifax, Nova Scotia, Canada. It was loaded with a cargo that included nearly 3,000 tons of various **explosives** and other flammable materials, chiefly **TNT**, **picric acid**, and **nitrocellulose**. Halifax and surrounding communities suffered extensive damage from the blast itself, the ensuing fires, and a blast-induced tsunami that swept inland from Halifax harbor. Estimates of the dead and injured were placed at approximately 2,000 and 9,000, respectively, but an accurate count proved impossible. The blast was estimated to have been equal to 3 kilotons of **TNT-equivalent**, making it the largest accidental explosion then recorded.

The *Mont-Blanc* had sailed from New York City on 1 December, after taking on a cargo of explosives and other provisions bound for Europe to aid the Allies in their wartime efforts. It arrived off Halifax on the evening of 5 December 1917, too late to enter the harbor prior to the closing of its antisubmarine nets. Around 8:30 a.m. the next day, the nets were lifted, and the *Mont-Blanc* entered the harbor. At the same time, an unloaded Norwegian vessel, the *Imo*, was departing the harbor for New York City, where it was to pick up relief supplies destined for Belgium. The channel is narrow, and the two ships passing in opposite directions collided around 8:45 a.m. The collision sparked a fire aboard the *Mont-Blanc*, whose crew attempted to extinguish the blaze. When their efforts proved futile, its captain ordered the ship to be abandoned. Taking two rowboats, in about 10 minutes they arrived safely on the Dartmouth side of the harbor, to the north and east of Halifax city. As they fled inland, they tried to sound the alarm regarding the cargo. Many ashore failed to understand the French-speaking crew, and the crew knew little English; consequently, few understood their warnings.

The situation was even worse on and around the *Mont-Blanc*. Although the captain and crew were aware of the extreme hazard posed by the cargo of explosives, most others were clueless. The needs of wartime secrecy limited knowledge by others regarding the true nature of the cargo, and the ship carried no warning flags to avoid being targeted by the Imperial German Navy. With the crew gone, the *Mont-Blanc* was adrift in Halifax harbor. It shortly wound up at one of the piers, where the fire jumped to the land and ignited munitions stored on the pier. Unmindful of the potential dangers, several ships rushed to aid the *Mont-Blanc*. With the flames having spread ashore, the Halifax Fire Department rushed to the scene. Residents of Halifax and its environs gathered along the shore or at windows of their homes and businesses to watch as the ship burned. All these actions would have tragic consequences, as they would add to the deaths and injuries.

At 9:06 a.m., the cargo detonated. Those people on or nearest to the ship were essentially vaporized, adding to the uncertainty over the number killed. The weather in Nova Scotia was cold, and many homes and businesses had fires going in coal- or wood-burning stoves. The shock wave from the blast overturned many of these stoves as well as oil-burning lamps, causing fires to erupt throughout the blast zone. The shock wave also broke windows for a vast distance from the harbor, killing or blinding many spectators who had gathered to watch the blazing ship.

The toll of dead and injured was long, but the heroism of Vince Coleman (1872–1917) kept it from being longer. A telegraph dispatcher for the Intercolonial Railway working in the rail yards near the harbor, Coleman learned of the dangerous cargo aboard the *Mont-Blanc* from a sailor who was fleeing the scene. Coleman knew that a passenger train was due to arrive momentarily, so he stayed at his post to telegraph a warning: "Stop trains. Munitions ship on fire. Approaching Pier 6. Goodbye." The train stopped outside Halifax, thus saving all aboard. Additionally, telegraph operators at other stations heard the message, enabling railway officials to dispatch aid to the scene quickly. Coleman perished at his post when the *Mont-Blanc* detonated, but he became a fixture in Canadian history for his heroism. Others beyond the railway workers responded with aid. For instance, the Boston Red Cross and the Massachusetts Public Safety Committee were quick to send assistance. Since 1971, the citizens of Nova Scotia have donated a large Christmas tree to the City of Boston as a token of thanks and remembrance.

The Halifax explosion was felt and heard in Charlottetown, Prince Edward Island, over 100 miles to the north, and in North Cape Breton, over 200 miles to the east. The *Mont-Blanc* disintegrated in the blast, as did other nearby ships that had come to its aid. A portion of its anchor, weighing over a half-ton, landed 2 miles away. While a precise count of those killed by the explosion is impossible, it is certain that more residents of Nova Scotia died from this single event than died in combat during all of the First World War. Although the **Texas City disaster** in 1947 and the **Oppau explosion** in 1923 would be larger in terms of loss of life, injuries, and property damage, the Halifax explosion continues to rank among the largest unplanned blasts. *See also* LAKE DENMARK NAVAL AMMUNITION DEPOT EXPLOSION; PORT CHICAGO DISASTER.

HARRIS, JOHN PALMER (1892–1982). American military officer credited with two key contributions that averted a shortage of **TNT** in the United States during the Second World War: (1) Anticipating the need for an abundant supply of toluene and (2) Identifying an improved process for converting toluene to TNT.

During the First World War, the United States and other nations suffered a shortage of TNT. A contributing factor to this shortage was the scarcity of toluene, an organic chemical that can be mixed with nitric acid and sulfuric acid to form TNT. Historically, toluene had been obtained both from coal as a by-product of coke ovens and, to a lesser extent, by extraction from illuminating gas. Following the First World War, a process for producing toluene directly from petroleum was invented in Germany. In 1927, Standard Oil of New Jersey licensed this process, allowing it to make toluene in the United States from petroleum. During the 1930s, Harris began negotiations with Standard Oil of New Jersey to use this process to produce toluene for the US military. At the time, Harris was an Army officer assigned to Picatinny Arsenal, New Jersey, and understood the critical importance of toluene to the production of TNT. In June 1940, the US Army contracted with Standard Oil of New Jersey for toluene produced from petroleum. By October 1942, Standard Oil of New Jersey was producing toluene at a rate of 65 million gallons per year. By contrast, total toluene production in the United States in 1918 was a mere 9 million gallons. This increase in toluene production capacity eliminated toluene availability as a factor in TNT production.

However, one other hurdle remained in order for TNT production to meet the expected needs of the war effort. The rate at which toluene could be converted to TNT in existing US facilities remained low. Either the rate had to be increased, or more physical facilities had to be constructed. Harris succeeded in increasing the production rate. His success stemmed from observations he made in 1941 at a TNT production facility in Beloeil, Quebec. Although his visit to Canada was planned, the stop at the Beloeil plant was added at the end of his trip to fill time before his train arrived to return him to the United States. Harris noticed that the Beloeil plant made TNT by adding toluene to a mix of nitric acid and sulfuric acid. In the United States, the opposite procedure was followed: the acids were added to toluene. Harris urged TNT producers in the United States to adopt this reverse process, but the producers were reluctant to change their methods. Eventually, the Canadian process was tried on a trial basis at the Keystone Plant, Meadville, Pennsylvania. The results of the so-called "reverse nitration" convinced the producers of its merits. TNT output could be tripled, and production lines designed to turn out 33,000 pounds of TNT per day were suddenly able to produce more than 100,000 pounds of TNT per day. This increase eliminated the need to use substitutes for TNT, as had occurred during the First World War, and the price of TNT dropped to half the then prevailing price.

Harris ultimately retired from the US Army as a colonel. In recognition of his contributions to TNT production, he was inducted into the US Army Ordnance Corps Hall of Fame in 1975.

HÄUSSERMANN, CARL (1854–1918). German chemist credited with observing the **explosive** nature of **TNT**. Häussermann was a professor of chemical technology at the Technische Hochschule Stuttgart (Technical University of Stuttgart), where he specialized in the chemistry of coal-tars, such as toluol, a coal-tar product that is high in toluene. In the course of his research, Häussermann prepared TNT and noted that it could be **detonated**. In 1889, he reported his observations on TNT in the scientific literature, prompting research into the potential of using TNT as a military and industrial explosive.

HEADACHE-FREE EXPLOSIVES. *See* NITROSTARCH.

HELLHOFFITE. Name given to a **Sprengel explosive**, nominally consisting of 28 percent **nitrobenzene** and 72 percent fuming nitric acid. Development is credited to Albert Hellhoff and Hermann Gruson (1821–1895) in the 1870s. They obtained patents for an artillery shell in which two components were loaded into separate cylinders within the shell, thus being an early type of binary munition. When the shell was fired, a setback caused the separate cylinders to rupture, allowing the nitrobenzene and nitric acid to mix during flight and to explode on impact with a target. Having the liquids in separate cylinders enhanced safety in handling and storage. Hellhoff and Gruson applied for patents in at least 12 countries between 1879 and 1882, including in the United States and the UK. In August 1885, trials were held in St. Petersburg, Russia, in which the performance of Hellhoffite shells could be compared with both **black powder**-filled shells and **nitroglycerin**-filled shells. While Hellhoff and Gruson took the results as showing comparable or superior performance on the part of Hellhoffite shells, their explosive failed to gain a following, largely because of the extreme corrosiveness of the nitrobenzene and, especially, the fuming nitric acid. Little biographical information is known about Hellhoff beyond his having been an artillery officer in the Prussian army. Gruson, however, was a well-known German engineer, industrialist, and inventor. His fondness for collecting cacti led to his having a species of cactus named in his honor: *Echinocactus grusonii*, a round, spiny plant better known as the mother-in-law's cushion.

HENNING, GEORG FRIEDRICH (1863–1945). German pharmacist credited with the first synthesis of **RDX**. Henning was engaged in the preparation of novel drugs when he synthesized RDX in the late 1890s. He was awarded German patent 104,280 on 15 July 1898 for a process for making a nitro compound from hexamethylenetetramine. Henning proposed no name for this compound but did provide its correct empirical formula ($C_3H_6N_6O_6$). Later work substantiated that what he claimed in his patent corresponds to RDX. In his patent claim, he anticipated that the substance would prove useful as a starting material for medicinal products. Such use apparently failed to materialize. The **explosive** nature of RDX would be recognized decades later by others, especially by **Edmund Ritter von Herz**.

HERCULES POWDER COMPANY. *See* POWDER TRUST.

HERZ, EDMUND RITTER VON (1891–1964). Austrian-born chemist and inventor of **explosives**. Edmund von Herz relocated from Vienna to Germany around the First World War, where he worked in Köln (Cologne) as an industrial chemist for various commercial explosives industries. He was a prolific inventor of explosives and made many notable contributions. In particular, he is the first person known to have received patents for the use of **RDX** as an explosive. He was granted British patent 145,791 in 1920 and US patent 1,402,693 on 3 January 1922, both covering its use as an explosive of the substance then called "cyclotrimethylene trinitramine." In his US patent,

Herz noted that RDX is odorless, tasteless, and nontoxic, unlike most other nitro compounds then in use as explosives. He noted that this property would prove useful during manufacture and loading, avoiding problems such as those seen with the **canary girls**. Moreover, he wrote that the "most surprising feature of this compound, however, is it [*sic*] extraordinary explosive and shattering power." Testing would prove this statement correct, as RDX has a **Relative Effectiveness Factor** of 1.60, one of the highest for any explosive in routine use.

Herz also discovered 5-nitrotetrazole and its various metal salts, such as mercuric 5-nitrotetrazolate, which became a replacement for **lead azide** as a **detonator**. He is credited, jointly with Hans Joachim Rathsburg (1888–1975), with the development of the SINOXID **primer** composition, which achieved considerable commercial success. His interests extended beyond conventional explosives to **pyrotechnics**, **signaling flares**, and smoke agents, and he is the eponym of the Herz smoke cartridges used as a means of controlling infestations by rodents, especially voles.

HEXAMETHYLENE TRIPEROXIDE DIAMINE (HMTD). White crystalline solid formerly used as a **detonator** in **blasting caps** and currently of concern as a **homemade explosive** (HME) associated with use or attempted use by terrorists. The preferred, systematic IUPAC name for hexamethylene triperoxide diamine (HMTD) is 3,4,8,9,12,13-hexaoxa-1,6-diazobicyclo[4.4.4]tetramine, and its chemical formula is $C_6H_{12}N_2O_6$. HMTD is an organic peroxide, similar to other **acetone peroxides**. HMTD is unstable, being sensitive to **friction**, heat, **shock**, and vibration. HMTD can be made from hexamine, hydrogen peroxide, and citric acid. Because these three ingredients are readily accessible, HMTD is of concern as an HME. Evidence of terrorist interest was seen from terrorist events, such as three bomb explosions during 17–19 September 2016 in New Jersey and New York that injured more than 30 persons. On 17 December 2024, the FBI executed a search of a property in Isle of Wight County, Virginia, where 155 **improvised explosive devices** were seized and a container of HMTD was retrieved from a refrigerator. The suspect of this search had no known, legitimate purpose for having explosives of any type and later pleaded guilty to illegal possession of a destructive device.

HEXOGEN. Synonym for **RDX**.

HIGH BLAST EXPLOSIVE (HBX). Ternary explosive comprised of **aluminum**, **RDX**, and **TNT**. Wax and lecithin are also present as **phlegmatizing agents**. Three commonly used formulations of high blast explosives are (1) HBX-1 (40 percent RDX, 38 percent TNT, 17 percent aluminum, and 5 percent D2 composition wax (a mix of 84 percent polyethylene, 14 percent **nitrocellulose**, and 2 percent lecithin); (2) HBX-3 (31 percent RDX, 29 percent TNT, 35 percent aluminum, and 5 percent D2 composition wax; and (3) **H-6**.

HIGH EXPLOSIVE (HE); HIGH VELOCITY EXPLOSIVE. An **explosive** that, when functioned, decomposes rapidly at a supersonic rate, resulting in **detonation** rather than **deflagration**. Examples of high explosives include **ammonium nitrate-fuel oil mixture (ANFO)**, **Composition B**, **mercury fulminate**, **nitroglycerin**, and **TNT**. *See also* EXPLOSIVE LENS; LOW EXPLOSIVE; LOW VELOCITY EXPLOSIVE.

HIGH-IMPULSE THERMOBARIC WEAPON. *See* FUEL-AIR EXPLOSIVE (FAE).

HIGHLY ENRICHED URANIUM (HEU). Uranium having a content of uranium 235 isotope of 20 percent or greater is referred to as highly enriched. When the uranium-235 content reaches or exceeds 90 percent, then the material is regarded as weapons-grade HEU or, simply, **weapons-grade material** and is suitable for use in a **nuclear explosive**. *See also* LOW-ENRICHED URANIUM (LEU).

HIGH-TEST PEROXIDE (HTP). Hydrogen peroxide (H_2O_2) that has been concentrated above 68 percent, especially in the range of 85–98 percent. High-test peroxide (HTP) behaves similarly to **concentrated hydrogen peroxide** in that both decompose to steam and oxygen, making them suitable for use as a **propellant**. Both can be purchased commercially, making them of concern if diverted to use in **homemade explosives**. Hydrogen peroxide becomes more stable as it is concentrated, especially as the concentration approaches 98 percent. This property has been exploited for military use, with HTP plus a suitable catalyst being used in rockets and **torpedoes**. Handling and storage of HTP can be hazardous, however, especially if it should leak and contact water in a confined space. This situation arose on 16 June 1955 when an HTP-fueled torpedo exploded aboard the British submarine HMS *Sidon* while it was submerged at port off the southern coast of England. The explosion caused the vessel to sink, although most crew escaped. However, 12 crew members and one member of a rescue party were lost. An HTP-fueled torpedo is suspected as the cause of the 12 August 2000 sinking of the Russian submarine *Kursk* (K-141). All 118 crew members perished, but proof that HTP caused the sinking is lacking.

HMTD. *See* HEXAMETHYLENE TRIPEROXIDE DIAMINE (HMTD).

HMX. White, crystalline substance used extensively as an **explosive**. The preferred, systematic IUPAC name for HMX is 1,3,5,7-tetranitro-1,3,5,7-tetrazocane, and its chemical formula is $C_4H_8N_8O_8$. HMX is often referred to as "octogen," reflecting its eight-membered ring structure.

HMX is among the most powerful **chemical explosives**. It has a **Relative Effectiveness Factor** of 1.66, the same as that for **PETN** and somewhat greater than that for **RDX**. HMX is chemically and structurally similar to RDX. HMX came into widespread use as an explosive following the Second World War, especially in certain **plastic explosives**, such as **PBX**. Some sources credit the discovery of HMX to work done in Germany in the 1930s, although published reports of its properties are lacking in the scientific literature until after the war. Various attributions are given for the abbreviation "HMX." These attributions include its being an abbreviation for His (or Her) Majesty's explosive, high-velocity military explosive, high-melting explosive, high-molecular-weight explosive, and high-melting unknown. HMX was originally isolated as an impurity formed during the synthesis of RDX in a method that came to be known as the **Bachmann Process**. Although the process was developed in 1941, details were withheld from the scientific literature until after the war for reasons of national security. In their 1949 *Journal of the American Chemical Society* article describing this process, Werner E. Bachmann and John C. Sheehan include a footnote in which they credit University of Toronto chemist George F Wright (1904–1976) with designating an unknown present in the RDX as "HMX." Bachmann and Sheehan go on to note that "HM = high melting" but fail to specify whether "X" refers to "explosive" or "unknown."

Their 1949 article is silent on the explosive nature of HMX. However, a footnote in the experimental section emphasizes the need for using a safety shield when determining the melting point of HMX, leading one to conclude that its explosive properties had been recognized. Wright's use of HMX to designate an unknown substance found in RDX arose from his work for the Canadian government, which was searching for improved methods of RDX production. This work took place in 1941, in coordination with the research being conducted by Bachmann, Sheehan, and others in the United States. Therefore, although the precise attribution for the abbreviation "HMX" is uncertain, the abbreviation clearly originated with Wright, who deserves credit for coining the abbreviation. *See also* KISTIAKOWSKY, GEORGE BOGDANOVICH (1900–1982).

HOLLOW CHARGE. *See* SHAPED CHARGE.

HOMEMADE EXPLOSIVE (HME). An **explosive** prepared from commercially available precursor chemicals but intended for illicit or illegal use. Examples of such precursor chemicals include **ammonium nitrate**, **concentrated hydrogen peroxide**, fuel oil, **icing sugar**, **nitrobenzene**, and **nitromethane**. Commercially available chemicals can be attractive to prospective bomb-makers because they might be acquired legally without arousing suspicion, as happened in the case of those responsible for the **Oklahoma City Bombing** and the **7/7 Terrorist Bombings**. Alternatively, such chemicals might be obtained by theft or diversion from legitimate users. *See also* BERNARD, SIMON FRANÇOIS (1817–1862).

HOOSAC TUNNEL. Railroad tunnel through the Berkshire Hills in western Massachusetts that served as an important link between Albany, New York, and northeastern US cities, such as Boston. Construction of the Hoosac tunnel marked the first large-scale use in the United States of **nitroglycerin** and **blasting caps** for tunneling showed the superiority of nitroglycerin over **black powder** for such purposes. Work on the tunnel began on 8 January 1851 and continued for the next quarter century. Progress lagged considerably until the use of nitroglycerin was introduced in 1868, when **George Mowbray** began its production in North Adams, Massachusetts, near the western portal to the tunnel. During the course of supplying nitroglycerin for the Hoosac tunnel, Mowbray discovered that nitroglycerin is less **shock**-sensitive when frozen than when liquid. This discovery enhanced the safety of transporting and storing nitroglycerin and had important consequences for increasing its use. The first train passed through the tunnel on 9 February 1875, although its official opening was delayed until 1 July 1876.

When completed, the Hoosac tunnel measured 25,081 feet (4.75 miles), making it the longest tunnel in the Western Hemisphere. It retained this distinction until the 1916 opening of the Moffat Tunnel in the Rocky Mountains of Colorado. In 1975, the American Society of Civil Engineers designated the tunnel a Historic Civil Engineering Landmark, noting that its construction witnessed the introduction of tunneling innovations such as the use of steam drills and nitroglycerin.

HOWARD, EDWARD CHARLES (1774–1816). English scientist credited with discovering fulminic acid and its salts, **mercury fulminate**, and silver fulminate. On 13 March 1800, Howard announced the results of his research at a meeting of the Royal Society. His intended goal was to achieve the first synthesis of what is now known as hydrochloric acid (HCl), but was then called muriatic acid. That research produced the unexpected discovery of the **fulminates** as well as detailed descriptions of their nature as **explosives**.

Howard's announcement, as well as his paper in the prestigious *Transactions of the Royal Society*, attracted immediate attention throughout Europe and was promptly translated into German and French. The significance of his work was recognized, too, when the Royal Society bestowed its Copley medal on Howard, making him the youngest person to have been so honored.

Howard, however, soon abandoned this area of research owing to violent explosions in his laboratory. When he poured sulfuric acid on mercury fulminate, an "explosion nearly at the instant of contact was effected; I was wounded severely, and most of my apparatus destroyed . . . I must confess I shall feel more disposed to prosecute other chemical subjects."

HTA-3. *See* TERNARY EXPLOSIVE.

HURRICANE. Code name for the first **nuclear weapon** tested by the UK. Hurricane, an **implosion device** having 7 kilograms of **plutonium** as its **explosive**, was **detonated** on 3 October 1952 with an estimated yield of 25 kilotons TNT-equivalent. The detonation took place in a lagoon within the Montebello Islands, Western Australia, and was an important step for the British effort to develop a nuclear weapon. Despite the participation of British personnel in the **Manhattan Project**, the US Atomic Energy Act of 1946 prevented the United States from sharing design data with other nations. This act denied the British access to their collaborative work done on **Gadget**, **Little Boy**, and **Fat Man** during the Second World War. Because Hurricane had been placed in the hold of the HMS *Plym* for this detonation, it allowed the effects of a ship-smuggled nuclear weapon to be assessed.

HYATT, JOHN WESELY. *See* NITROCELLULOSE.

HYBRID EXPLOSIVE. An **explosive** combining properties of a **chemical explosive** and a **mechanical explosive**. Unlike typical chemical explosives, the chemical reactions that occur in a hybrid explosive cause neither **deflagration** nor **detonation**. For this reason, a hybrid explosive is also called a "non-deflagrating, non-detonating explosive." The hybrid explosive uses a chemical reaction to create gases, which are then used as a physical force for purposes such as demolition, debris removal, drilling, quarrying, and tunneling. *See also* HYDROX.

HYDROGEN BOMB. A **thermonuclear bomb** using hydrogen as the **primary explosive**. The device functions by **fusion**, requiring temperatures of several million degrees. The explosive yield of a hydrogen bomb is equivalent to about 58 kilotons **TNT-equivalent** per kilogram (2.2 pounds) of hydrogen fuel (H-fuel); that is, 1 kilograms of H-fuel yields a 58-kiloton explosion. This yield is nearly three times that expected from an equivalent mass of **uranium** fuel, which might be used in a **fission**-type nuclear bomb. The first successful test of a hydrogen bomb took place on 11 November 1952, when the United States conducted a test called "**Ivy Mike**." The yield of the blast was estimated to be 10.4–12 megatons TNT-equivalent.

HYDROX. A **hybrid explosive** using chemical reactions to generate gases, which are then used as **mechanical explosives**. Hydrox is a trademark for a system developed in the United States around 1955. It uses solid sodium nitrite ($NaNO_2$), which mixes with solid ammonium chloride (NH_4Cl). On mixing, these two solids undergo a reaction forming sodium chloride (NaCl), water, and molecular nitrogen (N_2). Nitrogen is a gas, and the pressure exerted by this gas can be used as a heaving force (e.g., in mining to dislodge rock) and for debris removal, underwater demolition, drilling, and tunneling. *See also* CARDOX.

I

ICING SUGAR (IS). Finely ground sugar produced by milling granulated sugar into a powdered state; also known as confectioner's sugar and powdered sugar. Icing sugar (IS) is used commercially in industrial-scale food production and by home cooks, especially for icing or decorating cakes and other edibles. As with any fine powder of a combustible material, IS can undergo **dust explosions**, as shown by the 7 February 2008 explosion at the Imperial Sugar Company, Port Wentworth, Georgia, that left 14 dead and injured scores. This property of sugar has been exploited for use in **homemade explosives**. Examples include the 24 April 1993 Bishopsgate Building Bombing in London, which killed one person; the 9 February 1996 South Quay Bombing in London, which killed two; and the 15 June 1996 Manchester, England, Shopping Centre Bombing, which had no fatalities but caused extensive property damage. All three bombings were attributed to the Provisional Irish Republican Army (PIRA) and involved IS mixed with **ammonium nitrate**.

IMPLOSION DEVICE. A type of **nuclear weapon** whose design features a spherical device that contains a quantity of **fissionable material** slightly below **critical mass** at ordinary conditions of pressure and volume. At **detonation**, the device undergoes implosion, having its volume suddenly reduced by compression brought about by **chemical explosives**. This compression takes the fissionable material to a supercritical mass, and a **nuclear explosion** ensues. **Plutonium** has often been used as the fissionable material in such a design. **Fat Man**, the bomb dropped on Nagasaki, is an example of an implosion device. *See also* EXPLOSIVE LENS.

IMPROVISED EXPLOSIVE DEVICE (IED). A device incorporating **explosive** materials intended to result in an explosion. An improvised explosive device (IED) might be fabricated in a completely improvised manner, using **homemade explosives** or commercial explosives acquired by theft. Alternatively, they might be an improvised modification to one or more conventional military weapons. *See also* IMPROVISED NUCLEAR DEVICE (IND).

IMPROVISED NUCLEAR DEVICE (IND). A device incorporating radioactive material designed to result in the formation of a nuclear yield reaction. An improvised nuclear device (IND) may be fabricated in a completely improvised manner or may be an improvised modification to a **nuclear weapon**. *See also* IMPROVISED EXPLOSIVE DEVICE (IED).

INDIA. Nuclear weapon state (NWS). The nuclear program in India began in the mid-1940s, perhaps as early as March 1944. Basic nuclear research was directed by Homi Jehangir Bhabha (1909–1966), founder of the Tata Institute of Fundamental Research. This program was directed initially toward peaceful uses of nuclear energy, meant to boost national prestige while providing inexpensive electricity. International assistance from Canada and the United States, among others, provided India essential technology, materials, and know-how that could be diverted toward building a **nuclear explosive** device. Such a device, given the code name **Smiling Buddha**, was **detonated** on 18 May 1974 and was touted by India as a PNE demonstration. Evidence of the development of nuclear weapons emerged following the May 1998 nuclear tests. Known as Operation Shakti, the tests took place at the Pokhran Test Range, Rajasthan. At 3:45 p.m., local time, on 11 May, three devices were detonated simultaneously. Two days later, at 12:21 p.m., two more devices were detonated simultaneously. All five detonations took place underground. One device was a **fusion**-type; the other four were **fission**-types. Yields ranged from sub-kiloton to 45 kilotons **TNT-equivalent**. The tests were announced publicly by Indian prime minister Vajpayee on 13 May, when it was also announced that the tests were concluded. His announcement confirmed that India possesses nuclear weapons. *See also* BANGALORE TORPEDO; PAKISTAN.

INDIA SALTPETER. Synonym for **potassium nitrate**. Deposits naturally rich in potassium nitrate occur in various regions of the world, including the **guano** found in caves and areas largely inaccessible to humans. Such deposits are found in **India**, especially in the northeastern regions. During the period of British domination or rule over India, these deposits became economically important as a source of potassium nitrate for use in manufacturing **black powder** and, eventually, fertilizers. The deposits were harvested, then mixed with water, and evaporated, leaving an impure form of potassium nitrate known as "grough." The impurities in the grough ranged from 1 to 10 percent and included various chlorides, hydroxides, and sulfates, plus sand and vegetable matter. A rough measure of the purity was obtained by casting the grough into a cake, which was broken apart and examined. The impurities were identified qualitatively by their effect on the overall appearance, and the measure of purity thus obtained was referred to as "refraction of niter." Beginning in the late eighteenth century and continuing into the twentieth century, trade in grough was dominated by the British through their control of the deposits in India, and potassium nitrate came to be referred to in the **explosives** and fertilizer industries as India Saltpeter. The importance of India Saltpeter as a source of potassium nitrate diminished greatly with the development of an economical industrial process for **nitrogen fixation**.

INITIATOR. *See* PRIMARY EXPLOSIVE.

INSENSITIVE HIGH EXPLOSIVE (IHE). An **explosive** that is extremely insensitive to fire, **friction**, impact, **shock**, or vibration and that is difficult to **detonate** by accident. The detonation of the conventional explosives in the **hydrogen bombs** in an aviation accident over Palomares, Spain, in 1966 and during an incident in **Greenland** in 1968 prompted a search by the United States for safer conventional explosives. This search culminated in the development by 1979 of a replacement, which came to be called an insensitive high explosive (IHE). Examples of IHE used by the United States include **FOX-7** and **triaminotrinitrobenzene**.

INSOLUBLE NITROCELLULOSE. Nitrocellulose that is insoluble in an ether-alcohol mixture. All forms of nitrocellulose are soluble in acetone and **nitroglycerin**; therefore, the term "insoluble nitrocellulose" is inexact and problematic, despite its being widely used in reference to **explosives**. The ether-alcohol mixture contains two parts (volume/volume) of diethyl ether to one part 95 percent ethanol. The solubility of nitrocellulose in this mixture is highly dependent on the nitrogen content of the nitrocellulose. As the percentage of nitrogen increases, the solubility of the nitrocellulose decreases. For example, nitrocellulose containing around 10–11 percent nitrogen, such as **collodion**, **pyrocollodion**, and **pyroxylin**, is soluble in the ether-alcohol mixture, while **guncotton**, which has 13–14 percent nitrogen, is insoluble in this mixture. Therefore, guncotton is an example of insoluble nitrocellulose, while collodion, pyrocollodion, and pyroxylin are examples of soluble nitrocelluloses.

Use of the terms "soluble nitrocellulose" and "insoluble nitrocellulose" proved important to the British government in deciding a patent infringement claim lodged by **Alfred Nobel**. Nobel disputed the award of a patent to Sir **Frederick Abel** and Sir **James Dewar** for what became known as **Cordite**.

The government ruled against Nobel, noting that Nobel's British patent had referred to the use of "soluble nitrocellulose," thus failing to anticipate the use of insoluble nitrocellulose, as was the case for the **guncotton** in Cordite.

IRAN. State known to have pursued **nuclear weapons**. From 1989 to 2003, Iran followed the Amad Project, a scientific effort to develop nuclear weapons, most likely with assistance from the **A. Q. Khan** nuclear technology smuggling network. A goal of this project was to design, produce, and test five nuclear devices, each to have a potential yield of 10 kilotons **TNT-equivalent** and each capable of being integrated onto a missile. While the IAEA has asserted that the effort ceased in 2003, it is strongly suspected to have continued. Iran is not known currently to possess nuclear weapons, but is known to have the infrastructure and experience to quickly produce **weapons-grade uranium** at multiple facilities, if it chooses to do so.

ISRAEL. Israel is widely suspected of possessing **nuclear weapons**. For example, a 1968 report by the Central Intelligence Agency (CIA) of the United States concluded that Israel possessed nuclear weapons, but the evidence offered was confined to discussions between a senior CIA official and renowned US physicist **Edward Teller**. Various reports have surfaced speculating on the Israeli nuclear weapons stockpile, with estimates ranging as high as 400 warheads. The size and composition of such a stockpile, if it exists, remain uncertain, as Israel engages in a policy of opacity toward nuclear weapons. On 12 December 2006, Israeli prime minister Ehud Olmert, during an interview on German television, appeared to state that his country in fact possesses such weapons, but official confirmation is lacking. Israel is not a party to various arms control agreements dealing with nuclear weapons, such as the Treaty on Non-Proliferation of Nuclear Weapons and the Treaty on the Prohibition of Nuclear Weapons.

Israel received US assistance with its nuclear program in 1955 under the "Atoms for Peace" initiative, through which the United States provided Israel a small nuclear reactor under bilateral safeguards. Israel recognized that this reactor would fail to provide the technology needed for a **nuclear weapons** program, and its efforts to obtain a reactor capable of producing **plutonium** from the United States were rebuffed. Thereafter, Israel received assistance from France, particularly through **Bertrand Goldschmidt**, beginning no later than 1956. France assisted in constructing the Negev Nuclear Research Center, near Dimona, ostensibly as a research reactor. Construction took place in secret and outside the inspection regime established by the IAEA. By 1958, the United States was aware of the construction through overflights of the U-2 reconnaissance aircraft. In December 1960, the Israeli government acknowledged the existence of a nuclear center near Dimona but declared it was being used for peaceful purposes. The reactor at the center went critical on 26 December 1963. The 1979 **Vela Incident** was viewed by some as a nuclear weapon test conducted jointly by Israel and **South Africa**, although the participation of South Africa seems to have been ruled out subsequent to its 1993 revelations regarding its nuclear weapon program.

More substantive information on the Israel program, though, emerged in 1986 as a result of revelations from Mordechai Vanunu. He had worked as a technician at Negev Nuclear Research Center but quit that job in October 1985. In 1986, he provided details to a reporter for a British newspaper on what he purported to be nuclear weapon-related work being carried out at the center. Shortly thereafter, Vanunu, who had emigrated to Australia but was visiting London, was lured to Rome by an Israeli intelligence service agent. On arrival, he was abducted, drugged, transported to Israel, and convicted of revealing state secrets. Vanunu was released on parole in 2004 after serving 18 years of imprisonment. His parole requires that he remain in Israel and refrain from meeting with foreigners.

ITALIAN-AMERICAN GALLEANISTI BOMBING CAMPAIGN. Assorted bombings carried out in the United States in the second decade of the twentieth century by followers of Luigi Galleani (1861–1931), a dedicated anarchist. Arriving in the United States in 1901, Galleani joined the Italian immigrant workers' movement, shortly becoming one of its leaders. In 1903, he joined with two others to found *Cronaca Sovversiva* (*Subversive Chronicle*), an Italian-language newspaper that promoted insurrectionary anarchism as a means of addressing workers' grievances. In 1905, he published the bomb-making handbook ***La Salute* è *in Voi!*** His followers, referred to as Galleanisti, likely used this handbook when they pursued their bombing campaign, which was most intense in April and June 1919. Their homemade bombs were simple and effective. Opening a package released the contents of a vial of sulfuric acid; the acid contacted **mercury fulminate blasting caps**; the caps **detonated**, setting off sticks of **dynamite**, which were surrounded by metal slugs. The slugs served as **Shrapnel**, adding to the destructive potential of each bomb when it exploded. The source of supplies for these bombs is uncertain but likely they were diverted from industries where Italian-American immigrants worked and where **explosives** were used routinely for legitimate purposes, such as farming, mining, quarrying, and building roads.

Their bombing campaign is known to have included bombs sent in April and June 1919. It is speculated to have included the 22 July 1916 Preparedness Day bombing in San Francisco that killed 10 and the 16 September 1920 **Wall Street Bombing** that killed 50. In both cases, though, proof that the

Galleanisti were responsible is lacking. Abundant proof exists to tie them to the 1919 bombings. These bombings came in two waves. In the first wave, at least 36 bombs were mailed in April. The hoped-for delivery date was 1 May, to coincide with worldwide celebrations honoring various labor, radical, and revolutionary movements. One package was delivered early, however, and opened, severely injuring two. Remnants were taken to postal authorities, enabling many others to be interdicted successfully, and no loss of life was experienced. The second wave involved nine bombs, detonated nearly simultaneously in eight US cities on 2 June 1919. As with the April bombs, each bomb used dynamite surrounded by metal slugs. But these bombs were substantially larger than those sent in April, having up to 25 pounds of dynamite. Two deaths were attributed to the June bombs, one of which was of a bomb-maker. While human deaths, injuries, and structural damages were limited, the effects were considerable. Most of the intended recipients of the April and June 1919 bombs were businessmen and government officials, especially those tied to efforts to suppress the labor movement. The response by these officials and others in the United States included a series of illegal searches and seizures, arrests with insufficient cause, and detentions of suspected anarchists and other radicals. While such actions were harsh and may have involved wrongdoing by officials, they appear to have succeeded in keeping more bombings from being carried out by the Italian-American Galleanisti, with the sole possible exception of the Wall Street bombings in 1920. By that time, Galleani himself had been deported to his native Italy, along with eight of his Galleanisti.

IVY MIKE. Code name for the first successful **detonation** of a staged **thermonuclear** device. The detonation was carried out by the United States on 1 November 1952, taking place on Elugelab Island in the Enewetak Atoll of the Pacific Proving Grounds, an area that encompassed the Marshall Islands, Christmas Island (Kiritimati), and Johnston Island. It had an estimated yield of 10.4 to 12 megatons **TNT-equivalent** and used the "fission-fusion-fission" design concept of the Teller-Ulam configuration. The overall device, known by the code name "Mike," weighed 82 tons, including the refrigeration plant required to liquefy deuterium, a component of the secondary portion of the explosive, by cooling it to approximately -418° Fahrenheit (-250° Celsius). This bulk and its massive size meant Mike was not a deliverable **nuclear weapon**; rather, its importance was that the test successfully demonstrated the **boosted fission weapon design concept**. Ivy Mike obliterated Elugelab Island, creating a crater over a mile in diameter and 164 feet deep. The test also led to the discovery of the elements einsteinium and fermium, which were created by the detonation and later found in the debris from the blast. *See also* GEORGE.

J

JOE-1, JOE-4, JOE-111. On 29 August 1949, the Soviet Union conducted its first successful **detonation** of a **nuclear explosive** device. The Soviets referred to the device officially as the RDS-1 but gave it the nickname "First Lightning" (in Russian: *Pervaya Molniya*). The US government gave it the nickname "Joe," after Soviet leader Joseph Stalin. The nickname for this first test would later become "Joe-1" when the Soviets conducted further nuclear tests. The design was similar to that of **Fat Man**. The device had a **plutonium** core, and its explosion had a yield of approximately 22 kilotons (kt) **TNT-equivalent**. The Soviets conducted the test in secrecy, and the United States was unaware of the test at the time. However, a US detection system onboard weather reconnaissance aircraft provided the evidence that would both detect and date this detonation, informing the West that the Soviet Union had perfected a **nuclear weapon**. On 12 August 1953, the Soviets successfully detonated their first **thermonuclear** device, using a design of alternating layers of the **fission** fuel (**uranium**-235) and **fusion** fuel (lithium-6 deuteride) similar to that of **Alarm Clock**. Its estimated yield was 400 kilotons TNT-equivalent, an apparently low yield for such a device, calling into question whether it was truly a **hydrogen bomb**. The Soviets officially designated the device the RDS-4, but the US government dubbed this device "Joe-4," in line with designations used for prior Soviet nuclear devices. On 30 October 1961, the Soviet Union detonated a three-stage device, widely known as "**Tsar Bomba**," but dubbed "Joe-111" by the US government. The detonation was the most powerful nuclear weapon detonated to that time. *See also* KURCHATOV, IGOR VASILIYEVICH (1903–1960); SAKHAROV, ANDREI DMITRIYEVICH (1921–1989).

JOHN HALL & SON. Private English firm engaged in manufacturing **explosives** from 1812, principally through operations at Oare Works and Marsh Works, near Faversham, Kent, England. John Hall (1764–1836) began business as a smith working in Dartford, where he made processing machinery for the local **black powder** mills. In 1812, he purchased the Royal Gunpowder Factory, Faversham, becoming directly involved in making black powder at the nearby Oare Works and Marsh Works. When he died in 1836, his sons William and Peter inherited the gunpowder operations in Faversham. In 1846, they were inspired by the demonstrations of **guncotton** given at the Royal Arsenal, Woolwich, by **Christian Schönbein** and **Rudolf Böttger**. They quickly purchased the rights to British patent 11,407 for "Improvements in the Manufacture of Explosive Compounds." This patent was granted on 8 October 1846 to John Taylor of Adelphi, Middlesex, England, but was based on information communicated to Taylor by Schönbein. The purchase gave John Hall & Son the sole right to manufacture guncotton in Britain for three years. The company constructed a factory near Faversham and began the commercial production of guncotton, becoming the earliest known commercial producer of guncotton in the UK. The factory exploded on 14 July 1847, killing many, only 10 of whose remains could be identified. John Hall & Son ceased production of guncotton in the aftermath of the explosion, although it continued to make black powder at the Oare Works and Marsh Works. The explosion was significant in that no further attempts would be mounted to produce guncotton in Britain until around 1863. *See also* PATENT SAFETY GUN-COTTON COMPANY.

K

KA PROCESS. *See* BACHMANN PROCESS.

KATIE. Common name for the Mark-23 **nuclear weapon** designed for use on US Navy ships. In September 1953, the United States began developing a nuclear weapon to be used on the 16-inch guns of the US Navy's *Iowa*-class battleships. Katie was placed in the US stockpile in December 1956 and remained part of the nuclear stockpile until October 1962. Each shell was slightly over 5 feet long and weighed nearly a ton. Its explosive yield was estimated to be 15–20 kilotons **TNT-equivalent**. There is no record that any US battleship ever carried Katie on board while at sea.

KHAN, ABDUL QADEER (1935–2021). Pakistani physicist and engineer widely viewed as a founder of the **nuclear weapons** program in **Pakistan**. In 1976, he became director of the Engineering Research Laboratories in Kahuta, in essence becoming the director of the Pakistani nuclear weapons program. His contributions and leadership of the program led to the 1981 renaming of the laboratories as the A. Q. Khan Research Laboratories (KRL) in his honor. In addition to his work on the Pakistani nuclear program, Khan was the central figure in the so-called "A. Q. Khan nuclear technology smuggling network." The existence of the network and of his role in it came to light largely because of revelations following **Libya**'s decision in 2003 to renounce nuclear weapons. Those revelations prompted the Pakistani government to arrest Khan on 31 January 2004 for illegally transferring nuclear technology. On 4 February 2004, he appeared on Pakistani television, admitting his role and absolving the government and military of any involvement. The following day, Khan was pardoned by Pakistani president Pervez Musharraf but was placed under house arrest. He was released from house arrest by order of the Islamabad High Court on 6 February 2009. Khan died in Islamabad on 10 October 2021. *See also* IRAN; NORTH KOREA.

KHARITON, YULI BORISOVICH (1904–1996). Russian nuclear physicist. Khariton was among the founders of the Soviet **nuclear weapon**

program, having been invited by **Igor Kurchatov** in 1943 to join the program. His influence was greatest through his position as director of **Arzamas-16**, the secret nuclear weapon complex that he helped to establish in 1946. His staff at the institute included **Andrei Sakharov**. Early in his career, Khariton received permission to study abroad, joining the research group headed by Ernest Rutherford at Cambridge University, England, and working alongside **James Chadwick**. Khariton earned his doctorate in physics from Cambridge in 1928. Thereafter, he returned to the Soviet Union, working in the Moscow region at the Institute of Chemical Physics before joining Arzamas-16, where he served as director for 46 years. His importance to the nuclear weapons program is illustrated both by his being forbidden to travel by air and by his having a dedicated, private railroad train carriage at his disposal.

KIESELGUHR; KIESELGUHR DYNAMITE. *See* DYNAMITE.

KISTIAKOWSKY, GEORGE BOGDANOVICH (1900–1982). Ukrainian-born physical chemist who made important contributions to the development of **nuclear explosives**. Kistiakowsky fled Ukraine during the Russian Civil War, arriving in Germany, where he earned his doctorate. He emigrated to the United States in 1926. In 1930, he gained US citizenship and joined the Harvard University chemistry faculty, where he would spend the bulk of his career. He took a break, however, immediately prior to US entry into the Second World War to serve as technical director of the Explosives Research Laboratory near Bruceton, Pennsylvania. In this role, he oversaw research into **shaped charges**, led development of a process capable of producing **RDX** on an industrial scale, and occasionally responded to special requests. One such request was from the US government's Office of Strategic Services (OSS) for an explosive that could be smuggled safely past Japanese guards into mainland Asia and used for sabotage by Chinese resistance fighters. To fulfill this request, Kistiakowsky mixed **HMX** with flour, creating a somewhat edible product that was also an explosive. The mix was dubbed "Aunt Jemima," a trade name of a then-popular American brand of pancake flour. While the mix was edible if taken in tiny amounts, the preferred response was to cook something with it if it was found and its use was challenged. Fifteen tons of this mix were produced during the war, and none was ever discovered by the Japanese or their collaborators on the Asian mainland to be an explosive rather than genuine pancake flour.

In 1943, Kistiakowsky was appointed to the **Manhattan Project**, leading X Division in Los Alamos, New Mexico, where he was responsible for the successful development of **explosive lenses** for use in **implosion-type** nuclear explosives. He was among the witnesses to the 16 July 1945 detonation of **Gadget** at the **Trinity Test Site**. After the war, he returned to Harvard, departing during the Eisenhower Administration to serve in the White House on the president's Science Advisory Committee and as director of the Office of Science and Technology Policy.

KURCHATOV, IGOR VASILIYEVICH (1903–1960). Russian physicist widely seen as the father of the Soviet **nuclear weapon** program. In 1943, Soviet leader Joseph Stalin ordered a small-scale project to be started in the Soviet Union aimed at developing a nuclear weapon and placing Kurchatov in charge. His work culminated in the successful **detonation** of a weapon on 29 August 1949. Kurchatov was revered by the Soviet leadership for his work on nuclear weapons, earning him the honor of having his ashes buried in the Kremlin wall necropolis on Red Square following his death in 1960. *See also* ARZAMAS-16; JOE-1, JOE-4, JOE-111; KHARITON, YULI BORISOVICH (1904–1996); SAKHAROV, ANDREI DMITRIYEVICH (1921–1989).

L

LAFLIN & RAND. *See* DU PONT, LAMMOT (1831–1884); GUNPOWDER TRADE ASSOCIATION.

LAKE DENMARK NAVAL AMMUNITION DEPOT EXPLOSION. A thunderstorm rolled across northern New Jersey late on the hot and muggy afternoon of Saturday, 10 July 1926. At 5:15 p.m., a bolt of lightning struck Temporary Magazine Number 8 on Lake Denmark Naval Ammunition Depot. A fire ensued within the magazine, which contained around 800 tons of assembled munitions and bulk **explosives**, mostly **TNT**-filled shells left over from the First World War. The fire caused the contents of the magazine to explode. The shockwave from this explosion was felt 30 miles away and caused the clock in the tower of the First Methodist Episcopal Church in Rockaway Borough to stop.

Within 45 minutes, the fire had spread to the nearby Temporary Magazine Number 9 and Shell House Number 22, causing their contents to **detonate**. These initial blasts fed even more fires and explosions on the depot and the adjoining Picatinny Arsenal. Fire and explosions continued for three days. In the process, 187 of 200 magazines were destroyed, and many other structures were damaged heavily. Metal fragments, flaming timbers, pieces of concrete, and hot ash rained down over an area as far as 22 miles from the depot. Nineteen men lost their lives from the explosions or as they fought the fires, although fatalities may have been limited because fewer personnel were on site during the weekend. The dead included 12 US Marines, who had been stationed at the depot. Contemporary estimates of the damages placed the cost at around $47 million.

The US Navy, which was responsible for the depot, and the US Army, which was responsible for Picatinny Arsenal, immediately opened investigations into the events of 10 July and its aftermath. These investigations revealed that storage of assembled munitions and bulk explosives in the magazines exceeded what the investigators took to be reasonable storage limits. However, written guidance on such limits was lacking, leaving interpretation of reasonable limits to the judgment of local authorities.

An indirect factor contributing to the events at Lake Denmark Naval Ammunition Depot was the end of hostilities in the First World War, which caused assembled munitions manufactured in the United States but intended for Europe to pile up at storage sites along the Atlantic seaboard. Additionally, some surplus Allied munitions as well as some captured munitions were shipped from overseas to the United States, most often destined for Atlantic seaboard ports. All these surplus munitions needed proper storage. Lake Denmark Naval Ammunition Depot was one of many sites selected to provide such storage. The rush to off-load munition-laden ships and to remove surplus munitions from ports along the Atlantic seaport likely added to insufficient attention being paid to the cumulative effect of placing additional munitions and explosives in any one location.

The loss of life and widespread destruction prompted far-reaching changes to munition storage magazines. One notable change was the 1928 adoption by the US Army of the American Table of Distances for Storage of Explosives. This document provided written guidance on munition storage, and its effective use was thought to decrease the likelihood of another event such as that at the Lake Denmark Naval Ammunition Depot. It established the Armed Forces Explosives Safety Board, later renamed the Department of Defense Explosives Safety Board, to be responsible for determining the safe separation distances for stored munitions and bulk explosives. Other changes included the mandatory installation of grounded lightning masts to divert the electricity from any lightning strike away from the magazine contents, and the use of an arched, earth-covered design for above-ground storage magazines. *See also* HALIFAX EXPLOSION; PORT CHICAGO DISASTER.

LAND CLEARING CAMPAIGN. Movement in the United States in the decade following the First World War to use **explosives** to clear land of tree stumps, enabling the land to be used for crop acreage. By the early years of the twentieth century, decades of harvesting timber had left expanses throughout the continental United States dotted with tree stumps. Many of these stumps were large and rot-resistant. Because the tree stumps were obstacles to cultivation, these cutover lands posed challenges to anyone hoping to use them for agriculture. The end of the First World War and its promise of enduring peace brought a possible solution. Stocks of military explosives were deemed to be in excess of needs for peacetime gunnery practice or similar uses. By reformulating some of these explosives, they might prove useful for commercial uses such as stump removal. With such use in mind, a land clearing campaign began, focusing particularly on the cutover Great Lakes regions in Michigan, Minnesota, and Wisconsin. Initially, surplus **TNT** was used, reformulated in many cases as "Sodatol," a combination of 45 percent TNT and 55 percent **sodium nitrate**. By 1924, **Pyrotol**, a **DuPont** product made with surplus **smokeless powder**, had become available to farmers at no cost other than shipping and packaging. The University of Wisconsin College of Agriculture became engaged in the campaign, sending a truck on tour into 51 counties to promote land clearing using explosives. The truck was driven by Walter A. Rowlands (1896–1972), who called himself by the nickname "Pyrotol Pete." By 1928, supplies of Pyrotol were exhausted. DuPont promoted "Agritol," made with **dynamite** rather than smokeless powder or TNT, as a replacement. But by then, the land clearing campaign had ended. Overall, Wisconsin was the largest distributor of war surplus explosives for land clearing, followed by Minnesota, Washington, Michigan, Oregon, and North Carolina. In all, 63 million pounds of surplus explosives had been distributed for use nationally, primarily for use in stump removal. The effectiveness of the campaign relative to adding land for use in agriculture is uncertain. Falling crop prices in the late 1920s and a growing recognition of alternative uses for the cutover lands combined to foster more nuanced land-use decisions, with farming joining forestry, recreation, and residential as choices. *See also* BATH SCHOOL BOMBING.

LA SALUTE È IN VOI! **Italian-language handbook on homemade explosives and improvised explosive devices**. The title can be translated as either "Health is in You!" or "Salvation is Within You!" This 48-page handbook appeared in the United States in the first decade of the twentieth century and is associated with a group of Italian anarchists and radicals in the United States, which followed Luigi Galleani and his message of violence in the face of capitalists' oppression. The author or authors are unknown but are speculated to be Galleani and **Ettore Moliani**. The contents include sections on explosive materials, **dynamite**, **fuses**, **mercury fulminate**, and **nitroglycerin**.

The information contained in the handbook was available in other sources, such as chemistry texts and reference works. But such sources might have been inaccessible to the intended audience or might have been overly technical for this audience to use successfully in making explosives. Furthermore, the handbook was written as both an easily usable guide to explosives and as a political manifesto urging the audience to commit revolutionary acts against their bosses and other capitalists who profited at their expense. The message informing the manifesto was consistent with Moliani's known writings, which included *Guerra all'oppressore* (*War Against Oppressors*). Followers of Galleani—referred to as the **Italian-American Galleanisti**—are known to have been responsible for various bombings, especially ones carried out in 1918–1919. They are speculated to have had ties to other bombings, including the **Wall Street Bombing** of 1920. Definitive proof of their involvement in these other bombings is lacking. Furthermore, no bombings in the United States have been conclusively tied to the use of *La Salute è in Voi!* as a guide.

LAVOISIER, ANTOINE (1743–1794). French scientist and government official. He was originally named Antoine-Laurent de Lavoisier, but he shortened it to the less royalist-sounding Antoine Lavoisier in 1789 during the French Revolution. Lavoisier contributed significantly to the growth of chemistry and is often labeled the father of modern chemistry. In the context of **explosives**, he made important contributions that resulted in increasing the quality and quantity of **gunpowder**. These contributions arose in large part through his application of scientific principles to the study of explosions and explosives. The results had important consequences for the efforts of the colonies in North America to achieve their independence from Britain, both through the quantity and quality of gunpowder exported to the colonies during the American Revolution and by the example he set for E. I. **Du Pont**, one of his apprentices.

Lavoisier served as a founding commissioner on the Gunpowder and Saltpeter Administration (1775–1792). In this position, he brought a scientific approach to understanding the chemistry of gunpowder. Through his scientific discoveries, Lavoisier delineated the chemical reaction of exploding gunpowder as the reaction between a nitrate, present in the form of **potassium nitrate** in crude saltpeter, and carbon, present in the form of **charcoal**. This delineation allowed him to determine quantitatively that the optimal proportions of saltpeter and charcoal in gunpowder are approximately eight parts saltpeter to one of charcoal. Prior to the work of Lavoisier, such proportions were developed empirically and lacked a sound scientific basis. Lavoisier also reformed what had previously been an archaic, inefficient, and rudimentary process of harvesting and purifying potassium nitrate for use in making **black powder**. For example, he sent technicians to **India** to observe and report on their methods for harvesting **India Saltpeter**. In studying the natural formation of saltpeter, Lavoisier identified slow circulation of air and moisture through the soil as optimal conditions for saltpeter production. Accompanied by Jean-Baptiste-Paul-Antoine Clouet (1739–1816), he toured France, looking for sites matching these conditions of slow circulation of air and moisture. Based on their examination of the local geography, they identified the Touraine region of southwestern France as the prime candidate for natural saltpeter production.

Lavoisier studied methods used elsewhere in Europe for niter beds, which are raised enclosures used to produce a nitrate-rich efflorescence that might be harvested and purified to yield potassium nitrate. He used the term *nitriary* for such beds. As a result of his studies, he concluded that the methods used by the Swedes were superior and urged saltpeter farmers to adopt these methods. In 1777, he published a manual *Instruction sur l'établissement des nitrières et sur la fabrication du salpêtre* (*Instructions for establishing nitriaries* [niter beds] *and for making saltpeter*). It gave detailed instructions as well as diagrams on selecting a site suitable for operating a niter-bed, building pyramid-shaped saltpeter-collection beds, selecting rotting mater, urine, and other liquid wastes for use in the beds, and refining the saltpeter once it is harvested. This manual would place **saltpeter farming** on a more systematic and scientific basis.

Lavoisier also sought replacements for one or all of the ingredients in black powder. In this regard, he encouraged the research of his colleague **Claude Louis Berthollet**, who was experimenting with **potassium chlorate** as a possible substitute for potassium nitrate. The results of this experiment, however, proved disastrous when a milling operation accidentally exploded on 26 October 1788, killing one of Lavoisier's fellow commissioners from the Gunpowder and Saltpeter Administration as well as the sister of another commissioner. This tragedy caused Berthollet and Lavoisier to cease their work on the possibility of substituting potassium chlorate for potassium nitrate. *See also* GUNPOWDER SERVICE (*SERVICE DES POUDRES ET SALPÊTRES*).

LEAD AZIDE. White to yellow crystalline inorganic compound, having the empirical formula $Pb(N_3)_2$. Lead azide is usually handled and stored under water. It is extremely sensitive to **friction**, heat, **shock**, and vibration. These characteristics have made it useful as a **primary explosive**. German chemist Theodor Curtius (1857–1928) is credited with the discovery and preparation of lead azide in 1891. Shortly thereafter, German manufacturers of explosives studied it as a replacement for **mercury fulminate**. These studies were halted in 1893 when an explosion in a Spandau laboratory killed a man. The first reported successful use of lead azide in **explosives** was in a 1907 French patent, number 384,792, granted to François Hyronimus (1874–1952) of Société Française des Munitions de Chasse, de Tir et de Guerre (French Society of Ammunition for Hunting, Shooting and War). This patent was followed in 1908 by British patent 1,819, also granted to Hyronimus and titled "Improvements relating to the charge of ammunition primers."

Despite these technical developments, the adoption of lead azide in explosives lagged. This situation changed during the First World War when Italy declared war on Austria-Hungary on 23 May 1915. Italian army attacks on Gorizia, then part of Austria, jeopardized unimpeded access by the Central Powers to the mercury mines of Idria, situated on the eastern edge of Gorizia.

The only other operating European mercury mines were in Spain, a neutral party during the conflict. Consequently, the Central Powers took steps to develop safe handling techniques for producing, using, and storing lead azide so that it might serve as a replacement for mercury fulminate. Investigations during the 1920s and 1930s into improvements in lead azide led to methods for mixing it with any of various polysaccharides as a **stabilizer**. Lead azide treated in this manner is referred to as dextrinated lead azide (DLA), and it is this form of lead azide that has come into more common use. *See also* EXPLOSIVE BULLET.

LEAD BLOCK TEST. *See* TRAUZL LEAD BLOCK TEST.

LEBEL MODEL 1886 RIFLE. French 8-mm bolt-action rifle credited as the first military rifle designed to use a cartridge based on **smokeless powder**. The Prussian army had earlier introduced a type of smokeless powder in cartridges used by the **Dreyse needle-gun**, but this gun was not specifically designed for a smokeless cartridge. Prussian use of that cartridge during the **Franco-Prussian War** may have added to interest in developing a weapon specifically designed for a smokeless cartridge.

In April 1887, the French formally adopted the rifle that became known as the Lebel model 1886 rifle. It is named for Nicolas Lebel (1838–1891), who designed the cartridge. At the time, Lebel was a Lieutenant Colonel in the French Infantry and was a member, along with **Paul Vieille**, of the Commission for Repeating Firearms (*Commission des Armes à Répétition*), which had been tasked by the French government to exploit Vieille's newly developed **Poudre B**. Production of the Lebel model 1886 ceased in May 1920, by which time more than 2.8 million had been manufactured. It remained in use with the French Foreign Legion until the late 1930s.

The introduction of the Lebel model 1886 touched off an arms race as various nations sought to match or exceed its performance. Examples include the 1888 adoption by Germany of the Gewehr 88 or Model 1888 Commission rifle and the 1891 adoption by Italy of the Mannlicher-Carcano rifle and by Russia of the Mosin-Nagant rifle. In 1892, the United States selected the Krag-Jørgensen, a Danish-designed rifle that fired a smokeless cartridge known as the .30–40 Krag. In 1895, the UK adopted the .303 Lee-Enfield Mk I rifle, specifically designed to fire a smokeless cartridge. *See also* FRANCO-PRUSSIAN WAR; SPANISH–AMERICAN WAR (APRIL–AUGUST 1898).

LENK VON WOLFSBERG, WILHELM, FREIHERR (1809–1894). Austrian military officer who developed methods for preparing **guncotton**.

Although the methods were widely adopted, they failed to overcome the stability problems that had plagued guncotton and were eventually abandoned. The potential use of guncotton for military applications was demonstrated conclusively in 1846 by, among others, **Christian Schönbein** and **Rudolf Böttger**. In 1852, the Austrian government acquired the rights from Schönbein to use his process for manufacturing guncotton. That same year, the Emperor of Austria established a committee to investigate possible military uses of guncotton. He appointed Lenk, an officer in the Austrian army and a nobleman, to this committee, and thereafter Lenk heavily influenced its activity. Lenk immediately set about developing a more elaborate manufacturing process than that of Schönbein in an effort to enhance the stability of the final product. In 1853, the Austrian government built a factory at Hirtenberg to make guncotton using Lenk's process. Testing of this yarn-like product, however, yielded poor results. When used as a **propellant**, it would **detonate** too rapidly and with intense shock, often causing the projectile to burst within the gun barrel. It did function satisfactorily as a bursting charge, with **black powder** being used as the propellant. Despite the disappointing results, the Austrian government granted Lenk permission to communicate his results to other nations. In 1862 and 1863, he took out British patents on his manufacturing process, using the pseudonym of "Révy," apparently in an effort to hide his identity as an Austrian officer and nobleman. In 1863, he visited England and made presentations on his manufacturing process. These presentations caught the interest of Edward Prentice and his nephew, William Prentice. They adopted Lenk's process for use at their newly formed **Patent Safety Gun-Cotton Company**, Stowmarket.

On 20 July 1863, a magazine at Hirtenberg exploded accidentally, causing the Austrian government to abandon use of guncotton as a propellant, but permitting its continued use as a bursting charge. In 1865, a guncotton magazine at Steinfelder Heath, near Vienna, exploded accidentally. As a result, the Austrian government forbade all use and manufacturing of guncotton as of 11 October 1865. Lenk, however, prospered in his military career, despite the setbacks with guncotton and his advocacy for its use and manufacture, and in 1877 was promoted to full general.

LE ROSSIGNOL, ROBERT. *See* HABER-BOSCH PROCESS; HABER PROCESS.

LIBYA. State known to have pursued a **nuclear weapons** program. In 1975, Libya became a party to the **Treaty on Nonproliferation of Nuclear Weapons**. However, it violated this treaty and conducted a nuclear weapons program with assistance from the **A. Q. Khan** nuclear technology smuggling network. In December 2003, Libya renounced all weapons of mass destruction programs, and its nuclear weapons program was dismantled by the UK and the United States with oversight by the IAEA.

LINDE, CARL PAUL GOTTFRIED VON (1842–1934). German scientist, engineer, and inventor. Carl von Linde invented a continuous process for liquefying gases in large quantities. This invention formed the basis for the science of refrigeration, making possible basic scientific studies at very low temperatures. It also became the basis for an industry that Linde headed and that was centered around producing gases for industrial uses. Of special importance was Linde's development of the large-scale production of liquid oxygen, which could now be used as a fuel in industrial processes, such as high-temperature furnaces for manufacturing steel. In 1895, he introduced a novel **explosive** comprised of carbon black (a soot-like substance), packed into a porous bag, which was then dipped into liquid oxygen. This type of **Sprengel explosive** became known as an **oxyliquit** and saw extensive use during the construction of the Simplon Tunnel. This railway tunnel goes through the Alps and connects Brig, Switzerland, and Domodossola, Italy. When it opened in 1906, it was the longest railway tunnel in the world, remaining so until the 1982 opening of the Daishimizu Tunnel in Japan.

LIQUID EXPLOSIVE. Chiefly, an **explosive** in which all components are liquid. The term is also applied to, but infrequently, an explosive in which some but not all components are liquid, as is the case for the **Sprengel explosives**.

LIQUID OXYGEN EXPLOSIVE. *See* OXYLIQUIT.

LITTLE BOY. Initially, the nickname given to the first **nuclear weapon** successfully **detonated** in warfare; later, the nickname for bombs of a similar design. The original Little Boy was dropped over Hiroshima, Japan, by the United States on 6 August 1945. It used **uranium**-235 as the **fuel** for its **nuclear explosive**. The resulting nuclear detonation was estimated to be in the range of 16–20 kilotons **TNT-equivalent**. Little Boy was designed as a **gun-type nuclear device**. The final assembled device weighed approximately 4.5 tons, was 10 feet long, and was slightly over 2 feet in diameter. The design of the nuclear explosive called for nine washer-shaped rings to be launched by detonation of **conventional explosives** as a single projectile toward a target consisting of six discs arrayed over a polonium-beryllium **primary explosive**. When the holes in the washer-shaped rings slipped over the discs, they combined to form a supercritical mass, and a nuclear detonation ensued. Each ring was 6.25 inches in diameter, with a combined mass of approximately 84 pounds. Each target disc was 4 inches in diameter, with a combined mass of approximately 57 pounds. When the projectile was dropped over the target area, the total mass of uranium was 142 pounds. Studies of the detonation revealed that the explosion consumed slightly less than 2 pounds of the 142 pounds, for an overall efficiency of 1.2–1.4 percent. The Little Boy bomb was officially designated the Mark-1 (Mk-1).

Despite the successful detonation of the bomb over Hiroshima, the US military preferred the design of the **Fat Man** bomb, and production of that type of bomb overshadowed production of the Mk-1 type. A few Mk-1 bomb casings (perhaps 30) were constructed, but no nuclear explosive cores were made. By 1951, the Mk-1 Little Boy bomb was no longer in the US nuclear weapon stockpile. *See also* GADGET.

LOCKERBIE BOMBING. *See* PAN AM FLIGHT 103.

LOW-ENRICHED URANIUM (LEU). Uranium that has been processed to increase the uranium-235 content (i.e., "enriched") but that still contains less than 20 percent of that isotope. Further processing is required to make the material suitable for use as a **fuel** in a **nuclear explosive**. *See also* HIGHLY ENRICHED URANIUM (HEU).

LOW EXPLOSIVE; LOW VELOCITY EXPLOSIVE. An **explosive** that, when functioned, decomposes rapidly but at a subsonic speed, resulting in **deflagration** rather than **detonation**. Examples include substances used as **gunpowders**, **pyrotechnics**, and **propellants**. *See also* EXPLOSIVE LENS; HIGH EXPLOSIVE (HE); HIGH VELOCITY EXPLOSIVE.

LOW-FREEZING DYNAMITE. Dynamite treated in some manner so that the freezing point of **nitroglycerin** is lowered. The normal freezing point of nitroglycerin in dynamite is about 55°F (13°C). For many regions of the world and at many times of the year, the ambient temperature is at or below this temperature. If nitroglycerin should freeze following its incorporation into dynamite, it will separate out as an oil on melting. Nitroglycerin that accumulates in this manner—that is, by melting and separating out of the dynamite—has enhanced sensitivity to **friction**, **shock**, and vibration. Consequently, the separated material presents an extreme hazard for storage, transport, and handling. Lowering the freezing point extends the useful range of the dynamite and reduces the possibility that a hazardous condition might be created if the nitroglycerin should freeze.

In 1890, German chemist Alfred Wohl (1863–1939) suggested using mononitroglycerin and **dinitroglycerin (DNG)** to lower the freezing point of nitroglycerin. Wohl's research interests were primarily in fermentation chemistry, rather than in **explosives**, and he apparently failed to pursue the topic further. His suggestion was noted, though, and was reduced to practice by Anton Mikolajczak, who mixed DNG with nitroglycerin, producing dynamite having a freezing point significantly below 0°F (-18°C). In 1904, Mikolajczak published his experimental results in the scientific literature ["Neue Sprengstoffe" ("New explosives"), *Glückauf* **40** 629]. That same year, he obtained French patent 341,911 and British patent 27,706 for his invention of dynamite modified by the use of a mix of nitroglycerin and DNG, receiving US patent 798,436 for this invention on 29 August 1905. That invention came to be known as low-freezing dynamite. Mikolajczak is widely credited with its invention, both because of his early patents and because he openly published his results in the scientific literature. While DNG continued to see use in dynamite for many years, it has the disadvantage of being soluble in water. Consequently, it eventually fell into disuse. Mikolajczak's work was important, however, because it inspired other scientists to investigate related substances, especially other glycerin esters, as possible freezing-point depressants for nitroglycerin.

LULU. Common name given to a **nuclear weapon** developed and fielded by the United States for use as a depth charge. Officially designated the Mk-101, Lulu weighed 1,200 lbs. and had dimensions of 7.5 feet long and 1.5 feet in diameter at its widest. Approximately 2,000 were produced in five different models. Lulu's yield was estimated at 11 kilotons **TNT-equivalent**. Lulu was in service from 1958 to 1971 with the US Navy for antisubmarine warfare. *See also* SPECIAL ATOMIC DEMOLITION MUNITION (SADM); WE.177A.

LYDDITE. Name adopted by the British military to protect the identity of a **picric acid**-based **smokeless powder**. In response to the successful development of **mélinite** by the French, the British experimented with picric acid and various derivatives as a possible filling for munitions. In 1888, a mix most likely containing picric acid, **nitrobenzene**, and petroleum jelly was tested at a military artillery practice range near Lydd, in the south of England, near the English Channel coast. The testing proved the mix was a success, and it was dubbed "Lyddite" in honor of the nearby town. It also served as a code name to obscure its chemical composition.

Lyddite became a standard **explosive** for British munitions. It was used for the first time in combat on 1 September 1898 during the opening operations of the Battle of Omdurman in Sudan. The 37th Royal Artillery battery fired 50-pound Lyddite-filled shells from Ordnance Breech-Loading 5-inch howitzers into Omdurman. The firing continued the next day and succeeded in destroying portions of the city walls and the Mahdi's tomb. More extensive use of Lyddite occurred during the **Second Boer War** (1899–1902) and the First World War. However, it suffered from the drawbacks of all picric acid-based explosives in that it requires a powerful **detonator**, and the picric acid reacts with exposed metals in the shell lining to form unstable picrates. The need for a powerful detonator was made apparent during the Second Boer War when many Lyddite-filled shells failed to detonate owing to a faulty detonator. A further drawback to Lyddite was that it was sufficiently **shock-sensitive** that it tended to detonate on contact with a dense surface, such as armor plate. This surface detonation dissipated the impact of the explosive, limiting its effectiveness in penetrating armor and other defenses. *See also* EXPLOSIVE D; SHELLITE.

M

MACCLESFIELD BRIDGE DISASTER. *See* MAJENDIE, VIVIAN DERING (1836–1898).

MADRID TRAIN BOMBINGS. Ten **detonations** took place nearly simultaneously on commuter train cars during the morning rush hour in Madrid on 11 March 2004. These bombings killed 191 outright. Two more died later, with the last death attributable to the bombings occurring in 2014 to a victim who had remained in a coma for a decade. More than 1,800 were injured, and property damage was extensive. The event became known in Spain as "11-M" (11 de Marzo) after the date of the event.

The explosions occurred on trains at the Atocha, El Pozo, and Santa Eugenia stations and alongside the Calle Téllez station. The four trains shared the same line and were headed in the same direction. In addition to the 10 explosions, three more unexploded devices were recovered eventually. Two were found that same day and were **rendered safe** by Spanish **explosive ordnance disposal (EOD)** experts. The third one was found on 2 April and was determined to have failed to detonate owing to a lack of a functioning **primary explosive**. The recovery of these three devices enabled authorities to determine that commercial **explosives** had been used. Both **dynamite** and Goma-2 ECO were identified, with the latter being comprised of **ammonium nitrate**, **ethylene glycol dinitrate (EGDN)**, **nitrocellulose**, and dibutyl phthalate. Further investigation found that these explosives had been diverted illegally by an individual having legal access to explosives for mining operations. That individual had traded the explosives for drugs. He would be identified, prosecuted, and convicted for his role in the bombings, receiving a prison sentence in excess of 34,000 years.

Early reports, including statements by Spanish government officials, labeled the bombings the work of ETA, a Basque separatist group responsible for a campaign of violence over more than 30 years and claiming more than 800 deaths. Eventually, though, evidence pointed to Islamic fundamentalists. Although a person identifying himself as a spokesman for Al Qaeda claimed that this militant group was responsible, the link to Al Qaeda remains unverified. In October 2007, 18 Islamic fundamentalists of mainly North African origin and three Spanish accomplices were convicted for their roles in the bombings. The Madrid train bombings are the worst Islamist attack in European history since the **Pan Am Flight 103** bombing on the 21 December 1988, which killed 270.

MAGNESIUM. Silvery metal. Finely divided magnesium burns readily in air. This property has been exploited successfully in various **explosives**, especially in **pyrotechnics**.

MAINE. *See* SPANISH–AMERICAN WAR (APRIL–AUGUST 1898).

MAJENDIE, VIVIAN DERING (1836–1898). British engineer and Royal Artillery officer. Majendie pioneered the fields of **explosive ordnance disposal (EOD)** and **forensic analysis of explosives**. After serving as the assistant superintendent at the Royal Arsenal at Woolwich, he was appointed in 1871 chief inspector of explosives for the UK. He remained in this position until his death in 1898. In his role as chief inspector, he led the inquiry into the **Stowmarket Gun-cotton Explosion of 1871**. In doing so, he used methods and procedures that would become the basis for forensic analysis of explosives. The UK Defence Science and Technology Laboratory (DSTL) dates the founding of its Forensic Explosives Laboratory to Majendie's work on this 1871 inquiry. He investigated the Macclesfield Bridge Disaster, an explosion on 2 October 1874 of a **gunpowder**-laden barge on the Regent's Canal that killed three men and a boy as well as destroying the bridge. His findings from that investigation helped frame the UK Explosives Act of 1875. This Act amended and consolidated laws governing explosives, covering their import, manufacture, sale, storage, and transport. Among its provisions, the Act placed the burden on those handling explosives to take reasonable precautions to prevent unauthorized access. It banned the sale of explosives to those under age 16. It required retailers to be registered. It also established the position of government inspectors whose role included ensuring the provisions of the Act were being enforced. In recognition of his role in developing the Act, Majendie received the honor of being appointed a Companion of the Order of the Bath (CB). After a bomb was detonated at Victoria Station, London, on 26 February 1884, Majendie **rendered safe** an unexploded bomb by disabling its clockwork mechanism, showcasing the emerging discipline of EOD. In 1895, he was made a Knight Commander of the Order of the Bath (KCB) in recognition of his services.

MALPAS TUNNEL. First tunnel excavated entirely with the use of **explosives**. In this case, the excavation was done in 1679, and the explosive was **black powder**. The tunnel is on the Canal du Midi and runs beneath the d'Ensérune hill near Hérault, France. The work was conducted in secret at the direction of chief engineer Pierre-Paul Riquet and under the supervision of master mason Pascal de Nissan. Riquet directed that the work be done in secret owing to the risk of collapse, which he feared would harm the overall canal building project. The tunnel was completed in eight days and runs 165 meters, with an 8-meter arch.

MAMMOTH POWDER. A preparation of **black powder** having a large grain size; specifically, that which was developed by **Thomas J. Rodman** and **Lammot du Pont** in 1856. The powder resembled small pebbles rather than the fine powders typically in use at that time as **propellants**. The relatively large size of the powder caused it to be dubbed "mammoth." By the outbreak of the **American Civil War**, mammoth powder was a standard propellant for 15-inch guns and was used by both sides during that conflict.

MANHATTAN PROJECT. Albert Einstein signed a letter to US president Franklin D. Roosevelt on 2 August 1939. Sent at the urging of **Leo Szilárd**, who helped compose it along with fellow nuclear physicists **Edward Teller** and Eugene Wigner (1902–1995), the letter described how **uranium** could be used to create "extremely powerful bombs." After a two-month delay, Roosevelt set in motion what would prove to be a massive, multinational effort to develop **nuclear weapons**. To maintain secrecy, the effort to build the nuclear weapons was placed under the US Army Corps of Engineers. A member of this corps, Leslie R. Groves (1896–1970), was chosen to command the effort, which was placed under the Manhattan Engineering District (MED). Initially, the effort was given the benign-sounding cover name of the "Development of Substitute Materials Project." But its link to the MED led the effort to become known as the Manhattan Project.

The project was truly international, multidisciplinary, and geographically dispersed. Scientists, engineers, and others from Canada, France, Italy, the United Kingdom, and the United States contributed to the Manhattan Project, working out of more than 30 locations. Major sites involved were:

- Alamogordo, New Mexico, designated the **Trinity Test Site**, where the first successful **detonation** of a nuclear device, **Gadget**, would take place on 16 July 1945.
- Berkeley, California, where the Radiation Laboratory made fundamental contributions to nuclear physics.

- Chicago, Illinois, where Italian-born physicist Enrico Fermi led the team at the Metallurgical Laboratory that achieved the first successful sustained chain reaction. Among other scientists at this laboratory were **James Franck** and **Bertrand Goldschmidt**.
- Los Alamos, New Mexico, known as **Site Y**, where much of the research and design of nuclear weapons took place and where many of the eminent scientists worked, such as Edward Teller, **James Chadwick**, **Otto Robert Frisch**, **George Kistiakowsky**, and **Julius Robert Oppenheimer**, who served as its scientific director.
- Oak Ridge, Tennessee, known as **Site X**, which became home to the Headquarters, MED, as well as the Clinton Engineering Works, where **uranium** was enriched for use in weapons.
- Richland, Washington, known as **Site W** and home to the Hanford Site that produced **plutonium** for use as **fuel** in a nuclear explosive.
- Wendover, Utah, where the US Army Air Corps airfield was used by the **509th Composite Group** as part of "Project Alberta," where the aircrews practiced the unusual and difficult maneuvers associated with dropping a single, 5-ton bomb, then performing a nearly 360-degree turn so that both the path of the bomb might be followed following its release and its detonation might be captured on film. These maneuvers were used in combat when members of this group dropped the nuclear weapons **Little Boy** and **Fat Man**.

Although the Second World War ended with the formal surrender of Japan on 2 September 1945, the work of the Manhattan Project continued for another 16 months. In January 1947, the Manhattan Project ceased when its work was transferred to the newly established Atomic Energy Commission (AEC). *See also* AKERS, WALLACE ALAN (1888–1954); EXPLODING BRIDGEWIRE DETONATOR (EBW); PENNEY, WILLIAM GEORGE (1909–1991).

MASSIVE ORDNANCE AIR BLAST BOMB (MOAB). Large-yield non-nuclear **explosive** device, believed to have been the largest non-nuclear weapon ever **detonated** when it was originally tested in 2003. The Massive Ordnance Air Blast Bomb is often known by its acronym, MOAB, as well as its nickname, "mother of all bombs." Its official designation is the GBU-43/B, a bomb designed by the US Air Force and intended for delivery by a C-130 cargo aircraft. The bomb is reported to weigh over 21,000 pounds (10.5 tons), of which 18,700 pounds (nearly nine tons) is its explosive charge, **H-6**. On 11 March 2003, the bomb was tested successfully at Eglin Air Force Base, Florida. That test resulted in a detonation estimated to have been 11 tons **TNT-equivalent**, placing its yield at about 0.1 percent that of Little Boy, the **nuclear explosive** device detonated over Hiroshima, Japan, in 1945. The Massive Ordnance Air Blast Bomb is not known to have been used in combat. *See also* FATHER OF ALL BOMBS (FOAB); MASSIVE ORDNANCE PENETRATOR (MOP).

MASSIVE ORDNANCE PENETRATOR (MOP). Precision-guided bomb designed to destroy hardened, underground targets by penetrating up to 200 feet of unspecified material before **detonating**. The Massive Ordnance Penetrator is often known by its abbreviation, MOP, as well as its nickname, "bunker buster bomb." Officially designated the GBU-57A/B, the MOP was developed by the US Air Force. Each bomb is 20.5 feet long and 31.5 inches in diameter and weighs around 15 tons. Its size limits deployment to a single aircraft type, the B-2 strategic bomber. The **explosive** payload of the MOP is approximately 2.5 tons and uses a mix including **plastic-bonded explosives** and **polymer-bonded explosives** (PBX), optimized for detonation in confined spaces. The United States used the MOP during its 22 June 2025 strike against nuclear material enrichment facilities in Iran. *See also* FATHER OF ALL BOMBS (FOAB); MASSIVE ORDNANCE AIR BLAST BOMB (MOAB).

MAXIM, HIRAM STEVENS (1840–1916). Prolific American-born inventor, many of whose inventions had military applications. Raised in Maine, Maxim had a limited formal education before going to work as a coachbuilder, instrument maker, and draftsman. Despite his lack of training in engineering or science, he developed various inventions relating to incandescent lamps and electric light bulbs, beginning in the 1870s, for which he received patents. These patents and others led to various long-standing patent disputes with, among others, the famed American inventor Thomas Edison. In 1881, Maxim relocated to Europe, eventually settling in England, outside London. While he continued to pursue business ventures related to electricity and electric light, he began to turn his inventiveness to guns and ammunition. In 1883, he perfected what became known as the Maxim gun, which is widely viewed as the first successful fully automatic machine gun. This invention would soon become more important when the newly invented **smokeless powders** were used as filling for cartridges, permitting a gunner to continue sighting and firing his weapon without being engulfed in a cloud of smoke.

Maxim formed a company to produce and market this weapon and other inventions, which included **Maxim's Powder**, a smokeless powder for which he received British patent 16,213 in 1889 and US patent 430,212 in 1890. In time, his company became the Maxim-Nordenfelt Guns and Ammunition

Company, Ltd. In 1889, the British government granted a patent to Sir **Frederick Abel** and Sir **James Dewar** for a smokeless powder. This powder eventually became known as **Cordite** and was similar in composition to Maxim's Powder. Maxim challenged this patent. The challenge went to the House of Lords, which eventually ruled against Maxim. On 21 September 1894, Maxim was granted British patent 17,994 for a **multi-perforated smokeless powder**. This patent would be the basis for a long-running dispute with his brother, **Hudson Maxim**, who was granted various US patents for a similar invention. In 1896, Hiram Maxim sold this company to Vickers, Sons & Company, which named the combined firm "Vicker, Sons & Maxim." In 1900, Maxim became a British citizen and in 1901 was knighted for his various contributions, especially to the military. Thereafter, Maxim devoted most of his creative energies to enterprises outside **explosives**, including his novel-powered airplane and the "Captive Flying Machine," an amusement ride that became a staple of British fairgrounds. He would later claim to have been the first to achieve powered flight by a heavier-than-air craft, although this claim is not widely accepted. Maxim was granted 122 US patents and 149 British patents, for ideas ranging from electric lights to explosives. His accomplishments seem all the more remarkable in view of his limited formal education in engineering or science.

MAXIM, HUDSON (1853–1927). Prolific American inventor, chiefly of **explosives**. Born in Maine, Maxim joined his older brother, **Hiram Stevens Maxim**, in New York City, working in various manual-labor jobs for his brother's business, Maxim Gas Machine Company, as well as driving a wagon for "Howe's Great London Circus and Sanger's English Menagerie of Training Animals." Several months of this sort of work convinced Hudson to get an education so that he might enjoy more intellectual pursuits. In 1871, he enrolled at the Maine Wesleyan Seminary in Kent's Hill, which offered a classical education that included chemistry and physics. He also changed his name from "Isaac," which had been given to him at birth, to "Hudson," after a lawyer friend whom he admired. He failed to graduate, but he did pick up sufficient chemistry to begin his career as a printer, specializing in books on penmanship and, in 1883, inventing a method of color printing of newspapers.

In the early 1880s, he moved to England, where his brother Hiram had relocated, again joining his brother's firm, which by this time emphasized **explosives**, ammunition, and military hardware. Hudson would spend long intervals in England with his brother over the next two decades, although the brothers would eventually have a falling out from which they were never reconciled. During one of Hudson's early stints in England, he met a French artilleryman employed by his brother who provided a sample of a French version of **smokeless powder**—most likely **Poudre B**. At the time, the composition of this powder was a closely guarded secret of the French government, which regarded smokeless powder as giving it distinct military advantages. Applying his chemistry skills from the Maine Wesleyan Seminary, Hudson analyzed the sample and found that it was mostly **nitrocellulose**. In 1888, he returned to the United States to be the American representative of the Maxim-Nordenfelt Guns and Ammunition Company, Ltd., bringing some of the French smokeless powder with him. He immediately began experimenting with his own formulations of explosives.

When his contract with his brother's firm expired in 1891, he formed his own company, Columbia Powder Manufacturing Company, Farmingdale, New Jersey, with a goal of entering the **dynamite** business. The company failed in 1893, however. Maxim reorganized it as the Maxim Powder & **Torpedo** Company. Later renamed the Pneumatic Torpedo and Construction Company, it was closely associated with the design and construction of the **pneumatic guns** for the US military during the 1890s. This company proved successful and included a manufacturing plant near Farmingdale in a town that Maxim named for himself. Success came through the development of a smokeless powder, the **Maxim-Schüpphaus smokeless powder**, and, more importantly, of the **multi-perforated smokeless powder**.

On 1 November 1894, he experimented with **mercury fulminate**. Sprinkling some on the floor of his workshop but leaving the rest in his left hand, he struck a match, intending to **detonate** the mass on the floor. Instead, the flame was carried to his hand and exploded. His left hand was torn off at the wrist. He got the wound dressed, went home, and returned to work the next day. In 1895, he received four US patents related to multi-perforated smokeless powder. These patents conflicted with one granted in England to his brother, Hiram, in 1894, contributing to Hiram's perception that Hudson had stolen the idea during his stay in England. In 1898, **DuPont** purchased the rights to the Maxim-Schüpphaus smokeless powder and to the various multi-perforated smokeless powders and gave Hudson a long-term contract as a consultant. With various renewals, that contract remained in effect until his death in 1927.

His earnings from DuPont made him wealthy, and he used this wealth to fund various endeavors, such as the development of "**Maximite**" (a successful **picric acid**-based explosive) and "**Stabillite**" (most likely a failure), and the development of a **propellant** for torpedoes, which he called "**motorite**." His expertise in explosives was recognized by the US government, which appointed him to the Civilian Advisory Board (later renamed the Naval Consulting Board). Chaired by famed American inventor Thomas Edison, the Board acted as a link between American industry and the armed services.

One outgrowth of this service was the appointment of Hudson as chairman of the Committee on Ordnance and Explosives, which advised the military throughout the First World War. Also in 1915, he lent his name to the Maxim Munitions Corporation, which was most likely a bogus effort to exploit his name and notoriety to pump up otherwise worthless stock. By 1916, the company was touting a scheme to adopt the "Enricht process" for converting water into a fuel to replace gasoline. The process was the brainchild of Louis Enricht (1844–1923) and became a notorious scam. Hudson promptly recognized it as such, resigning as a director of the Maxim Munitions Corporation. He followed up by sending circulars to newspapers announcing his decision and his lack of association with Maxim Munitions Corporation. Despite these actions, he spent the next decade fending off lawsuits brought by people claiming they had been swindled by the firm.

These lawsuits failed to interrupt his creative work on other developments, which included two food products based on soybeans: "Maxicream" and "Maxim-Feast." The first was a sandwich filling intended to rival peanut butter; the latter, a multipurpose food suitable for canning and marketed as a source of nutrition for starving populations. Neither product found success, nor did his invention of a new game, "The Game of War," which was intended as an improvement on chess. He continued his self-funded work on novel propellants for **torpedoes**, with several patents being issued following his 1927 death and with all rights being granted royalty-free to the US government. It is uncertain whether any of these patents were drawn upon by the US Navy.

MAXIMITE. A **picric acid**-class **explosive** reportedly 50 percent more powerful than **dynamite**. Invented in 1901 by **Hudson Maxim**, Maximite consisted of about 88 percent **picric acid** and 12 percent **1-nitronaphthalene**. The composition may have been influenced by a sample of **Poudre B** that Maxim had obtained covertly through a French source employed by his brother, **Hiram Stevens Maxim**, in England. Hudson Maxim claimed Maximite was very resistant to the **shock** associated with both the firing and the penetration of a shell or projectile, while having the explosive power of pure picric acid. Preliminary trials by the US Army were sufficiently favorable that 75,000 pounds were purchased for use in **torpedoes** and armor-piercing shells. However, the premature explosion of a torpedo shell containing 140 pounds of Maximite when being fired from a 12-inch mortar caused extensive damage but no loss of life. This experience caused the US military to discontinue all use of Maximite, adopting **Explosive D** in its place. Hudson Maxim claimed that Maximite had been adopted officially and was in use by the US military until it was replaced by Explosive D in 1906, although his claim is at variance with the official record.

MAXIM-SCHÜPPHAUS SMOKELESS POWDER. First **smokeless powder** to be invented in the United States and the first to be adopted for use by the US military. In 1893, **Hudson Maxim** and his partner at the Maxim Powder & **Torpedo** Company, **Robert C. Schüpphaus**, developed a mixture of **guncotton** and **nitroglycerin** that proved superior to other samples submitted to the US military for testing. The mixture was comprised of 9 percent nitroglycerin, 1 percent urea, and 90 percent a mix of 92 percent guncotton and 8 percent soluble **pyroxylin**. The final product was extruded as long cylinders that were perforated axially with small holes. In 1898, **DuPont** acquired the rights to this formulation from Maxim and Schüpphaus and set about producing it for the US Army at a facility in Carney's Point, New Jersey. The Maxim-Schüpphaus smokeless powder was distinctive in two regards. First, the amount of nitroglycerin was small, compared to that seen in other smokeless powders then in use, such as **ballistite** and **Cordite**. This feature produced a benefit in terms of the life of the gun barrel, because nitroglycerin and its products of combustion were corrosive toward the lining of the barrels. Lessened amounts of nitroglycerin meant lessened wear on the barrel. Second, the manner in which the Maxim-Schüpphaus smokeless powder was perforated enabled absolute control over the rate at which the powder burned. It was this latter feature that proved particularly important to the future development of smokeless powders and became the basis for a long-standing dispute between Hudson Maxim and his brother, **Hiram Maxim**. *See also* MULTI-PERFORATED SMOKELESS POWDER.

MAXIM'S POWDER. Any of various formulations of **smokeless powder** comprised of **nitroglycerin**, **nitrocellulose**, and castor oil and developed by **Hiram Stevens Maxim** and his firm, the Maxim-Nordenfeldt Guns & Ammunition Company, Ltd., of England. According to one account, castor oil was included in the mix to reduce the temperatures produced on **detonation**. An alternative explanation is that the residual castor oil following **detonation** helped lubricate the gun barrel and protect against erosion of the barrel lining. In either case, the inclusion of nitroglycerin in Maxim's Powder presented operational problems owing to the high temperatures obtained on detonation and the aggressive behavior of nitroglycerin and its combustion products toward the lining of the gun barrel. Beginning in 1889, the US War Department tested various smokeless powders, ranking Maxim's Powder as the most promising of a dozen or so varieties of powder that were being considered. However, the United States failed to adopt any of the formulations of Maxim's Powder for military use, eventually adopting the **Maxim-Schüpphaus smokeless powder**. This choice may have been influenced by the desire of the US government to ensure a domestic, rather than foreign, supply of smokeless powder, as well as by the problems seen when nitroglycerin was used in a smokeless powder. *See also* MAXIM, HUDSON (1853–1927); MULTI-PERFORATED SMOKELESS POWDER.

MCCARTY, ROBERT. *See* DICKINSON, CHARLES S. (1809–1874).

MEAD, SAMUEL HOLMES, JR. (1848–1875). American inventor of an **explosive bullet**. Mead received US patent 133,714 on 10 December 1872, for an explosive bullet. In his application, he stated that he did not claim an explosive bullet as such, noting that "bullets charged with **fulminating** compounds have been in use for many years, notably by General [John] Jacobs in India and by Von Dreyse, the inventor of the Prussian **[Dreyse] needle-gun**." Rather, he asserted that his invention dealt with the metallic cartridge, which was fitted within a hollow bullet and in which the explosive charge might be contained. Mead described his invention as a combination of a breech-loading metallic cartridge, with a hollow bullet containing fine-grain **gunpowder**, and a small, primed metal cartridge shell. On penetrating flesh or striking bone, the bullet explodes, giving a shock which usually causes the instant death of the animal struck. Before Mead could arrange to manufacture and market his invention, he learned that US Army Brigadier General Montgomery C. Meigs (1816–1892) had invented something similar. Mead approached Meigs and proposed that they share this invention. Although no record exists that Meigs had invented something similar, he agreed to the proposal. By 1874, the "Mead-Meigs shell," also known as the "Safety Explosive Bullet," was on the market. Mead, though, had little time to enjoy any income or fame that his invention might have brought. On 20 May 1875, while packing his gear for what a newspaper report called a "shark-shooting frolic" off Martha's Vineyard, Massachusetts, Mead accidentally discharged his rifle. A Mead-Meigs bullet flew into his forehead, exploded, and ended his life.

MECHANICAL EXPLOSIVE. An **explosive** in which a sudden increase in pressure arises from physical processes rather than chemical reactions. In a mechanical explosive, there may be a change of state (e.g., from liquid to gas) but there is no change in chemical composition. An example of a mechanical explosive in an industrial setting is steam bursting through a weakness in a pipe or container. An example in a more domestic setting would be steam causing an intact potato to rupture while being heated in a microwave oven. In both examples, water is acting as a mechanical explosive. Accidents involving mechanical explosives can have catastrophic consequences, as was the case for the disaster that struck the steamboat ***Sultana*** in April 1865. However, mechanical explosives themselves are of limited practical use in warfare or in other situations where **chemical** and **nuclear explosives** might find use. "**Cardox**" is an example of a mechanical explosive that has seen extensive use in mining, demolition, and other industrial operations. A second example is compressed air, which was tested extensively by the United States as a **propellant** in **pneumatic guns** saw limited use during combat in the Spanish–American War, but ultimately proved a failure. However, compressed air proved successful as a means of launching **torpedoes**, both from the decks of surface ships and from submarines, and compressed air came to be used extensively for this purpose. *See also* HYDRID EXPLOSIVE.

MÉLINITE. Family of **picric acid**-based **smokeless explosives** invented circa 1884 by **Eugène Turpin** and adopted shortly thereafter by the French for military use. Turpin dubbed his invention "mélinite" because of its color, which he thought resembled honey, from the Greek μέλι (*honey*). The original composition of mélinite was a mix of 70 percent picric acid and 30 percent **pyroxylin** that was dissolved in acetone. This composition was later changed to pure picric acid, which was melted and then poured directly into a shell. It was allowed to solidify, with a channel being created for a **fuze** to be inserted. The French tested this composition extensively in 1892 and found it safe against **shock**, heat, and accidental rupture of its contents. Thereafter, this original composition, dubbed "Mélinite D," and various other compositions of mélinite were adopted for use by the French in **torpedo** shells and as a demolition explosive. These other compositions included Mélinite/O (for "ordinaire"), which was 99.7 percent picric acid and 0.3 percent **trinitrocresol**, and Mélinite/P (for "paraffine"), which was 88 percent picric acid and 12 percent paraffin. Turpin was accused by the French government of attempting to sell the secret of mélinite to foreign governments, especially Germany. He was convicted, sentenced to four years imprisonment in 1891, and was released in 1893.

MENDELEYEV, DMITRI IVANOVICH (1834–1907). Russian chemist and inventor of **pyrocollodion**. Mendeleyev is remembered chiefly for his contributions to the development of the Periodic Table of the Elements, especially his unique contribution of predicting the properties of elements that were yet to be discovered. He made many other contributions to chemistry and was recognized in Russia during his lifetime for his brilliance both as a theoretician and on practical problems.

His reputation in this latter regard accounted for his being asked by Ivan Mikhailovich Chel'tsov (1848–1904) to assist in developing a **smokeless powder** for use by the Russian military. In 1890, Chel'tsov, the sole chemist working for the Russian School of Marine Officers, St. Petersburg, and Mendeleyev set off for England and France in an attempt to learn what these nations had developed. By this time, Mendeleyev enjoyed an international reputation for his scientific accomplishments, and that reputation gained him access to important officials and laboratories in England and France. Arriving first in England in June 1890, they visited with **Frederick Abel** and **James Dewar** and witnessed a field demonstration of **Cordite**. Moreover, they were given a small sample to take with them. France proved less cooperative, as the French government zealously protected the formula for its smokeless powder, **Poudre B**. Eventually, the Russians did succeed in visiting the Central Laboratory in Paris, where its director, **Émile Serrau**, provided them a two-gram sample of Poudre B.

Returning to Russia, Mendeleyev directed the analysis of their samples and shortly determined that the French material contained a mixture of types of **nitrocellulose**, with one type being soluble in a mixture of alcohol and ether and the other type being insoluble in this alcohol-ether mixture. He further determined that the insoluble type of nitrocellulose was unevenly distributed in the final product. Although the final product was considered by the French to be satisfactory for their purposes, Mendeleyev recognized that a type of nitrocellulose that was uniformly soluble in the alcohol-ether mixture would be inherently homogeneous and, hence, would be superior to Poudre B. In January 1891, Mendeleyev informed officials with the Russian Navy, which had commissioned the work, that he had developed a novel form of nitrocellulose having the desired uniform solubility in an alcohol-ether mixture. He dubbed this new substance "pyrocollodion" and gave its chemical formula as $C_{30}H_{38}O_{25}(NO_2)_{12}$. He calculated that this formula represented the ideal product because its products of combustion were entirely gaseous, leaving no residue and having no visible combustion products.

Although this material appeared ideal for military use, it would not be exploited successfully by the Russians. Instead, rivalry between the Ministry of the Navy, which favored its adoption, and the Ministry for War, which did not, impeded full-scale industrial production. Eventually, a plant capable of producing 100 tons per year was built and operated. But the output of this plant proved inadequate to meet the demands for smokeless powder following the eruption of the First World War. Consequently, Russia was forced to purchase supplies from foreign sources, particularly the United States, where various firms were producing smokeless powder based on Mendeleyev's invention of pyrocollodion.

MERCURY FULMINATE. White to grayish yellow, bitter-tasting crystalline powder. The preferred, systematic IUPAC name for mercury fulminate is mercury (II) oxidoazaniumylidynemethane, and its chemical formula is $C_2HgN_2O_2$. Mercury fulminate is most likely the earliest **high explosive** whose discovery and preparation were reported in the scientific literature. It is also the first high explosive to be used by terrorists—in this case, an attempt in 1858 to **assassinate Napoléon III**. Mercury fulminate is sensitive to **friction**, heat, and **shock**, and these properties have accounted for its use as an initiator in **blasting caps**, **percussion caps**, and **primers**. Credit for the discovery of mercury fulminate is shared by Pierre Bayen, who prepared it in 1775, and **Edward Charles Howard**, who prepared it in 1799 and reported its synthesis and other properties in 1800. Similarly, **Edwin Goode Wright**, through his publication in 1823, is credited with introducing its use as a primer for firearms. In 1857, British Army major John Jacobs (1812–1858) proposed using it as the explosive charge in an **explosive bullet**. Preparation, handling, and storage of mercury fulminate are hazardous. It deteriorates when stored under conditions of high temperature or high humidity, making it unsuitable for storage in tropical environments. Its use in explosives has been largely supplanted by the use of other substances, beginning with **lead azide** and **lead styphnate** in the early twentieth century. *See also* BERNARD, SIMON FRANÇOIS (1817–1862); COPPER FULMINATE; FULMINATE; FULMINATING GOLD.

MIKE. *See* IVY MIKE.

MILLS BOMB. A handheld or rifle-delivered **grenade** with deep grooves intended to help with handling. Its design is credited to British inventor William Mills (1856–1932), whose Mills Munition Factory, Birmingham, England, produced 75 million for use during the First World War. Mills was granted a patent in the UK in 1915, but his design became the subject of a protracted dispute over patent infringement and priority of invention. The typical design of a Mills bomb called for an iron body having an explosive fill and a **detonator**, with the outside featuring a hand-lever striker held in check by a securing pin. Fills included **Amatol** (1915–1920s), replaced thereafter by **Baratol** or **TNT** and indicated by a green band around the body, and **Ammonal**, indicated by a pink band. In 1922, Mills was knighted in recognition of the importance of his work.

MIXED OXIDES OF NITROGEN. *See* NITROGEN TETRAOXIDE.

MOAB. Acronym for **Massive Ordnance Air Blast Bomb**. In addition, MOAB is commonly associated with an acronym for "Mother of All Bombs." This latter association arose as a back-formation from MOAB and its original association with Massive Ordnance Air Blast Bomb. *See also* FATHER OF ALL BOMBS (FOAB); MASSIVE ORDNANCE PENETRATOR (MOP).

MODIFIED CORDITE. *See* CORDITE.

MOLINARI, ETTORE (1867–1926). Italian chemist and anarchist. Molinari devoted much of his career to working as an industrial chemist and was widely regarded as one of the most esteemed industrial chemists of his time. At various times, he supplemented his industrial career by teaching general, industrial, and inorganic chemistry classes at the high school, technical school, and college levels. In 1895, he was one of the founders of the Society of Chemistry of Milan. In 1915, he was named the director of the chemistry laboratory of the Italian Society of Producers of **Explosives**. He was considered the greatest expert on explosives in Italy during the first decades of the twentieth century and was credited with perfecting a process for manufacturing **TNT** during the First World War when access to supplies from outside Italy became difficult. In addition to his work in industrial chemistry, Molinari was a dedicated anarchist, arousing suspicions that he used his expertise in explosives to assist other anarchists in creating **homemade explosives** and **improvised explosive devices**. In 1891, for example, while in London, he worked to have instructions for making explosives published in the French magazine *L'International*. He maintained his ties to anarchist groups throughout his industrial chemistry career, becoming a founding member of the Union of Italian Anarchists in 1919. The combination of his extensive knowledge of explosives, his experience as a teacher, and his sympathy for Italian anarchist causes fueled speculation that he wrote or helped write ***La Salute è in Voi!***, a handbook on explosives imagined to have been used by Italian-American anarchists. The message informing the manifesto within this handbook was consistent with Molinari's known writings. No proof of his tie to this handbook has emerged, however, and Molinari professed that while he was a committed anarchist, he rejected violence. *See also* ITALIAN-AMERICAN GALLEANISTI BOMBING CAMPAIGN.

MONOBELS. Class of **permissibles** containing approximately 80 percent **ammonium nitrate**, 10 percent **nitroglycerin**, and 10 percent carbonaceous materials, such as wood pulp, flour, and **sawdust**. Monobels and **Carbonites** were the first explosives to be certified by the US Bureau of Mines as permissibles, meaning their use was judged safe for mining operations. *See also* FIREDAMP.

***o*-MONONITROTOLUENE; *p*-MONONITROTOLUENE.** *See* DETECTION AGENT.

MOONLIGHTER. Person **shooting** oil wells, especially one attempting to avoid paying either for the services of the **Roberts Petroleum Torpedo Company** or royalties to that firm for the concept of using **explosives** to improve the flow of wells. The term arose because persons engaged in such illegal practices often conducted their shoots on moon-lit nights in an effort to avoid detection by agents of the Roberts Petroleum Torpedo Company. Moonlighting was widespread during the 1860s and 1870s, particularly in the oil fields of New York, Ohio, and Pennsylvania. These areas had many wells that had become unproductive owing to the harvesting of the oil, but could be restored to productivity with shooting. Three factors spurred the development of moonlighting. First, oil well operators were angry over what they perceived as excessive fees charged by the Roberts Petroleum Torpedo Company for its services. Second, explosives and explosive devices could be made in isolated locations or in equipment that could be moved as needed, making detection difficult. Third, even if the explosion was detected and the explosive device was located, attributing the explosion and device to a particular manufacturer proved difficult. The Roberts Petroleum Torpedo Company aggressively pursued those individuals and firms it believed to be infringing on its patents, leading to the **Great Law War**. The need for moonlighting diminished after 1883 with the expiration of the patents held by **Edward A. L. Roberts** and his company.

MOP. Abbreviation for **Massive Ordnance Penetrator**.

MORDECAI, ALFRED (1804–1887). American military officer credited with introducing scientific methods to the development of military munitions. Mordecai was educated at the US Military Academy, West Point, New York, where he graduated first in his class of 1823. In 1836, he was appointed commander of Frankfort Arsenal, Philadelphia, Pennsylvania, launching a quarter-century association with ordnance development and testing. Mordecai wrote the first ordnance manual for the US military, *The Ordnance Manual for the Use of Officers of the Army*, which was published in 1841. He also conducted experiments with artillery and **explosives**, including the first documented testing for the military of what would become known as **smokeless powders**. In 1845, he released *Report on Experiments on Gunpowder,* *made at Washington Arsenal in 1843 and 1844*. This publication and his subsequent reports were translated into French and German, earning Mordecai an international reputation for his methodical, scientific approach.

In April 1855, Mordecai was dispatched as an official observer during the Crimean War. His observations were released in the *Military Commission to Europe in 1855 and 1856. Report of Major Alfred Mordecai*, which came to be regarded as a classic work in American military science. A native of Warrenton, North Carolina, Mordecai had family living in the southern states when the **American Civil War** erupted in April 1861. Therefore, he requested to be assigned to a posting in California so that he would avoid being engaged directly in the fight against his native South. When his request was denied, he resigned from the military, ending his career in military ordnance. Thereafter, he worked on assorted railway engineering projects, mostly for the Pennsylvania Railroad. *See also* RODMAN, THOMAS JACKSON (1816–1871).

MOTHER OF ALL BOMBS (MOAB). Nickname for the **Massive Ordnance Air Blast Bomb**.

MOTORITE. A **propellant** for **torpedoes**. In 1901, **Hudson Maxim** received a US patent for a combination of **nitroglycerin** and **guncotton**, which he described as a "motive-power combustible for automobile torpedoes, designated 'motorite'." At the time, the term "automobile torpedoes" was used for what would later be termed "self-propelled torpedoes." During 1902 to 1904, the US Navy tested motorite and other propellants, ultimately selecting a compressed air system. A major factor working against the selection of motorite was its inclusion of nitroglycerin, which was considered unsafe. The Navy, however, purchased the rights to motorite from Maxim, perhaps to retain the rights to manufacture it if further improvements should warrant, and perhaps to prevent the rights from being acquired by foreign governments or companies.

Despite the failure of his original design for motorite, Maxim continued to work on propellants for torpedoes. He also continued to use *motorite* as the name for each variant he developed, adding some confusion to efforts to establish the precise composition of motorite. By 1907, his design had evolved such that an explosive mixture was used to heat water, creating steam that was used to drive a turbine engine in the torpedo. This design is contained in one of the last patents granted to Maxim (US patent 1,512,354, "Liquid composition for driving self-propelled torpedoes" issued 21 October 1924), which used a mix of **ammonium nitrate**, **dinitroglycerin**, ethanol, and water. Maxim devoted at least three decades to his search for a propellant for torpedoes and claimed he spent $75,000 of his own funds to support his research. However, there is no evidence that any of his designs were ever adopted for military use.

MOWBRAY, GEORGE MORDEY (1814–1891). Chemist and inventor credited with innovations in the production and use of **nitroglycerin**. Born in England and educated as a chemist in France and Germany, Mowbray began his career manufacturing and dispensing drugs in England. In 1848, he was appointed editor of *The Chemical Times and Journal of Pharmacy*. His business interests extended outside manufacturing drugs, however, and included construction, plumbing, and window glazing. When these interests failed in 1852, he was forced to declare bankruptcy. This experience ruined his health, and in 1853, he embarked on a long sea voyage at the recommendation of his physician. Mowbray disembarked in California, where he found work as a surgeon, druggist, and operator of lime kilns. About 1858, he moved to New York City, again working in manufacturing and dispensing drugs. Following the 1859 discovery of oil in Pennsylvania, Mowbray moved to Titusville, where he built and operated a refinery. In 1866, he placed an advertisement in *Scientific American*, inviting "parties requiring nitroglycerine [*sic*] in quantity—say 100 lbs. per day—to correspond with this subscriber, who has devised a new method for its manufacture, reducing the cost as well as the risk to a minimum." His interest in producing nitroglycerin might have been stimulated by its use in oil well **torpedoes** to increase oil flow. Additionally, his career in drug manufacturing and sales is likely to have contributed to his knowledge of nitroglycerin, owing to its original use as a headache remedy under the name **glonoin**. His advertisement in *Scientific American* prompted a reply, dated 1 July 1867 and written on behalf of the commissioners of the Troy and Greenfield Railroad and of the **Hoosac tunnel**. In their reply, the commissioners invited Mowbray to come to North Adams, Massachusetts, and furnish nitroglycerin for blasting the tunnel, which was being built through the Berkshire Mountains.

Mowbray accepted the invitation. Arriving in North Adams on 29 October 1867, he immediately began building a nitroglycerin production plant of his own design. By the time blasting for the Hoosac tunnel was completed in 1875, Mowbray had delivered over one million pounds of nitroglycerin. At the same time that he was supplying nitroglycerin to the Hoosac tunnel, Mowbray was supplying nitroglycerin to other enterprises. One such delivery resulted in a serendipitous discovery having important consequences for the storage, transport, storage, and use of nitroglycerin. During the winter of 1867–1868, Mowbray supplied nitroglycerin to be used to clear ice obstructing a dam near Deerfield, Massachusetts, to the east of North Adams. The engineer-in-charge of operations at the dam, William Philander Granger (1834–1903), loaded several containers of liquid nitroglycerin onto a horse-drawn sleigh for transport 9 miles through snow over the Berkshire Mountains. Each container was wrapped in a blanket to insulate it against the cold. However, the sleigh overturned during the journey, and the nitroglycerin containers lost their insulating blankets and rolled off into the snow. When the containers were recovered, the contents were discovered to be frozen. Up to this time, nitroglycerin was imagined to be more sensitive to **shock** when cold than when warm. Solid nitroglycerin was thought to be especially dangerous. Despite this discovery and the possible danger, Granger continued his journey. Arriving at the dam, he had each container placed into a hole previously prepared in the ice. Each container was fitted with a **blasting cap**. When Granger attempted to **detonate** the containers, they failed to explode, although the blasting caps functioned as designed. Once Mowbray learned of this experience, he correctly deduced that nitroglycerin is less sensitive to shock when frozen than when liquid. Thereafter, he proceeded to freeze nitroglycerin for storage and transport. Others learned of this discovery, and the freezing of nitroglycerin was widely adopted, contributing to increased safety in its storage, transport, and use.

Mowbray engaged in a long-running patent dispute with **Alfred Nobel** and others over nitroglycerin. Mowbray insisted that the substance he produced and sold was different chemically and physically from the substance or substances covered by the Nobel patents. In his view, his substance was "trinitroglycerin" (the term he always used in referring to his product) rather than merely "nitroglycerin." Specifically, he argued that his product contained three nitro groups (actually, three nitrate esters) for each molecule of glycerin, the parent compound, whereas nitroglycerin might contain one, two, or possibly three such groups. Mowbray pointed to chemical and physical differences between his "trinitroglycerin" and the substance produced following the procedures covered by the Nobel patents. Mowbray's arguments may have had a basis in fact, in that his process may have resulted in pure nitroglycerin (i.e., glycerin trinitrate) rather than a mix of mononitroglycerin, **dinitroglycerin**, and trinitroglycerin produced by other processes. However, his arguments proved to be futile. He was judged to be in violation of the Nobel patents and forced to cease production. Thereafter, he helped set up the American Zylonite Company, which specialized in making commercial goods based on **nitrocellulose**. He also worked as a consultant to the explosives businesses of **Hiram Maxim**.

MULE-BORNE BOMBING. In February 1862, Union Captain James "Paddy" Graydon (1832–1862) enacted a scheme to disrupt Confederate forces in New Mexico during the **American Civil War**. At the time, the Irish-born Graydon, a longtime resident of the New Mexico territory, was serving in the 1st New Mexico Volunteer Cavalry. Confederate forces were engaged in a campaign aimed at cutting Union lines of communication between Fort Craig, where Graydon and others were stationed, and Santa Fe, where Union forces had their headquarters. Graydon's plan was to send **explosive**-laden mules into the Confederate picket lines, **detonate** the explosives, and charge his cavalry through the gap created in their defenses. Late on 20 February 1862, he selected two mules and placed **fused** 24-pounder howitzer shells on their sides. The fuses were lit, and the mules urged forward. They balked, turned around, and headed into Union lines, where the explosives detonated. The only casualties were the mules. But the noise and excitement of the explosions spooked herds of beef cattle and horses within the Confederate lines. They stampeded and ran toward the Union lines, denying the Confederates some of their much-needed food and transport. While Graydon's plan failed to work exactly as planned, it had a good outcome from the perspective of the Union forces. Moreover, the use of mule-borne explosives was a harbinger of similar tactics to be seen a half-century later with the horse-drawn cart at the **Wall Street Bombing** and further escalation to the vehicle-borne terrorist bombings such as the **Oklahoma City Bombing** and the **Bali Bombing**.

MULTI-PERFORATED SMOKELESS POWDER. Shape of **smokeless powder** in which the powder is extruded as rods formed to have holes running lengthwise (i.e., axially). The presence, size, and duration of these holes enable the rate of burning of the powder to be controlled. Such control is an important feature in **explosives**, especially when developing larger caliber guns. Experience has shown that an overly rapid rate of burning could cause the gun barrel to rupture. Similarly, an unduly slow rate of burning could limit the **propelling** force imparted to a projectile, thereby reducing its range. These basic concepts had been demonstrated prior to the development of smokeless powder, as witnessed by the known usefulness of **corning** for **black powder**. A more scientific approach had been taken by **Thomas Rodman** and **Lammot du Pont** in their development of **Mammoth Powder**, which demonstrated the importance of both the size and shape of a powdered explosive. The lesson of that importance for both black powder and **brown powder** likely influenced the development of smokeless powder.

Two individuals—one working in England and the other in the United States—filed for patents related to multi-perforated smokeless powder at nearly the same time. The fact that the two individuals were brothers—**Hiram Stevens Maxim** and **Hudson Maxim**—may have contributed to a long-standing dispute that each had pirated the idea from the other. On 21 September 1894, Hiram Maxim was granted British patent 17,994 for multi-perforated smokeless powder. A half-year later, on 30 April 1895, Hudson Maxim was granted US patent 538,618 for apparatus, processes, and dies for making explosive rods or grains. This last patent was of particular importance because it covered not only the concept and design of multi-perforated smokeless powder but also the equipment used in its production. On 4 June 1895, Hudson was granted three US patents (540,326; 540,327; and 540,328) for inventions quite similar to the one covered by British patent 17,994 held by his brother, as all dealt with the shape and size of multi-perforated smokeless powder. In 1898, **DuPont** purchased the four patents from Hudson Maxim and made him a consultant, an arrangement that endured for the rest of his life and that brought him considerable wealth. These patents and the services of Hudson Maxim contributed to the dominance that DuPont enjoyed during the First World War and thereafter as the preferred provider of smokeless powder to the US military.

In October 1910, Hiram Maxim wrote US president William Howard Taft. In his letter, Maxim claimed to be the sole inventor of multi-perforated smokeless powder. He also claimed to be the original inventor of all smokeless powders containing **nitroglycerin**, such as those powders in use by the US Navy. His claims prompted an official inquiry jointly by the Navy and the Army. Hiram Maxim's claims were weakened by the fact that none of the smokeless powders used by the Navy contained nitroglycerin. Additionally, the work done by Rodman, Lammot du Pont, and others had demonstrated the critical nature of size and shape for explosive powders. This work meant that the concept of size and shape as critical to an explosive had already been established, thereby limiting the grounds for claiming originality in an invention to those dealing with equipment, machinery, and the procedures for controlling the size and shape of the finished product. Hudson Maxim's patents seemed to prevail in this matter, and Hiram Maxim's claims were dismissed by the US military. However, the fact that Hudson Maxim had resided for long periods in England, often living with his brother and collaborating with him on various projects related to explosives, gave ample opportunity for each brother to expropriate ideas from the other. Therefore, there is considerable uncertainty as to the relative contributions made by Hiram Maxim versus those of Hudson Maxim with regard to multi-perforated smokeless powder.

MUNROE EFFECT. *See* SHAPED CHARGE.

N

NEEDLE-GUN. *See* DREYSE NEEDLE-GUN.

NEUMANN EFFECT. *See* SHAPED CHARGE.

NEUTRON BOMB. A type of **enhanced radiation weapon** in which the predominant radiation produced is in the form of neutrons.

NEW ORLEANS. "Protected Cruiser" that saw limited but important service during the **Spanish-American War**. In March 1898, the United States purchased two cruisers from Brazil, both to bolster the US fleet in anticipation of the conflict with Spain and to avoid having the ships be bought by Spain. Both ships were under construction in England by Armstrong, Mitchell, and Company. Brazil had named them *Amazonas* and *Almirante Abreu*, but the US Navy renamed them *New Orleans* and *Albany*, respectively. Both ships were officially designated "Protected Cruisers" but came to be called Elswick Cruisers after the English town where they were built. The *New Orleans* was commissioned shortly after its purchase, but the *Albany* would not join the fleet until 1900, well after the end of the conflict. Both ships carried 6-inch naval guns that used **smokeless powder**, unlike the **black powder** and **brown powder** otherwise in use by US military forces. Immediately on being commissioned, the *New Orleans* sailed from England, arriving in time to join other ships of the fleet in the Caribbean. It fired its guns on 6 and 16 June in action against the coastal defenses of Santiago de Cuba. The difference in performance between smokeless powder and black or brown powder was dramatic. The modest amount of smoke produced on firing the guns of the *New Orleans* ensured its gun crews maintained a clear view of the target and could adjust their firing accordingly, in contrast to the clouded view suffered by the crews of other ships. This experience spurred interest by the US military in replacing black and brown powder with smokeless powder. However, the smokeless powder used by the *New Orleans* and, later, by the *Albany* had to be purchased from England, which was costly and left the US dependent on a foreign supplier. Therefore, the US military undertook development of a domestic supply of smokeless powder, ultimately adopting **Explosive D** and the **Maxim-Schüpphaus smokeless powder**.

NITER; NITRE. Term for nitrates, chiefly **potassium nitrate**. *See also* AMERICAN CIVIL WAR (1861–1865); DU PONT, LAMMOT (1831–1884); NITER AND MINING BUREAU; NITRE AND MINING BUREAU.

NITER AND MINING BUREAU; NITRE AND MINING BUREAU. Independent agency in the government of the Confederate States of America. Its activities are emblematic of measures taken historically to assure a reliable supply of **potassium nitrate** for **black powder** production; however, the scale of its operations is distinctive. Formed in April 1863 by an act of the Confederate Congress, the **Niter** and Mining Bureau subsumed the mission and personnel of the Niter Bureau of the Ordnance Department, which had been formed one year earlier. Many of the personnel of the Bureau carried military ranks and wore uniforms. Its mission was to ensure an adequate domestic supply of various metals, such as lead and gold; **sulfur**; and **nitrates** (chiefly in the form of potassium nitrate). Sulfur and nitrates were of particular importance, as these were needed by the Confederacy for manufacturing black powder. By most accounts, the Bureau succeeded, although some of its more ingenious measures, such as the organized collection of human wastes from chamber pots and latrines, were of dubious value and became the subject of crude humor and ridicule. Despite the various shortages suffered by the Confederacy during the **American Civil War**, black powder supplies appear to have been ample. In fulfilling the needs of the Confederacy for the raw materials needed for black powder, the Bureau oversaw an extensive effort to harvest nitrates from caves and mines. This effort is impressive for its scope and success. Nitrates accumulate in caves and mines, appearing as white powders that effloresce on the ceilings, walls, and floors. This nitrate-rich efflorescence is commonly referred to as "peter dirt," and those who harvest the nitrates are called "**peter monkeys**." In both cases, "peter" refers to **saltpeter**, a synonym for potassium nitrate.

The chemistry of nitrate formation is complex, but the process of harvesting peter dirt is straightforward: It can be swept off the walls and dug from the floors. The Confederacy divided its territory into more than a dozen districts, with each exploiting various caves and mines as potential sources of peter dirt. The result was that sources of potassium nitrate were scattered throughout the Confederacy, and the loss of a single cave or mine had no appreciable impact on the overall supply. But the Confederacy proved vulnerable to its weak transportation system, and these weaknesses became more pronounced as the conflict entered its last two years. Therefore, the lack of reliable shipments of peter dirt or the refined potassium nitrate caused the Confederacy to look to other sources for nitrates. It was at this juncture in the conflict that serious attention was turned to two natural sources of nitrates: human feces and human urine. The Niter and Mining Bureau was tasked with exploiting these sources. It did so by three methods. Military latrines were organized and placed under guard so that the human wastes might be collected by the Niter and Mining Bureau. Independent operators harvested human wastes from outhouses still in use and, especially, from those that had been abandoned, with the harvested materials being sold to the Niter and Mining Bureau. Finally, a public appeal was launched by the Confederate government for households to stockpile their "chamber lye" (a polite term for human wastes, especially those collected in chamber pots) so that it might be turned over to specially outfitted Niter and Mining Bureau wagons for collection. All these wastes were then used to establish niter beds. Such beds had been in use for centuries as a source for nitrates. The difference in this case was one of scale, as the Niter and Mining Bureau organized their use on a massive scale. Despite its scale of operations, it is doubtful whether niter beds had proven to be a significant source of nitrates by the time the conflict ended.

NITER-BED. *See* LAVOISIER, ANTOINE (1743–1794); NITER AND MINING BUREAU; SALTPETER.

NITRAMINE. *See* TETRYL.

NITRATE. The radical $(NO_3)^-$ as well as compounds containing this radical. A familiar substance containing the nitrate radical is nitric acid (HNO_3), which is a powerful **oxidizer**. Nitrates are important in preparing **explosives** and in producing compounds used in explosives. For instance, potassium nitrate is a salt of nitric acid and is used directly in making **black powder** and other explosives. Similarly, **dinitroglycerin** and **nitroglycerin** are esters formed from the reaction between nitric acid and glycerin.

NITRIC OXIDE. *See* NITROGEN TETRAOXIDE.

NITROBENZENE. Pale yellow oil having an almond-like odor and used as a **fuel** in certain **explosives**, such as the **Cheddites**, **Hellhoffite**, **Lyddite**, **Panclastite**, and **Rack-a-Rock**. The chemical formula for nitrobenzene is $C_6H_5NO_2$. Nitrobenzene has many legitimate industrial uses, especially as a precursor in the manufacturing of drugs and dyes. Such uses account for its availability for diversion to illicit use in **homemade explosives** (HME). For example, its mixture with **ammonium nitrate** is referred to in some literature on HME as ANNIE (ammonium nitrate and nitrobenzene **improvised explosive**).

NITROCELLULOSE. Any substance formed by nitrating cellulose. Nitration is typically done by mixing the cellulose with either nitric acid (HNO_3) or a mixture, such as sulfuric acid (H_2SO_4) and **potassium nitrate**. A more proper term is "cellulose nitrate," although "nitrocellulose" has gained wide acceptance, especially with regard to **explosives**.

The degree to which cellulose is nitrated depends on the time and temperature used during mixing as well as the ratio of nitrating agent to cellulose. Controlling these conditions enables the nitrogen content of the resultant nitrocellulose to be controlled. Increasing the nitrogen content (i.e., the percentage of nitrogen in the overall product) decreases the solubility of the product in an ether-alcohol mixture. The resultant products are often referred to as "soluble nitrocellulose" and "**insoluble nitrocellulose**." Nitrocellulose, having a nitrogen content of 13–14 percent, is known as **guncotton**. Both soluble nitrocellulose and insoluble nitrocellulose have seen wide use in explosives, especially in **propellants** such as **Ballistite** and **Cordite**.

French chemist **Henri Braconnot** might have been the first scientist to prepare nitrocellulose, although his 1833 publication on his experiments failed to note its explosive nature. In 1838, **Théophile-Jules Pelouze** noted the flammable and potentially explosive nature of nitrocellulose and suggested its use in **fireworks**. By the late 1840s, many scientists in Europe were exploring the use of some form of nitrocellulose in explosives. **Christian Schönbein** was especially active in promoting and demonstrating the use of guncotton as a propellant and as a **blasting agent**. In 1870, **Friedrich Volkmann** invented the first **smokeless powder** made solely of nitrocellulose, although his invention failed to gain wide use. In the early 1880s, French chemist **Paul Vieille** experimented with both soluble nitrocellulose and insoluble nitrocellulose, using these as the basic ingredients in **Poudre B**, the first smokeless powder to experience widespread use. Nitrocellulose was used widely for various industrial applications outside of explosives. In 1862, British scientist Alexander Parkes (1813–1890) used nitrocellulose as the basis for a substance that could be shaped and molded when heated but would retain its shape on cooling. Parkes dubbed this product "Parkesine," which is believed to be the first synthetic thermoplastic. His attempt at commercializing his invention failed, however. His associate, Daniel Spill (1832–1887), had greater success with a similar product that he dubbed "xylonite," which was used to make imitation coral jewelry. In 1869, American inventor John Wesley Hyatt (1837–1920) patented a similar product, which he dubbed "Celluloid," and which came to be used widely in the manufacture of billiard balls and dental plates and as a flexible base for photographic film. The chemistry of celluloid is believed to have inspired **Alfred Nobel** in his invention of **Ballistite**.

NITROGEN FIXATION. Process of converting molecular nitrogen (N_2) from the air into a compound that is useful for biological or industrial processes. Nitrogen fixation occurs both in natural systems and by modern industrial methods. Natural systems that fix atmospheric nitrogen include lightning discharges, which convert nitrogen into nitric acid (HNO_3), and certain bacteria in soil, which convert nitrogen into ammonia (NH_3) and then into **nitrates** that can be taken up directly and used by plants. The primary industrial process is the **Haber-Bosch process** or derivatives of that process, which converts atmospheric nitrogen into ammonia, which is used as a raw material in manufacturing various chemicals, including **explosives**.

NITROGEN TETROXIDE (NTO). Low-boiling liquid used as an **oxidizing agent** in **propellants** and in **Sprengel explosives**. Nitrogen tetroxide (NTO) has the chemical formula N_2O_4. Therefore, NTO is more properly called "dinitrogen tetraoxide," but "nitrogen tetroxide" has proved a more enduring name. Its potential use in explosives was noted in 1873 by **Hermann Sprengel** in an article he published in *The Journal of the Chemical Society*. Shortly thereafter, the German military experimented with NTO for use in a **torpedo**. French chemist **Eugène Turpin** proposed the use of NTO in **panclastites**, which he purported to be a novel class of explosives, but are actually a type of Sprengel explosive. NTO found wider acceptance as a propellant, and its use for that purpose continues. In some instances, NTO is mixed with nitric oxide (NO). This mixture is referred to as "mixed oxides of nitrogen" (MON) and finds application in rocket **fuels**.

NITROGLYCERIN (NG). Pale yellow, viscous liquid used widely in industrial and military **explosives**. The preferred, systematic IUPAC name for nitroglycerin is 1,2,3-propanetriol trinitrate. However, "nitroglycerin" (also spelled "nitroglycerine") has proven an enduring and popular name for this chemical. Medical references to nitroglycerin often use the term "glyceryl trinitrate." Nitroglycerin has the chemical formula $C_3H_5N_3O_9$.

Ascanio Sobrero announced his discovery of nitroglycerin in 1847, naming it "pyroglycerine." He called attention to its explosive nature and warned of difficulties in handling it. In 1848, Russian chemist Nikolai Nikolayevich Zinin (1812–1880) demonstrated the explosive nature of nitroglycerin to one of his pupils, 15-year-old **Alfred Nobel**, and to Alfred's father, Immanuel. This demonstration might have inspired the Nobel family to pursue further research into nitroglycerin. Zinin advocated the use of nitroglycerin as a military explosive during the Crimean War (1853–1856). Specifically, he oversaw the production of nitroglycerin-filled grenades for the Russian military sometime after March 1854. These grenades failed to be adopted for general use, perhaps because the **shock-sensitive** nature of nitroglycerin made their handling and storage unsafe. Sobrero's warnings proved accurate, as the early history of nitroglycerin production, transport, and use is filled with accidental explosions, including ones that claimed the lives of **Lammot du Pont** and of Alfred Nobel's younger brother Emil Oskar Nobel (1843–1864). Such accidents caused many communities to ban nitroglycerin and forced producers to move to remote locations, as was the case for Alfred Nobel in selecting the initial production sites for both **Nitroglycerin AB** and **Alfred Nobel & Co**. In 1866 or 1867, Alfred Nobel discovered that liquid nitroglycerin could be absorbed into a solid, such as charcoal or diatomaceous earth, thereby creating dynamite. This discovery greatly enhanced safety and expanded the production and use of nitroglycerin.

Learning of Sobrero's discovery and of his method of preparing nitroglycerin, other chemists set about duplicating his work. In behavior that was typical of the mid-nineteenth century, some chemists chose to sample their work. As a result, nitroglycerin was reported to have a sweet, burning taste to cause an intense headache and to increase heart rate. This latter property was soon put to use, and dilute preparations of nitroglycerin were offered as cardiac stimulants, often under the name **glonoin**. Use of nitroglycerin as a vasodilator has long been recognized as a legitimate use, but nitroglycerin is also recognized as highly toxic by ingestion, inhalation, and skin absorption. *See also* DINITROGLYCERIN (DNG); MOWBRAY, GEORGE MORDEY (1814–1891).

NITROGLYCERIN AB. Firm founded in Sweden on 28 November 1864 by **Alfred Nobel** to produce **nitroglycerin**. It was the first firm to produce nitroglycerin on an industrial scale. Production of nitroglycerin and other **explosives** originally took place at Vinterviken, east of Stockholm. The original owners were Nobel, his father, Immanuel (1801–1872), Carl Wennerström (1820–1893), and Johann Wilhelm Smitt (1821–1904), a wealthy and well-respected Swedish businessman who served as chairman of the firm until his death. Alfred Nobel's holdings in Nitroglycerin AB were liquidated following his death in 1896, ending any direct involvement by the Nobel family in the operation of the firm. In 1965, Nitroglycerin AB changed its name to Nitro Nobel.

NITROGLYCOL. Synonym for **ethylene glycol dinitrate**.

NITROGUANIDINE. White crystalline solid that is a **high velocity explosive**. Nitroguanidine has been used extensively in **triple-base powders**. It is relatively **insensitive** to **shock** and **friction**. The temperature of explosion for nitroguanidine is relatively low. This low temperature helps to make the explosion practically flashless, especially in comparison with **single-base** and **double-base powders**. The first use of nitroguanidine in military explosives is credited to the Germans. During the First World War, they used a mix of 50 percent nitroguanidine, 30 percent **ammonium nitrate**, and 20 percent paraffin as a **propellant** for trench mortars. Following the war, its use in triple-base powders, such as **Cordite N**, became common.

The first synthesis of nitroguanidine is credited to Léon Jousselin, a French chemist. Jousselin published his results in the scientific literature in 1877 and 1879, reporting on experiments in which he treated guanidine nitrate with fuming nitric acid. He incorrectly identified the product, however, as nitrosoguanidine. This error was corrected in 1891 by Italian chemist Guido Pellizzari (1858–1938), who reported the correct chemical composition of the product formed on mixing guanidine nitrate and fuming nitric acid. The chemical formula for nitroguanidine is $CH_4N_4O_2$. Other names used for nitroguanidine, especially in the context of explosives, are petrolite and picrite.

NITROMETHANE (NM). Colorless, oily liquid having a slight fruity odor. Its chemical formula is CH_3NO_2. Nitromethane is used extensively for legitimate industrial purposes, especially as a precursor chemical. It is also used legitimately as a **fuel** or fuel additive in motorsports and by hobbyists. It has been used illicitly in **homemade explosives**, especially in combination with **ammonium nitrate**. This combination, sometimes referred to as **ANNM**, was one of the **explosives** used in the 1995 **Oklahoma City Bombing**.

1-NITRONAPHTHALENE. Yellow crystalline solid used as a **fuel** in certain **explosives**, such as the **Cheddites** and **Maximite**. 1-Nitronaphthalene has the chemical formula $C_{10}H_7NO_2$.

NITROPENTA. *See* PETN.

NITROSTARCH. Explosive formed by the nitration of starch. Nitrostarch is not a discrete compound; instead, it is a mixture of various nitric acid esters of starch having varying degrees of nitration. Starch is a polymeric form of glucose, as is cellulose. The general chemical formulas for starch and cellulose are the same.

Nitrostarch was discovered in 1833 by **Henri Braconnot**, who prepared it by treating potato starch with concentrated nitric acid. On adding water, a reddish-orange substance precipitated. Braconnot also discovered **nitrocellulose**, which he prepared in a similar manner but using cellulosic materials rather than starch. For the next two decades, nitrostarch and nitrocellulose were believed to be the same material. In 1855, however, French chemist Antoine Béchamp (1816–1906) published results claiming to distinguish between these two related materials. By 1862, he had succeeded in isolating nitrostarch. Thereafter, nitrostarch was recognized as a substance separate from nitrocellulose. Nitrostarch is less sensitive to **shock** and **friction** than nitrocellulose and **nitroglycerin**, but it is less powerful as an explosive and is slower burning. It is more sensitive to shock and friction than **TNT** but, again, is less powerful as an explosive. Its sensitivity to shock and friction and its highly flammable nature posed barriers to its production and use as an explosive. The United States appears to have been the only nation to have adopted its use widely, with production being most brisk during the First World War when nitrostarch was used as a substitute for TNT.

Interest in the United States in the use of nitrostarch can be dated to 1888, when James Lawrence (1854–1924) experimented with nitrostarch as an additive in Volney's Powder. On 2 February 1904, British chemist Arthur Hough was granted US patent 751,076 for "Nitrated carbohydrate," which used corn starch. He used the process covered by this patent as the basis for a plant constructed in Lake Junction, New Jersey, and proceeded to manufacture nitrostarch for incorporation into **non-freezing dynamite**. An explosion on 18 February 1908 destroyed this plant, killing Arthur Hough's brother Joseph and two others. Thereafter, the plant was abandoned. Other manufacturing ventures proved more successful. In 1905, Nyalite, a non-freezing dynamite-containing nitrostarch, was introduced and stayed on the market briefly. In 1907, DuPont introduced its Arctic brand of non-freezing dynamites, and production of these continued through the First World War. During the First World War, the Trojan Powder Company of New York introduced Trojan Explosive, which was a combination of nitrostarch, **ammonium nitrate**, **sodium nitrate**, and **charcoal**. This mixture enjoyed brisk international sales to Britain, France, and Italy during the war as well as domestic sales following US entry into the war. Trojan Explosive was used as a fill in trench mortar shells and in hand and rifle grenades. With the end of the war, the need for substitutes for TNT diminished, as did demand for Trojan Explosive, although Trojan Powder Company remained in business, with manufacturing plants in California, Colorado, and Pennsylvania.

Nitrostarch enjoyed some popularity in the United States for industrial explosives, such as those used in demolition, mining, quarrying, and tunneling. In particular, nitrostarch is the basis for what are promoted as headache-free explosives. Exposure to nitroglycerin causes severe headaches, but no such side effect is associated with exposure to nitrostarch.

NITROXYLIN. *See* SCHULTZE POWDER.

NOBEL, ALFRED BERNHARD (1833–1896). Swedish-born chemist, inventor, and philanthropist. Nobel was a prolific inventor, receiving 355 patents during his lifetime, covering such diverse subjects as an improved gas meter (Britain, 1857) to purification of pig iron (Britain, 1879). However, his discovery of **dynamite** circa 1866–1867 is among his best-known inventions and became the basis for his fortune.

On 24 November 1864, Nobel established Nitroglycerin AB in Vinterviken, Sweden, to produce **nitroglycerin**. The prior year, he had invented a **detonator** consisting of a wooden plug that could be inserted into a container of nitroglycerin. The plug contained a small amount of **black powder**. When the plug was ignited, it detonated the nitroglycerin. This detonator is considered the first practical means of safely detonating nitroglycerin. He improved this detonator with his 1865 invention of a **blasting cap**, consisting of a small metal cap containing **mercury fulminate** that could be exploded by either **shock** or moderate heat. Inspired in part by the 3 September 1864 explosion during nitroglycerin production in Sweden that took the life of his brother, Emil, and others, Nobel sought a means of converting nitroglycerin into a form having enhanced safety in handling, storage, transport, and use. By 1866, Nobel had established the firm Alfred Nobel & Co. in Krümmel, Germany, producing nitroglycerin for sale on the European continent. During 1866–1867, his research at his firm in Krümmel, Germany, led to the discovery that when nitroglycerin was mixed with a solid, such as charcoal or sand, the resultant mixture proved relatively insensitive to shock. Furthermore, the mix was moldable and could be drilled safely. Nobel dubbed the mix *dynamite* and soon patented it in the UK and the United States. Russian historians would later claim that Nobel got the idea of mixing nitroglycerin with an absorbent, solid material from Nikolai Zinin, a Russian chemist with whom Nobel had studied. By the 1880s, Nobel had established a laboratory in Paris. While working there around 1886, he developed **Ballistite**, obtaining French and British patents on this invention soon thereafter. He offered his invention to the French government, which declined. After he signed a contract with the Italian government for its use of Ballistite, he was accused in France of treason and of having spied on **Paul-Marie-Eugène Vieille**, the developer of **Poudre B**. Eventually, the French government banned Nobel from conducting any further research in France, prompting him to move to

San Remo, Italy. He would maintain San Remo as his base of operations until his death on 10 December 1896.

His inventions occasionally brought him into conflict with others claiming priority of invention. For instance, in the 1870s, Nobel successfully challenged Carl Dittmar for infringing upon Nobel's patent on dynamite with his claimed invention of **dualin**. Similarly, he succeeded in defending against **George Mowbray**'s claims that Mowbray's forms of **nitroglycerin** and related substances differed from those covered by patents held by Nobel and others. Nobel failed, though, in pursuing his claim that Ballistite was the same substance as **Cordite**, for which Sir **Frederick** Abel and Sir **James Dewar** had been granted a patent.

His scientific and engineering skills were matched by his business acumen. Within 10 years of inventing dynamite, Nobel had become a shareholder in or co-owner of 16 explosive-manufacturing firms in 14 countries. He acquired great wealth from these business ventures. His 1895 will provided that much of his wealth be used to establish a foundation that would award prizes in chemistry, literature, physics, physiology or medicine, and peace. This foundation took on his name, as did the prizes, which continue to be seen as among the most prestigious awards one might receive in their respective fields. *See also* BLASTING GELATIN; DIPHENYLAMINE; PELOUZE, THÉOPHILE-JULES (1807–1867).

NOBEL'S SAFETY POWDER. *See* DYNAMITE.

NOBLE, ANDREW (1831–1915). Scottish scientist and industrialist noted for his contributions to the study of **explosives**. Noble graduated from the Royal Military Academy, Woolwich, remaining in the military until 1860, when he joined the firm of W. G. Armstrong & Company, Elswick, England. He remained there for the rest of his career, becoming chairman in 1900. He carried out a long collaboration with Sir **Frederick Abel** on the behavior of **black powder**. Noble developed novel methods for measuring the pressures produced in a gun barrel when an explosive **detonated** and for measuring the acceleration of a projectile as it traveled through the barrel. These methods advanced the understanding of the rapid reactions taking place when an explosive detonates, thus helping transform the development of explosives from a largely empirical enterprise to a more scientific discipline. In 1893, Noble was knighted for his contributions to science and the military and was created Baronet of Ardmore and Ardardan in 1902. *See also* ARMSTRONG, WILLIAM GEORGE (1810–1900).

NON-DEFLAGRATING, NON-DETONATING EXPLOSIVE. *See* HYBRID EXPLOSIVE.

NON-FREEZING DYNAMITE. Dynamite containing solid **explosive** materials as a replacement for **nitroglycerin**. Because these substances are solid, they are unaffected by exposure to low temperatures and, hence, cannot freeze, as is the case for nitroglycerin. In ordinary dynamite, liquid nitroglycerin is absorbed into some solid material, such as Kieselguhr. At low temperatures, the nitroglycerin can solidify; when the temperature rises, the nitroglycerin re-liquefies, but, in doing so, it is significantly less stable than previously. Therefore, a non-freezing dynamite has the advantage of being significantly less susceptible to changes in stability induced by exposure to low temperatures. *See also* LOW-FREEZING DYNAMITE; NITROSTARCH.

NONPROLIFERATION TREATY. *See* THE TREATY ON NONPROLIFERATION OF NUCLEAR WEAPONS.

NORTH KOREA. Nuclear weapon state (NWS). North Korea has a nuclear weapons program, asserting its need for a nuclear deterrent because the United States threatened the use of nuclear weapons during the Korean War (1950–1953). Development of this program was most likely facilitated through the **A. Q. Khan** nuclear technology smuggling network. The nation declared publicly on 10 February 2005 that it possesses nuclear weapons. On 9 October 2006, the United States detected a seismic event in North Korea, strongly suggesting a nuclear **detonation**, taken to be that nation's first successful detonation. That same day, North Korea announced that it had conducted a successful test of a nuclear device. Experts estimated the **explosive** yield from this detonation to be between 500- and 1,000-tons **TNT-equivalent**. This yield is considered low, fueling speculation that the device suffered from **preinitiation**.

North Korea has continued to conduct underground tests and, to date, is known to have conducted six detonations of nuclear devices, with its most recent detonation occurring on 3 September 2017. North Korea claimed that its fourth test, conducted on 6 January 2016, involved a **hydrogen bomb**, but this claim has been met with skepticism. The North Korean stockpile is estimated to be around 50 nuclear warheads, although such estimates are difficult to verify owing to the reclusive nature of the North Korean regime. *See also* THE TREATY ON NONPROLIFERATION OF NUCLEAR WEAPONS; WORLDWIDE NUCLEAR DETONATIONS.

NUCLEAR EXPLOSIVE. An **explosive** that uses radioactive materials as the primary **fuel** for the explosion. This fuel undergoes either a nuclear **fission** or a **fusion** reaction, resulting in the release of an enormous amount of energy. Fission is the process of splitting the nucleus of an atom into two or more lighter nuclei. Certain isotopes of **uranium** and **plutonium** are used for this purpose, as they readily undergo fission after absorbing a neutron. Fusion is the process of combining two light-nuclei atoms into a single heavier nucleus. The light-nuclei atoms typically used in this process are deuterium and tritium, which are isotopes of hydrogen.

The first successful man-made nuclear explosive was **Gadget**, which was **detonated** successfully on 16 July 1945 at the **Trinity Test Site**, New Mexico. This 5-ton device exploded with a blast of about 20 kilotons **TNT-equivalent**. The radioactive material used as the fuel for Gadget was plutonium. The same fuel was used in **Fat Man**, the second nuclear explosive to detonate successfully when used in war. It was dropped over Nagasaki, Japan, on 9 August 1945. Three days previously, on 6 August 1945, **Little Boy** detonated over Hiroshima, Japan, becoming the first nuclear explosive to be detonated successfully in war. It used uranium as the fuel.

NUCLEAR SUPPLIERS GROUP (NSG). *See* SMILING BUDDHA.

NUCLEAR WEAPON. Nuclear explosive device intended for hostile use against an adversary or in retaliation.

NUCLEAR WEAPON BAN TREATY. *See* TREATY ON THE PROHIBITION OF NUCLEAR WEAPONS.

NUCLEAR WEAPON-FREE ZONE (NWFZ). A geographical area in which the development, stationing, testing, and use of **nuclear weapons** are prohibited. The underlying assumption in establishing a nuclear weapon-free zone (NWFZ) is that excluding such weapons from a region is more likely to succeed than is eliminating or controlling such weapons after they have been introduced. This assumption has spurred six NWFZs recognized by the United Nations General Assembly (UNGA): in Africa, for the Antarctic region, in Central Asia, in Latin America, in Southeast Asia, and for the South Pacific. In addition, Mongolia has adopted national legislation, establishing its status as an NWFZ. This Mongolian legislation is recognized by the UNGA. Three other international instruments place prohibitions on, among other things, nuclear weapons, although the language specific to establishing an NWFZ is absent. These three are the Moon Agreement, which pertains to the moon and other celestial bodies; the Outer Space Treaty, which deals with all regions beyond the planet earth, including the moon and other celestial bodies and, as such, overlaps to an extent with the Moon Agreement; and the Seabed Arms Control Treaty, which covers the seabed, ocean floor, and subsoil thereof beyond a 12-mile territorial zone.

NUCLEAR WEAPONS ELIMINATION DAY. Observance held annually on 26 September, promoting the cause of nuclear disarmament. Officially known as the International Day for the Total Elimination of **Nuclear Weapons**, this observance was established on 30 December 2013 by UN General Assembly Resolution 68/32. According to that resolution, the day "provides an occasion for the world community to reaffirm its commitment to global nuclear disarmament through enhancing public awareness and education about the threat posed to humanity by nuclear weapons."

NYE COMMITTEE. US Senate investigation into the practices of American munitions industries during the First World War and subsequently, including makers of **explosives**. While the official name of this committee was the "Senate Special Committee on Investigation of Munitions," it became known for its chairman, Gerald Nye (1892–1971), a Republican US senator from North Dakota. It met for 16 months, beginning on 4 September 1934, and attracted wide coverage in the press and with the public.

In assuming the role of chairman of the seven-member committee, Nye promised, "when the Senate investigation is over, we shall see that war and preparation for war is not a matter of national honor and national defense, but a matter of profit for the few." His promise set the tone for Nye's agenda, consistent with his isolationist views. The investigation was prompted in part by a perception that the munitions industries had unduly influenced the decision by the US government to enter the war in 1917 and, as a consequence, had reaped enormous profits. The work of the committee gained a sense of urgency owing to political developments in Africa, Asia, and Europe, including the rearming of Germany in contravention of the Treaty of Versailles, and Japan's occupation of Manchukuo on the Asian mainland. Such developments spurred fears that the United States would be dragged into another foreign war by these same "merchants of death," the epithet Nye used to describe the armaments industries.

The Nye Committee conducted 93 hearings and questioned more than 200 witnesses prior to issuing its report ("The Nye Report") on 24 February 1936. The committee found "that the international commercial interests of such large organizations as **DuPont** and Imperial Chemical Industries may precede in the minds of those companies the importance of national policy as described publicly by the foreign office or the [US] State Department." The committee went on to note that "considerations of commercial interest were apparently foremost in the rearming of Germany beginning in 1924 and in the sale of a process which could be used to manufacture cheaper munitions in Japan in 1932, shortly after Secretary of State Stimson had taken steps to express the disapproval of this Nation for Japan's military activities in Manchukuo." The report highlighted efforts by DuPont to thwart negotiations during the Geneva Arms Control Conference of 1925 and noted the statement by a senior War Department official that "it was vastly more important [for the War Department] to encourage the DuPont Co. to continue in the manufacture of **propellants** for military use, than to endeavor to protect secrets relating to the manufacture."

Despite the hearings and its report, the Nye Committee may have had little influence on the conduct of government. It failed to achieve the goal of its chairman to nationalize the munitions industry. While three Neutrality Acts were adopted by Congress during the 1930s, one was passed prior to the Nye Committee, and the other two were passed after the committee was disbanded. However, the hearings alarmed DuPont, whose corporate officials worried that the revelations and denunciations of its business practices would harm its public image. DuPont sold a few products directly to the public, but its executives deemed corporate image important. The Nye Committee, in its view, had tarnished that image. In response, DuPont engaged an advertising agency to develop a public relations campaign to enhance its image. In October 1935, DuPont went public with its slogan: "Better Things for Better Living Through Chemistry." At first, use of this slogan was confined to a radio show sponsored by DuPont, "The Cavalcade of America." Later, it graced printed and television ads and continued in use through the mid-1980s.

O

OCTOGEN. Synonym for **HMX**.

OCTOL. *See* BINARY EXPLOSIVE.

OKLAHOMA CITY BOMBING. The explosion that rocked Oklahoma City and shocked the nation took place at 9:02 a.m., local time, Wednesday, 19 April 1995. A vehicle—later determined to be a Ryder rental van—laden with **explosives detonated** beneath the day-care center of the Alfred P. Murrah Federal Building. The blast damaged or destroyed around 300 buildings in addition to the Murrah building. Moreover, it killed 168 people, including 19 children. One rescue worker died of injuries received from falling debris during the recovery effort. Several hundred were injured. Property damage was estimated at nearly $700 million.

An extensive investigation ensued, during which 28,000 interviews were conducted by law enforcement personnel, 43,000 investigative leads were pursued, and 3.5 tons of evidence were taken. The FBI led the investigation, dubbing it "OKBOMB." The investigation succeeded in identifying the two primary perpetrators, Timothy McVeigh and Terry Nichols, and several other individuals who aided and abetted their scheme. McVeigh and Nichols, both US citizens, held extremist ideologies marked by an intense hatred of the US government. Their hope, apparently, was that the bombing would embolden like-minded persons to commit similar, violent acts, culminating in the toppling of the government. In this regard, they failed. Examination of evidence from the scene and taken elsewhere enabled the design of their **vehicle-borne improvised explosive device (VBIED)** to be reconstructed. McVeigh and Nichols placed 13 55-gallon barrels containing **homemade explosives** aboard the Ryder van. Nine barrels contained a mix of **ammonium nitrate** and **nitromethane (ANNM)**. The remaining four contained **ammonium nitrate-fuel oil mixture (ANFO)**. Each barrel contained about 500 pounds of explosive, for a total mass of 6,500 pounds. Time-delayed **fuzes** enabled McVeigh to park the van, then escape the scene, departing Oklahoma City in a getaway vehicle previously stashed near the Murrah building. The fuzes set off **blasting caps**, which set off 350 pounds of an explosive stolen from a quarry. The sequence of explosive charges culminated in the **detonation** of the barrels of ANNM and ANFO.

McVeigh and Nichols were tried separately for assorted crimes. McVeigh, viewed as the originator of the bombing plot, was convicted and sentenced to death on 2 June 1997. He was executed on 11 June 2001. Nichols was convicted as well and was sentenced to 161 consecutive life sentences, each without the possibility of parole.

OPERATION SHAKTI. *See* INDIA.

OPHORITE. Explosive manufactured from **potassium perchlorate** and a metal powder and typically used as an ignitor. British chemist Jocelyn Field Thorpe (1872–1940) is credited with developing the original formulation for ophorite during the First World War and with coining the name *ophorite*, which is a play on the four oxygen atoms in potassium perchlorate ("o four"). This original formulation consisted of a mix of nine parts magnesium and 13 parts potassium perchlorate. Later, other pyrophoric metals were used, especially **aluminum**. When he developed ophorite, Thorpe was the chair of organic chemistry at the Imperial College of Science and Technology, London, where his group was devoted to assisting the British government on problems related to the war. Ophorite saw application in early 1916 as an ignitor for thermite when it was used as an incendiary. Using ophorite proved advantageous in that it could be ignited by means of a **fuze** and it burned rapidly. Ordinary ignitors required a **detonator** in addition to the fuze and burned more slowly, failing to melt the thermite completely. Ophorite saw extensive use by the British during the First World War, and Thorpe's contribution was recognized in 1917 when he was designated a Commander, Order of the British Empire (CBE).

OPPAU EXPLOSION. Early on the morning of 21 September 1921, a tower silo containing a mix of ammonium sulfate and **ammonium nitrate** exploded at an Oppau, Germany, facility belonging to the chemical firm BASF. The official death toll was given as 561, with more than 2,000 injured. The blast destroyed most buildings in Oppau, leaving about 6,500 residents homeless and creating a crater some 19 meters deep. The noise of the explosion was reportedly heard in Munich, more than 300 kilometers away. Various estimates have been given for the size of the explosion, ranging as high as 4.0–4.500 kilotons (kt) of **TNT-equivalent**. More conservative estimates place it at 1–2 kilotons TNT-equivalent.

Investigating and reconstructing the explosion were hampered by the deaths of all persons having any direct knowledge of the work underway in the silo. The silo is known to have contained some 4,500 tons of the ammonium sulfate/ammonium nitrate mix at the time of the explosion. Both products were marketed by BASF for sale as fertilizer, although ammonium nitrate had been used during the First World War in various **explosives**. By contrast, ammonium sulfate [$(NH_4)_2SO_4$] is a white or grayish crystalline solid that is stable and is unlikely to have played any meaningful role in the explosion. It is also known that this mix routinely forms a near-solid cake within the silo. This caking occurred due to the presence of the ammonium nitrate, which is hygroscopic (i.e., it absorbs water and, in this case, turns from a fine powder into a rock-like mass). The cake proved difficult to remove from the silo, and workers would resort to using pickaxes to loosen it. This procedure carried the danger that the mass of the mix would collapse precipitously while the workmen were pounding away with their pickaxes, burying them inside the silo. To ease the process of removing the ammonium sulfate/ammonium nitrate mix, the workmen would place small charges of **dynamite** within the mix. Test firings of this procedure proved effective, and an estimated 20,000 charges had been fired safely prior to the events of 21 September.

A likely explanation for the explosion is that it was accidental, owing to the presence of a high proportion of ammonium nitrate in the region of the silo where the dynamite was **detonated**. This explosion is viewed as one of the largest industrial accidents. The root cause of the explosion also speaks to the overconfidence derived when an inherently unsafe procedure (in this case, using dynamite to loosen the mix) is conducted repeatedly without an adverse reaction. *See also* HALIFAX EXPLOSION; PORT CHICAGO DISASTER; TEXAS CITY DISASTER.

OPPENHEIMER, JULIUS ROBERT (1904–1967). American physicist. Oppenheimer made seminal contributions to theoretical physics but is perhaps best known for his work as scientific director of the Los Alamos Laboratory during the Second World War, overseeing the successful effort to develop **nuclear weapons**. After completing undergraduate studies in the United States, Oppenheimer continued his studies in the UK and Germany. Returning to the United States in 1927, he held various faculty positions prior to a May 1942 invitation to take responsibility for fast neutron calculations in support of an effort to develop nuclear weapons. When the **Manhattan Project** was established, Leslie Groves, its head, asked Oppenheimer to serve as scientific director of the Los Alamos, New Mexico, laboratory that would be the main site for weapons design, research, and development. In June 1945, he chaired an advisory panel tasked to share its views on whether and how nuclear weapons should be used on Japan. The panel reported on 16 June that "we can propose no technical demonstration [of a nuclear weapon] likely to bring an end to the war; we see no acceptable alternative to direct military use."

The **detonation** of **Gadget** on 16 July 1945 proved that the work overseen by Oppenheimer had succeeded, as did the detonations of **Little Boy** and **Fat Man** over Japan the following month. After the war, Oppenheimer opposed developing the **hydrogen bomb**, and his opposition caused some authorities to question his loyalty to the United States. On 21 December 1953, his security clearance was suspended, pending the outcome of further investigations. During those investigations, both **Edward Teller** and Leslie Groves offered testimony casting doubt on Oppenheimer's reliability and suitability for a security clearance. The situation was made worse by his past association with Communists and his support of Communist causes. In 1954, the suspension was made permanent, and his work on nuclear weapons came to an end. *See also* RADIOLOGICAL DISPERSAL DEVICE (RDD); TRINITY TEST SITE.

OXIDIZER. Any substance that readily donates oxygen atoms in a chemical reaction. In the context of **chemical explosives**, an oxidizer contributes to the rapid combustion of the **fuel**, with the result being the sudden release of pressure and the burning that are characteristic of either **deflagration** or **detonation**.

OXYLIQUIT. A **Sprengel explosive** comprised of a **fuel** that has been impregnated with liquid molecular oxygen, which acts as the **oxidizer**. German engineer Carl von Linde (1842–1934) is credited with inventing oxyliquits in the 1890s. Linde, who developed the first successful large-scale methods of liquefying oxygen and other gases, presented the concept of an oxyliquit in a scientific journal article published in 1899. Early oxyliquits used **sawdust** or carbon black as the fuel, and a paper cartridge containing the fuel was dipped into liquid oxygen shortly before use. The Germans used oxyliquits extensively during the First World War, in part due to domestic shortages of **nitrates**. However, by the Second World War, **nitration fixation** technology had advanced to the point where ample supplies of nitrates were available, and oxyliquits fell out of use as military explosives. According to one scheme, oxyliquits were designated according to their performance relative to other, more well-known explosives. The designation "D Oxyliquit" was used for an oxyliquit having performance roughly equivalent to that of **dynamite**, "A Oxyliquit" was used for one performing roughly like that of **ammonium nitrate**, and "P Oxyliquit" was used for one performing in a relatively slow manner similar to that of **black powder**. An oxyliquit is also referred to as a liquid oxygen explosive (LOX).

P

PAKISTAN. Nuclear weapon state (NWS). The beginnings of the Pakistani **nuclear weapons** program can be traced to January 1972, when the Pakistani prime minister Bhutto established Project 706 and set a deadline of the end of 1976 for having a weapon. The perceived urgency for having nuclear weapons likely arose from the 1971 loss of the territory that had been East Pakistan. The 18 May 1974 **detonation** of **Smiling Buddha** by **India** added to the sense of urgency. Progress, however, lagged, and the 1976 deadline remained unmet, owing partly to technical difficulties with **uranium** enrichment at the research laboratories in Kahuta, near Rawalpindi. In July 1976, **Abdul Qadeer Khan** was named to direct the work at the Engineering Research Laboratories, and the deadline was revised to 1984. His efforts succeeded, as demonstrated by five detonations of nuclear explosives on 28 May 1998 in the Ras Koh hills, Chagai district, Balochistan. Two days later, a sixth device was detonated in the Kharan desert, Balochistan. Khan claimed that the 1984 deadline was met in that Pakistan had the capability by then to detonate a **nuclear explosive**, although the actual detonation would not occur until 14 years later. Pakistan has not conducted any further nuclear weapon detonations beyond the one of 30 May 1998.

PAN AM FLIGHT 103. On 21 December 1988, Pan American World Airways flight 103 **exploded** while flying over the UK. The flight had originated in London and was destined for New York City. All 243 passengers and 16 crew aboard the Boeing 747 aircraft died as a result of the explosion, which caused the plane to come apart in flight. A large section of a wing landed with explosive force in Lockerbie, a town in southern Scotland, killing 11 and bringing the death toll to 270. An extensive investigation ensued, including testing for residues of explosives and reconstruction of recovered portions of the aircraft. Results revealed that the explosion resulted from a bomb rather than from some mechanical malfunction of the aircraft or other accidental cause. The bomb itself was believed to have been a **plastic explosive** concealed in a Samsonite suitcase carried within the forward luggage compartment of the aircraft. Based on tradecraft seen in other bombings and attempted bombings, the explosive was widely believed to have been **Semtex**. The operation was blamed on the work of two Liberian nationals and a Maltese merchant, although a broader conspiracy was imagined. On 15 August 2003, the Libyan ambassador to the United Nations submitted a letter to the UN Security Council in which he formally accepted responsibility for the actions of its citizens in relation to the bombing. The Libyan government paid over $2 billion in compensation to the families of victims. The destruction of Pan Am Flight 103 spurred the development and adoption of the **Convention on the Marking of Plastic Explosives for the Purpose of Detection**.

PANCLASTITE. Class of **Sprengel explosives** comprised of **nitrogen tetraoxide** (NTO) and a suitable **fuel**, such as **carbon disulfide** or **nitrobenzene**. French chemist **Eugène Turpin** claimed to have invented this class of explosives in the early 1880s, but a report published in 1873 by **Hermann Sprengel** in *The Journal of the Chemical Society* casts doubt on that claim. In his report, Sprengel called attention to the use of NTO as an **oxidizing agent** and noted the need for a suitable **fuel** to create an explosive. He failed to specify the possibility of using carbon disulfide or nitrobenzene as a fuel. However, his 1886 "Note of so-called 'Panclastites'," published in *The Journal of the Society of the Chemical Industry*, disputed Turpin's right to claim this class of explosives as his invention. The French used a form of Panclastite in aviation bombs during the First World War, calling their composition "**anilite**." Turpin coined the name "Panclastite" from the classical Greek meaning "all broken" [*παυ* (*all*) + *κλατος* (*broken*)], a reference to what he asserted was the destructive powder of this class of explosives.

PANPUSHKO, SEMYON VASILY'EVICH (1856–1891). Russian chemist and military officer. Panpushko was educated in chemistry in both Russia and Europe and was a graduate of various Russian military academies, where much of his studies involved artillery and, particularly, **explosives**. In the early 1880s, he spent four months in Germany learning modern methods of chemical analysis at the Wiesbaden Agricultural Institute, where he studied in the analytical chemistry laboratory of Carl Remigius Fresenius (1818–1897). In September 1887, the Russian government funded Panpushko to travel abroad for one year, which he spent visiting chemical laboratories, military academies, and **gunpowder** factories in Austria, Belgium, England, France, Italy, and Germany. This year-long trip exposed Panpushko to European practices regarding the production and military use of **nitroglycerin** and **picric acid**. On his return, he was assigned to the Mikhailovsky Artillery Academy, St. Petersburg, where he wrote a book detailing these practices. This book was well received by his superiors, and he was decorated by the artillery academy and instructed to assist in constructing a plant to produce picric acid. In 1890, he was sent to France for six weeks to learn about the production and military use of **guncotton**. On his return to the artillery academy, he continued to take an active role in experiments with novel munitions, gaining a reputation as a fearless and possibly reckless experimenter. His experiments would have tragic consequences. On 28 November 1891 (old style), Panpushko and three colleagues were killed when a six-inch shell filled with picric acid **detonated**. An inquiry revealed that someone was using a hammer to hit a wrench that was being used to screw on the bottom of the shell. Picric acid dust had accumulated in the threads of the screw and detonated from the **shock** or vibration of the hammer blows. Shortly thereafter, a monument was erected on the main artillery practice ground at the artillery academy in honor of his achievements and his sacrifice.

PASLEY, CHARLES WILLIAM (1780–1861). British military officer in the Royal Engineers and demolitions expert credited with many innovations, including one of the earliest uses of electric **detonators** to clear submerged obstacles to navigation. Around 2:05 p.m. on 23 October 1839, Pasley detonated 2,320 pounds of **gunpowder** contained in lead-sealed barrels beneath the waters of Spithead off Gilkicker Point, Hampshire, England. The subsequent explosion broke apart the HMS *Royal George*, which sank accidentally on 29 August 1782. On 17 May 1844, he oversaw a similar detonation of the HMS *Edgar*, which sank in 1711, also at Spithead. By removing these underwater obstacles, the two **explosions** enabled ships to resume using Spithead safely for anchorage. In both cases, the gunpowder was detonated by means of a platinum wire connected to a battery. At Pasley's command, the circuit running from the battery to the wire was closed. When the circuit was completed, the current heated the wire, causing the gunpowder to detonate. In recognition of his assorted services to his country, Pasley was knighted in 1846.

PATENT SAFETY GUN-COTTON COMPANY. Private British firm engaged in manufacturing **explosives**. Established around 1861 by the Prentice family, the Patent Safety Gun-Cotton Company of Stowmarket, Suffolk, England, initially made **guncotton** by the method of **Freiherr Wilhelm Lenk von Wolfsberg**. Lenk was an Austrian military officer who promoted a process that he claimed improved on the method of **Christian Schönbein**. Guncotton production using this process began in 1863, but an explosion in September 1864 caused the managers to abandon the process. In 1865, the factory was rebuilt and restarted, this time using a process devised by Sir **Frederick Augustus Abel**. This time, manufacturing operations proceeded successfully, and the company expanded its activities to include another factory, located in Penrhyndeudraeth, Gwynedd, North Wales, that produced an explosive comprised of guncotton, starch, and India rubber. But disaster struck on 11 August 1871 when one of three guncotton storage magazines in Stowmarket spontaneously **detonated**. That explosion touched off explosions in the other two magazines. Damage was extensive, and 28 people were killed. The victims included Edward Prentice (1838–1871), whose gold watch was all that was ever found of his remains, and his nephew, William Ridley Prentice (1847–1871), newly returned from studying abroad. In 2014, a memorial was emplaced in the Stowmarket Old Cemetery, honoring the 23 victims who are buried there.

Despite this explosion, the company rebuilt, although the Prentice family divested their interest in the firm. By 1881, the company had been renamed "The Explosives Co., Ltd.," changing its name again in 1885 to "The New Explosives Co. Ltd." It would go on to become a leading manufacturer of explosives for military, mining, and sporting purposes.

PBX. Abbreviation variously attributed to (1) plastic-bonded **explosive**, (2) polymer-bound explosive, (3) pressed bonded explosive, and (4) pressed plastic-bonded explosive. Although subtle differences might exist among these four terms, all refer to an explosive that has been modified by the addition of a polymeric material that functions as a **plasticizer**. The quantity of plasticizer is small, typically less than 10 percent of the total mass. Polymers used for this purpose include Kel-F (trademark for a family of fluorocarbons), polystyrene, and Viton (trade name for a family of vinylidene fluoride/hexafluoropropylene copolymers). In addition to having the desirable characteristics generally associated with a **plastic explosive**, many types of PBX have the added benefit of being able to be machined at room temperature into complex shapes, such as those associated with **explosive lenses**.

PEACEFUL NUCLEAR EXPLOSIONS (PNE) TREATY. A peaceful **nuclear explosion** (PNE) is one conducted for non-military purposes, such as large earthmoving operations during the construction of a canal or dam. The United States conducted PNEs during the 1960s and 1970s as part of the **Plowshare Program**. One example is the Sedan nuclear test, carried out at the Nevada Test Site and intended to stimulate natural gas production. However, the effort was abandoned as impractical. The Soviet Union carried out nearly 240 nuclear tests under programs designated as "nuclear explosions for the national economy," a term taken to be equivalent to PNE. **India** characterized its first successful nuclear explosion, **Smiling Buddha**, as a PNE. Concern over regulating such PNEs stimulated development of the Treaty between the United States of America and the Union of Soviet Socialist

Republics on Underground Nuclear Explosions for Peaceful Purposes, known familiarly as the PNE Treaty. It places limits on the size of nuclear explosions conducted for peaceful purposes. Any single PNE cannot have a yield exceeding 150 kilotons (kt) **TNT-equivalent**, but aggregate explosions totaling 1.5 megatons (mt) TNT-equivalent are permitted if each individual explosion can be verified to be less than 150 kilotons TNT-equivalent. The PNE Treaty prohibits explosions exceeding 1.5 megatons TNT-equivalent for any purpose and in any manner. The treaty was signed by the United States and the Soviet Union in Moscow on 28 May 1976 and entered into force on 11 December 1990. *See also* PROJECT CHARIOT.

PELOUZE, THÉOPHILE-JULES (1807–1867). French chemist; credited as the first person to note the flammable nature of a form of **nitrocellulose**. The son of an industrial chemist, Pelouze started off studying and practicing pharmacy. In 1831, he moved to Paris, taking a position with the Ecole Polytechnique and, later, with the Collège de France. In the mid-1830s, he learned of the work by **Henri Braconnot** on the nitration of starch and decided to conduct further studies into the chemistry of "xyloidine," as Braconnot had termed the product obtained by mixing concentrated nitric acid with starch. In 1838, Pelouze published his results in the form of a memoir. In his publication, Pelouze noted that xyloidine ignited at 180°C, leaving practically no residue. He reported, too, that he had dipped paper in nitric acid. When the paper was washed, it was parchment-like and impervious to moisture. Noting the flammable nature of this product, Pelouze recommended its use in fireworks. Because the product was flammable, he called it "**pyroxylin**," from the Greek *πῦρ* (*fire*) and *ξυλον* (*wood*). In 1846, Pelouze published additional results regarding his research on nitrocellulose materials. In his publication, he described the differences between xyloidine and pyroxylin. Pelouze stated that xyloidine is soluble in concentrated nitric acid, while pyroxylin is not, and that xyloidine is **shock-sensitive** and highly flammable, while pyroxylin is less sensitive to shock and is much less flammable, requiring temperatures of 175°–180°C to catch fire. His teachings apparently stimulated others to pursue the field of **explosives**. His pupils included **Ascanio Sobrero**, who would go on to discover **nitroglycerin**, and **Alfred Nobel**, who made many important contributions to the field of explosives. *See also* SCHÖNBEIN, CHRISTIAN FRIEDRICH (1799–1868).

PENNEY, WILLIAM GEORGE (1909–1991). British mathematician noted for his research into **blast effects** and who played a leading role in the development of **nuclear weapons** by the UK. During the Second World War, he oversaw the weapon development efforts of Canada and the UK He headed the UK delegation working on the **Manhattan Project** in New Mexico, where he concentrated on calculating the optimal height for **detonating Little Boy** and **Fat Man**. As a passenger aboard the B29 *Big Stinks*, Penney witnessed the detonation of Fat Man on 9 August 1945 over Nagasaki, Japan. After the war, he was appointed head, High Explosives Research, the name given to the effort by the UK to develop its own nuclear weapon arsenal. This effort became necessary when the United States refused to continue sharing nuclear weapon information. His work culminated in the successful detonation of **Hurricane** on 3 October 1952. Penney was honored for his contributions, which included being knighted in 1952 and being created a life peer in 1967, for which he took the title Baron Penney of East Hendred.

PENNIMAN, RUSSELL SYLVANUS (1853–1914). American chemist and inventor of **explosives**. Penniman began his career as an assistant to **George M. Mowbray**, where he worked in North Adams, Massachusetts, on the production of **nitroglycerin**. In 1885, Penniman invented a process for formulating **ammonium nitrate** as granules or prills. This process had important consequences for the use of ammonium nitrate as an explosive. Ammonium nitrate naturally absorbs water from the environment. As a result, ammonium nitrate degrades over time as it is exposed to the air, lessening its usefulness as an explosive. Penniman patented a process that mixed ammonium nitrate with a small amount of petroleum jelly, paraffin, or similar hydrocarbon, giving the ammonium nitrate a coating that protects it from moisture. That same year, Penniman and his associate John C. Schrader received a patent for the use of coated ammonium nitrate in explosives. These two patents became the basis for the industrial use of ammonium nitrate in **dynamite**. Penniman went on to become a senior executive for **DuPont**. Shortly after Penniman's death in 1914, DuPont began construction of a dynamite production plant near Williamsburg, Virginia, on the York River, and named the plant in his memory. Before the plant went operational, however, the United States entered the First World War, and the government took over the site for use as a shell loading facility. With this change in ownership of the plant and its grounds, a nearby town constructed to house workers took Penniman as its name. The population of Penniman, Virginia, dropped dramatically after the war when ammunition-related jobs disappeared. During the Second World War, most of the town property was taken over by the US Navy for use as government-owned storage and shipping facilities.

PENTAERYTHRITOL TETRANITRATE. *See* PETN.

PERCUSSION CAP. A container filled with a **shock-sensitive** substance that **detonates** on being struck. This detonation is typically used to initiate the ignition of another, typically much larger **explosive**. Percussion caps are routinely small and made of metal, especially copper, although early percussion caps included ones made of pasteboard. In a strict sense, the paper caps used in toy cap pistols and cap guns are percussion caps, although these caps are not used to detonate another explosive. By the 1860s, percussion caps had been integrated into a cartridge containing the main explosive charge and projectile (bullet) and became known as **primers**. Individual percussion caps are obsolete for modern weapons, but they are available for use, especially for historic firearms.

Establishing a precise date for the invention of the percussion cap is problematic. Most likely, the idea grew out of the detonator lock system pioneered by Scottish clergyman **Alexander Forsyth** in the late eighteenth century. Early percussion caps used a mixture of **charcoal**, **potassium chlorate**, and **sulfur**. The 1800 scientific report by **Edward Charles Howard** on the preparation of silver fulminate and **mercury fulminate** stimulated interest in using these substances, which are significantly more shock-sensitive than those that were used initially in percussion caps. Credit for inventing the percussion cap is ascribed to various individuals, including the English gunsmith **Joseph Egg** and the French gunsmith François Prélat (?–1859), who was granted a patent in July 1820 for the percussion cap. Although a strong case can be made that the right to claim priority of invention belongs to **Joshua Shaw**, the idea for the percussion cap may have occurred to various persons, several of whom reduced the idea to practice between 1810 and 1822.

PERCUSSION LOCK. *See* FORSYTH, ALEXANDER JOHN (1769–1843).

PERFORATED-CAKE POWDERS. *See* RODMAN, THOMAS JACKSON (1816–1871).

PERKINS, JACOB (1766–1849). American-born inventor, mechanical engineer, and physicist, whose many inventions included a **steam gun**. While working in the United States, Perkins developed a process that permitted steel plates to be substituted for copper plates in fine engraving and printing. About 1819, he moved to England in an effort to pursue this process as a commercial venture and remained there for the rest of his life. On 15 May 1824, he received British patent 4,592 for a steam gun. His design used steam directly to propel a projectile through a gun barrel. He built at least one gun using this design, and he displayed this gun for many years at the Adelaide Gallery, London. Dubbed "Mr. Perkins' Extraordinary Steam Gun" by the press, the gun attracted considerable interest from the public and specialists from various nations and inspired *Steam, A Poem*, which appeared in *The London Mirror* in February 1825:

Five hundred balls per minute shot,
Our foes in fight must kick the beam;
Let Perkins only boil his pot,
And he'll destroy them all by steam.

Despite the attention and demonstrations conducted for delegations from various nations, Perkins failed to receive any orders for his steam gun. Some 12 years after his death, his design was copied by the US Navy when it built a prototype steam gun during the **American Civil War**. Testing proved the gun had inadequate accuracy for military use, and the concept was abandoned. *See also* DICKINSON, CHARLES S. (1809–1874).

PERMISSIBLES. Explosives deemed safe for use in mines, especially in coal mining operations. The term is used chiefly in the United States, but similar terms are used by other nations and include "permitted explosives" (UK), "**explosifs antigrisouteux**" (France), "**explosifs S.G.P.**" (Belgium), and **Schlagwettersichere Sprengstoffe** (Germany).

The atmosphere of coal mines often contains methane (CH_4), other flammable gases, and coal **dust** in sufficient quantities to render it explosive. Any combustion source, such as a spark or a flame, can touch off an explosion in the presence of such mixtures. Many mining disasters can be traced to the combination of air, coal dust, and **firedamp**, as methane and other flammable gases came to be called by miners. The situation became sufficiently dire in the late 1870s that the UK, France, and Germany established "Firedamp" Commissions to study the matter and to recommend methods of reducing the risk of unplanned explosions in mines. These commissions identified that the flame accompanying the intentional use of explosives for blasting contributed to a follow-on unplanned secondary explosion. The British commission, under the leadership of Sir **Frederick Abel**, suggested placing a water cartridge atop any explosive placed in a borehole within the mine. This suggestion prompted similar methods of extinguishing any flame that might emit from the borehole when the planned, primary explosion occurred. These measures included using wet sand and moist moss and proved effective in many situations. But the search continued for a more scientifically based and predictable solution. That search led to the development of **nitroglycerin**-based explosives having salts that naturally contain "waters of hydration," such as sodium carbonate decahydrate ($Na_2CO_3 \cdot 10H_2O$) and magnesium sulfate heptahydrate ($MgSO_4 \cdot 7H_2O$). The addition of the hydrated salts lowered the temperature of the explosion, reducing the possibility of an unplanned secondary explosion. Such explosives were dubbed **wetter dynamites** or **grisoutites**. An accompanying development was that of the **Carbonites**, a group of dynamites having nitroglycerin plus some sort of carbonaceous material (such as wood pulp, **sawdust**, or flour) and **potassium nitrate** or sodium nitrate. The incorporation of the carbonaceous material and nitrate had the desired effect of reducing the temperature of the explosion, but this reduction in temperature arose from incomplete burning of the ingredients. Therefore, the use of Carbonites was accompanied by the release of carbon monoxide (CO), an asphyxiating gas whose presence posed a danger in all but the best ventilated areas.

A more satisfactory solution emerged with the development of explosives containing a large percentage of **ammonium nitrate**, whose **detonation** took place at a relatively low temperature and without the complication of a build-up of carbon monoxide emissions. Examples of such ammonium nitrate-based permissibles include **Ammonite**, **Grisoudynamite**, and **Monobel**. More recently, the successful development of **Cardox**, a **mechanical explosive**, and **Hydrox**, a **hybrid explosive**, added to the list of permissibles that might avoid mining disasters arising from unplanned, secondary explosions.

PERTUISET'S POWDER. Explosive used as the filling for **explosive bullets**. French hunting enthusiast Eugène Pertuiset (1833–1909) is credited with inventing this powder, for which he received US patent 78,322 on 26 May 1865 and British patent 2,837 in 1867. His invention came about because of his disappointment in the performance of explosive bullets developed by Louis-François Devisme (1806–1886), which contained **mercury fulminate** as their explosive charge. According to Pertuiset, Devisme's bullets failed to fell big game—such as lions and elephants—immediately. Instead, a wounded animal could wander some distance before dropping dead. Pertuiset found this situation inconvenient for his purposes, and he set about trying other possible alternative fillings, relying on a former employee of Devisme for help. Eventually, he succeeded with a two-part mixture. One part consisted of **potassium chlorate** in an intimate mix with approximately one percent animal charcoal (carbonized bones), with charcoal serving as a **stabilizer**. The second part consisted of **sulfur** mixed with approximately 12.5 percent "sporting gunpowder," the precise identity of which is obscure. The two mixtures were combined, then added as a filling within a bullet. The bullet would **detonate** from the heat generated from contacting an animal target, either a hard one, such as bone, or a soft one, such as tissues and organs. Massive wounds ensued from the fragmentation and shock, causing the animal to collapse immediately.

With his invention realized, Pertuiset went on a promotional tour. An early stop was Vincennes, France, where a rapid-firing, multi-barrel gun—a mitrailleuse—was loaded with his explosive bullets and fired at live cavalry horses, tethered to limit their movement and to prevent escape. The resulting slaughter of these horses was gruesome, but the French government declined to place an order for any of his powder or bullets. Soon after France, Pertuiset visited Russia, where his audience included Tsar Alexander II. The demonstration must have been impressive, because the Tsar presented Pertuiset an emerald and diamond ring as a token of appreciation. But others were less enamored by the demonstration. In particular, Prince Alexander Gorchakov, the Russian foreign minister, was sufficiently disturbed by what he had witnessed that he convened a meeting of representatives from 19 nations, including Russia, that resulted in the **Saint Petersburg Declaration of 1868**. This arms control agreement banned military use of explosive bullets under 400 grams (14 ounces).

Meanwhile, Pertuiset continued his tour, visiting an Austro-Hungarian Empire naval base in Pola as well as Constantinople, Cairo, Florence, and Rome. A late stop on his tour was London. On 3 August 1870, he was taken to a barn for condemned horses. His first assigned target was a horse deemed irreclaimably vicious and consigned to be destroyed. The weapon to be used with his explosive bullet was a revolver having a .450 caliber cartridge containing Pertuiset's Powder. A single shot was fired into the horse's head. A small hole on its forehead was all that was visible, but the horse was felled immediately. Subsequent examination of the horse's skull revealed a 7-inch by 6-inch hole within the skull cavity, with the brain utterly destroyed. Further tests were conducted on a second horse and, finally, a third horse, with the same result each time. His tour having ended, Pertuiset returned to his home in France. By this time—1870—explosive bullets had been banned as a weapon of war. He appears to have contented himself with the use of these bullets for his original target: big game. His fondness for this pursuit was captured in the 1881 oil painting by Édouard Manet, *Pertuiset, le chasseur de lions* (*Pertuiset, the Lion-Hunter*).

PERSON-BORNE IMPROVISED EXPLOSIVE DEVICE (PBIED). An **explosive** worn on or carried by an individual. Examples of person-borne improvised explosive devices (PBIED) include a backpack, as was seen in the **Bali Bombing** and the **7/7 Terrorist Bombings**; shoes, as was attempted by the **Shoe Bomber**; underwear, as was attempted by the **Underwear Bomber**; and a vest, as was seen in the **Sri Lanka Easter Bombings**. Explosives used for such devices range from **homemade explosives** to repurposed military explosives. The device itself might include metal objects, such as ball bearings, nails, screws, or slugs, to serve as **Shrapnel**, thus adding to the destructive potential of the device once it **detonates**. Most often, the person wearing or carrying the explosive is killed in the process of **detonating** the explosive.

PERU. *See* WAR OF THE PACIFIC (1879–1884).

PETER MONKEY. Individual who harvests cave soil for use in extracting **potassium nitrate**. The term was used, in particular, during the **American Civil War** for individuals working for the Confederate **Niter and Mining Bureau**. "Peter" is a shortening of "**saltpeter**," a term for potassium nitrate. Cave soil is rich in nitrates resulting from animal droppings, especially those of bats, and has been used for centuries as a source of nitrates. The work environment associated with harvesting cave soil is cold, damp, and dark. But it proved attractive for men during the American Civil War in lieu of being conscripted into the military. Consequently, early in the war, the ranks of peter monkeys were mostly adult, white males. Eventually, this loss of manpower to the Confederate military caused service as a peter monkey to cease being a means of avoiding conscription. This situation necessitated a new source of manpower. One alternative source that was tried was using enslaved men. This alternative proved unattractive to their enslavers, because the enslaved men would gain their freedom if Union forces should overrun their area of operations. A second alternative that was tried was using white males under conscription age. This alternative proved attractive, as these young men or boys were both sufficiently agile and small enough to handle the difficult working environment associated with caves. Consequently, by the later stages of the war, most peter monkeys were boys ages 16 and under.

PETN. White, solid substance used extensively as an **explosive**. PETN is an abbreviation for pentaerythritol tetranitrate, which is one of the names used for this substance. Other names used for PETN include nitropenta and pentrite. The preferred, systematic IUPAC name for PETN is 1,3-dinitrato-2,2-*bis*(nitratomethyl)-propane, and its chemical formula is $C_5H_8N_4O_{12}$.

PETN is among the most powerful **chemical explosives**. It has a **Relative Effectiveness Factor** of 1.66, the same as that for **HMX** and somewhat greater than that for **RDX**. PETN has been used extensively in **plastic explosives**, such as **Semtex**. It was reported in the scientific literature in 1891 by **Bernhard Tollens** and Peter Wigand, as part of their basic research into aldehydes and polyalcohols. Its explosive nature was recognized shortly thereafter; however, it was adopted in the mid-1890s for production as an explosive by the German firm United Rhenish-Westphalian Powder Works (Vereinigte Rheinisch-Westfälische Pulverfabriken). Germany used PETN in military munitions during the First World War, although other nations adopted it for military use much later.

PETROLITE. *See* NITROGUANIDINE.

PHLEGMATIZING AGENT. A substance added to an **explosive** as either a desensitizer or a **stabilizer**. The addition of a desensitizer or a stabilizer can enhance safety in handling the explosive by making the final product less **shock-sensitive** or less likely to undergo further reactions that could produce corrosive or otherwise harmful substances. Furthermore, a desensitizer can reduce the rate of combustion of the explosive, which is desired in certain applications, such as to reduce the maximum blast pressure experienced in a gun barrel when a **propellant** is **detonated**. An early example of a phlegmatizing agent is Kieselguhr, the diatomaceous earth into which **nitroglycerin** was absorbed, forming **dynamite**. As shown by the history of dynamite, researchers learned the importance of substituting a purely inert material, such as Kieselguhr, with a phlegmatizing agent that could also participate in the explosion, most often by being a **fuel**, such as charcoal, **sawdust**, and starch. Other examples of phlegmatizing agents include various waxes, such as beeswax, and paraffin oils and jellies, such as Vaseline.

PICRATOL. *See* BINARY EXPLOSIVE.

PICRIC ACID. Yellow crystalline solid that can **detonate** when subjected to **friction**, heat, **shock**, or vibration. Picric acid is the common name for 2,4,6-trinitrophenol and has the chemical formula $C_6H_3N_3O_7$. Irish-born **alchemist** and chemist Peter Woulfe (1727–1803) is credited with the earliest reporting in the scientific literature of the discovery of picric acid. His 1771 paper in the *Philosophical Transactions of the Royal Society of London* described a yellow substance obtained when he treated indigo with nitric acid. Woulfe observed that this substance might prove useful as a dye but failed to comment on its **explosive** nature. This substance would be rediscovered various times, including in the 1790s by Jean Joseph Welter (1763–1852), a French chemist and dye-maker. Welter prepared picric acid by treating silk with nitric acid. He also noted the intense bitter taste of the product, which came to be known as *jaune amer de Welter* (*yellow bitter of Welter*). Moreover, Welter reported in the scientific literature in 1799 that the bitter, yellow solid yields a potassium salt when treated with potassium hydroxide, and that this potassium salt explodes on heating. This potassium salt would later be shown to be potassium picrate.

In the early 1800s, chemists demonstrated that the substances prepared by Woulfe and Welter were the same, and it was given the name "carbazotic acid." Around 1830, French chemist Jean-Baptiste-André Dumas (1800–1884) dubbed this substance "picric acid" because of its intense bitter taste, from the Greek πικρός (*bitter*). His colleague, Auguste Laurent (1808–1853), correctly identified the chemical nature of picric acid as a trinitrophenol and reported his results in the scientific literature in 1841. Later research would establish its structure precisely as the 2,4,6-trinitrophenol isomer. Thereafter, picric acid continued to be investigated as a potential dye, although it proved unsatisfactory owing to a lack of color fastness.

The prospect of using picric acid as an explosive went unexplored until the 1880s, when French chemist **Eugène Turpin** developed **mélinite** for the French military. Mélinite was used in French artillery shells as a bursting charge. Shortly thereafter, other nations became aware of this novel explosive in use by the French military, which attempted to keep its chemical nature a secret. That attempt failed, in part owing to the efforts of Turpin to exploit his development by obtaining patents on it outside of France. Over the next 20 years, various nations introduced explosive mixtures in which picric acid was the primary component. Two of the best examples are **Lyddite**, which was used by the British, and **Shimose**, which was used by the Japanese.

Although picric acid was widely adopted during the 1880s, 1890s, and 1900s for military use, it suffered from two major defects. First, because it is highly sensitive to shock, it tends to **detonate** immediately on contact with a dense surface, such as the armor-plating on a ship. This immediate detonation means the explosive energy is often dissipated harmlessly outside a structure or with minimal damage. A preferred result is for the detonation to be delayed slightly so that the munition can penetrate and discharge its explosive effects within the structure rather than on its exterior. Second, picric acid readily forms salts ("picrates") when in contact with metals. These salts often are more shock-sensitive than is the picric acid, creating a bursting hazard for those handling the munition. The formation of these picrates also corrodes the lining of the munition.

PICRITE. *See* NITROGUANIDINE.

PLASTIC-BONDED EXPLOSIVE. *See* PBX.

PLASTIC EXPLOSIVE. An **explosive** modified by the addition of a **plasticizer**, resulting in a mass that is both sufficiently soft to be molded and capable of hardening into the desired fixed form. Some sources cite **blasting gelatin** as an early example of a plastic explosive. In this case, the plasticizer was **collodion cotton**. Other sources cite **guhr dynamite** as the earliest plastic explosive. In this case, **nitroglycerin** was both the explosive and the plasticizer. A more technically precise definition of plastic explosive is offered by the Technical Annex to the **Convention on the Marking of Plastic Explosives for the Purpose of Detection**. That definition provides that a plastic explosive is one that (a) is formulated with one or more **high explosives** which, in their pure form, have a vapor pressure less than 0.0001 pascals (Pa) at 25°C; (b) is formulated with a **binder**; and (c) is, as a mixture, malleable or flexible at normal room temperature.

Compared to an explosive lacking a plasticizer, a plastic explosive offers several advantages. First, it can take on a desired shape, such as a **sheet** or a specialized injection-molded form. Second, it is more **stable** against degradation. Third, it has greater mechanical strength. Finally, it is less sensitive to **shock** and **friction**. Prominent examples of plastic explosives include **Composition C4**, **Detasheet**, and **Semtex**. *See also* PBX.

PLASTIC EXPLOSIVES CONVENTION. *See* CONVENTION ON THE MARKING OF PLASTIC EXPLOSIVES FOR THE PURPOSE OF DETECTION.

PLASTICIZER. Any substance incorporated into the formulation for an **explosive** so that the **grains** of that explosive remain elastic and avoid becoming brittle. Brittleness in the grains of an explosive can adversely affect its **deflagrating** and **detonating** properties by changing the speed at which it burns. The inclusion of plasticizers is especially important with **nitrocellulose**-based explosives. Substances that have been used for this purpose include camphor; phthalates, such as di-*n*-octyl phthalate, which impart good resistance to water and oil; and citrates, such as tri-*n*-butyl citrate, which are more biodegradable than phthalates. Both di-*n*-octyl phthalate and tri-*n*-butyl citrate have been used in some formulations of **Semtex**. **Nitroglycerin** can also be both an explosive and a plasticizer, as was the case for **guhr dynamite**, which some sources attribute as the first **plastic explosive**.

PLOWSHARE PROGRAM. Name given to an investigation into using **nuclear explosives** for peaceful purposes. The use of *plowshares* was an illusion to a Biblical verse in which the Prophet Isaiah urges the people to beat their swords into plowshares (Isaiah 2:4). During the Plowshare Program, the United States set off 27 nuclear **detonations** to test the use of nuclear devices for extracting natural gas and oil and for earthmoving. These detonations took place between 10 December 1961 and 17 May 1973 in Colorado, New Mexico, and Nevada. The Soviet Union conducted a similar program, called "Nuclear Explosions for the National Economy." The Soviet program was both larger in scope and longer in duration. From 1965 to 1989, the Soviets conducted 156 tests, involving at least 240 nuclear detonations. *See also* PEACEFUL NUCLEAR EXPLOSIONS (PNE) TREATY; PROJECT CHARIOT; SMILING BUDDHA.

PLUTONIUM. A man-made element, with atomic number 94 and symbol "Pu." There are no stable **isotopes** of plutonium; hence, all isotopes are radioactive. The most significant radioisotopes in the context of **nuclear weapons** are plutonium-239 and plutonium-240. Plutonium-239 is a **fissile material** and is the primary isotope used as a **fuel** in US nuclear weapons. Personnel assigned to the **Manhattan Project** referred to plutonium-239 as "49 Metal." This designation served as a code name and was derived by combining the last digit of the atomic number for plutonium (94), the last digit of the isotope (239), and the fact that plutonium is a metal. Plutonium-240 has a high rate of spontaneous fission, and its presence increases the chances of **preinitiation**. For this reason, the plutonium used in nuclear weapons normally contains 7 percent or less plutonium-240. *See also* CRITICAL MASS; FAT MAN; GADGET; IMPLOSION DEVICE; SIGNIFICANT QUANTITY (SQ); SPECIAL NUCLEAR MATERIAL (SNM); WEAPONS-GRADE MATERIAL.

PNEUMATIC GUN. Weapon design in which compressed air is used as a **mechanical explosive**. The United States tested such guns extensively, beginning in 1883. Both the US Army and the US Navy took an interest in pneumatic guns, which were designed to use compressed air as a **propellant**, launching a projectile containing **dynamite** as the explosive charge. At the time this testing began, the United States lacked a successful formulation for a **smokeless explosive** that might be used as a propellant. **Black powder** and **brown powder** burned too rapidly and with too much force to be safe for use with dynamite-laden projectiles, which were then something of a novelty. Additionally, the heat generated on **detonation** of the propellant risked **detonating** the dynamite. Use of compressed air seemed promising, and testing proved the force of the compressed air could be adjusted to permit safe delivery of dynamite projectiles.

In 1886, the US Navy began work on the warship ***Vesuvius***, which would enter the fleet in 1890 with three 15-inch pneumatic guns. When the submarine, the *Holland*, entered the fleet on 12 October 1900, its armaments included two pneumatic guns, although one of these guns would later be removed. Around this same time, the US Army purchased three types of pneumatic guns, which were installed for testing and evaluation at Fort Hancock, near Sandy Hook, New Jersey. On 23 August 1894, the guns were demonstrated successfully, and in October 1895, the Army began installation of its first operational battery of pneumatic guns, guarding the Pacific Ocean entrance to San Francisco at Fort Winfield Scott, California. Construction was completed in 1898. Additional shore batteries with pneumatic guns were installed at Hilton Head, South Carolina, and Fishers Island, New York, during 1901 and 1902. Immediately prior to the **Spanish–American War**, the Army purchased 16 **Sims-Dudley dynamite guns**, which used a **hybrid explosive** system rather than purely mechanical force to compress air. Some of these guns, as well as the *Vesuvius*, saw use in combat during the Spanish–American War.

The earliest use of a pneumatic gun in combat, however, may have been during the Brazilian Naval Revolt of 1893–1894, although the record is unclear. At the outbreak of the revolt, the Brazilian government encouraged New York City industrialist Charles R. Flint (1850–1934) to assemble a squadron of six ships to aid in suppressing the revolt. One of these ships was christened the *Nictheroy* and was outfitted with a single 15-inch pneumatic gun capable of propelling a dynamite-containing munition. The ships were piloted and crewed by Americans, who were inducted into the Brazilian Navy upon the arrival of the ships in Brazil in March 1894. Its pneumatic gun apparently was fired two or three times. The first occasion was in November 1893, four months prior to the arrival of the *Nictheroy* into Brazilian waters. At the time, the ship was undergoing trials at sea. A dummy munition was fired, followed by the firing of a dynamite munition at a designated target near St. Thomas Island in the Caribbean Sea. The dynamite munition overshot the target, and no clear view was possible either of its landing or of its detonation. The second firing took place in mid-March 1894 when the *Nictheroy* neared positions held by the rebel naval forces in Guanabara Bay. A single dummy munition was fired off Ilho do Pai, southeast of Rio de Janeiro and south of Guanabara Bay. That shot demonstrated the accuracy of the gun, as the dummy round landed harmlessly near its intended target. But the shooting of the gun rendered it disabled. Thereafter, the gun was removed from the ship and was reported to have been mounted at Forteleza de Santa Cruz da Barra, which guards the eastern entrance to Guanabara Bay. There is no record of further use of this gun in combat. When Brazil failed to compensate Flint for the squadron he had assembled, he managed to sell some of the ships to the US Navy. One of those sold to the Navy was the *Nictheroy*. Minus its pneumatic gun, it was rechristened the *Buffalo*. Flint would later become one of the founders of Computing-Tabulating-Recording Company, the forerunner of IBM.

Interest in pneumatic guns waned by the turn of the century; however, as various formulations of smokeless explosives became available. Furthermore, pneumatic guns had undeniable shortcomings in terms of the relatively complex apparatus required to compress air and deliver that compressed air to the gun tube. The smokeless explosives proved superior to compressed air, black powder, and brown powder as propellants and to dynamite as explosive charges in the projectile. Work on pneumatic guns projecting dynamite-laden charges had ceased by the first decade of the twentieth century. The four shore installations in California, New Jersey, New York, and South Carolina were scrapped in 1904.

In reality, the pneumatic gun was an aerial projector rather than a cannon, and the dynamite projectile was considered a **torpedo**, a term that then meant a munition intended to damage a ship by concussive forces transmitted through the water. Consequently, other terms used for the pneumatic gun included "torpedo projector," "dynamite gun," "dynamite torpedo gun," and "pneumatic dynamite torpedo gun." *See also* ZALINSKI, EDMUND LOUIS GRAY (1840–1909).

POLYMER-BOUND EXPLOSIVE. *See* PBX.

PORT CHICAGO DISASTER. At 10:18 p.m. local time on 17 July 1944, an **explosion** erupted on the pier at the Port Chicago Naval Magazine, Contra Costa County, California. Moments later, a fire was observed on this same pier. Less than 10 seconds later, a much larger explosion followed, this one on a nearby munitions carrier, the Liberty Ship SS *E. A. Bryan*. The blast obliterated the *Bryan*, lifted the Liberty Ship SS *Quinault Victor* out of the water and ripped it apart, destroyed a railroad train, and capsized a US Coast Guard fire boat, causing it to sink. The explosion and ensuing fire left 320 dead, and nearly 400 others suffered injuries, many of them permanent. Property damage was extensive, and the town of Port Chicago, adjacent to the port, was nearly wiped off the map. Subsequently, a Naval Board of Inquiry was convened to identify what caused these explosions. Its findings failed to identify the precise cause but implied that mistakes made by enlisted men, most of whom were Black, were likely to blame.

Other twentieth-century disasters involving the accidental detonation of explosives caused greater loss of life or inflicted greater property damage, such as the **Halifax explosion** and the **Texas City disaster**. The Port Chicago disaster is distinctive, however, for its aftermath, especially for the unequal treatment of those personnel at or near the scene based solely on their race and for the lax attitude among selected US Navy personnel toward safe handling of munitions. For example, it emerged during the Board of Inquiry that there was no system for officers and enlisted men to become familiar with safety regulations for handling munitions. Also, supervisors and other superior officers engaged in competitions to see which group of munition handlers could load the most tonnage per hour, regardless of the risks inherent in speeding up loading operations. In the wake of the disaster, officers and senior enlisted men, all of whom were white, were given hardship time off, while the munition handlers, all of whom were black, were ordered to return to work. In August 1944, 258 black enlisted men refused to continue loading munitions until they had received training on safety procedures. Their refusal was treated by senior naval authorities as a failure to obey orders. In the face of this position, 208 returned to their munition handling duties without further training. Despite their willingness to return to duty, these 208 were charged with disobeying orders and were convicted of such at a court-martial. The remaining 50 were subjected to a court-martial in which they were charged with mutiny. All were convicted and received harsh sentences.

Eventually, two of these 258 men had their records cleared. On 17 July 2024, US secretary of the Navy Carlos Del Toro announced full exoneration of the remaining 256 men. His announcement noted that significant legal errors occurred during the courts-martial, that the defendants were tried together despite conflicting interests, and that the defendants were denied a meaningful right to counsel. *See also* OPPAU EXPLOSION.

POTASSIUM CHLORATE. Colorless or white crystalline solid having the chemical formula $KClO_3$. Potassium chlorate was discovered by French chemist **Claude Louis Berthollet** in the late eighteenth century and was originally called "Berthollet's salt" in his honor. Because potassium chlorate is a strong **oxidizer**, it is used in **explosives**, especially in **percussion caps** and **primers**. However, Berthollet's attempt in 1788 at using it as a replacement for **potassium nitrate** in **black powder** ended tragically, with an explosion that killed two colleagues. By the mid-nineteenth century, various inventors had patented methods of reducing the sensitivity of potassium chlorate to **friction** and **shock** by coating it with a liquid, such as castor oil, linseed oil, or turpentine, or a wax, such as paraffin. These coatings made possible the widespread use of potassium chlorate-based explosives, such as **Cheddite**. Potassium chlorate was also used in **Sprengel explosives**, especially **Rack-a-Rock**. *See also* POTASSIUM PERCHLORATE.

POTASSIUM NITRATE. Strong **oxidizing agent** used as an ingredient in many **explosives**. Historically, potassium nitrate was the oxidizer in **black powder**, and its availability was often the critical step in making black powder. It is a white or colorless crystalline powder and has the chemical formula

KNO_3. Names used for potassium nitrate include niter, nitre, and **saltpeter**. *See also* AMERICAN CIVIL WAR (1861–1865); DU PONT, LAMMOT (1831–1884); INDIA SALTPETER; SODIUM NITRATE.

POTASSIUM PERCHLORATE. Strong **oxidizing agent** used as an ingredient in many **explosives**, **percussion caps**, **primers**, **propellants,** and **pyrotechnics**. It is a colorless crystalline solid and has the chemical formula $KClO_4$. Its use in solid propellant rocket boosters has largely been supplanted by **ammonium perchlorate**. *See also* OPHORITE.

POTENTITE. Smokeless powder containing approximately 60 percent **guncotton** and 40 percent **potassium nitrate**. Potentite was made by the Potentite Company, Ltd., at a factory in Melling, England, near Liverpool, and was similar in properties to **Tonite**. The factory was established in 1873, and the company operated as a subsidiary of the **Cotton Powder Company, Ltd.**, Faversham, which made Tonite. The properties and uses of Potentite were similar to those of Tonite.

POUDRE B. First successful **smokeless powder**, invented in 1884 by French physicist **Paul-Marie-Eugène Vieille**. This powder was a dense, white colloidal mass. It was prepared by mixing **insoluble nitrocellulose**, which Vieille designated CP_1, and soluble nitrocellulose, designated CP_2. These two types of nitrocellulose were treated with an ether-alcohol mix, kneaded to a stiff jelly consistency, and cut into squares and dried. Later, the material was pressed through a die to form strips or cords. It came into nearly immediate use by the French military and continued to be used for several decades until replaced by superior formulations. The white color of the finished product accounted for the name given to it by Vieille: "Poudre Blanc," from the French *blanc* (*white*), which was later shortened to "Poudre B." It came to be called "white powder" in the United States, in particular. This white color distinguished it from conventional **black powder**, which was called "Poudre Noir," from the French *noir* (*black*). Other names used for Poudre B included "Vieille powder" (in honor of its inventor), "Poudre Blanc Nouvelle" (signifying its new or novel nature), and "BN Powder," where "BN" stood for "Blanc Nouvelle."

POUDRE SD. French term for a **solventless powder**. "SD" is an abbreviation for *sans dissolvent*, French for *without solvent.*

POWDER TRUST. Popular term used in the early 1900s for various American manufacturers of **explosives** whose business holdings were widely seen as a conspiracy in restraint of trade. On 31 July 1907, the US Department of Justice filed an antitrust suit against these manufacturers, alleging violations of the Sherman Antitrust Act of 1890. Although various parties were named in the suit, the main target was **DuPont**, which by this time controlled the majority of manufacturing in the United States of **black powder**, **dynamite**, and **smokeless powder**, either directly or through a complex network of partial ownership of other manufacturers. The Justice Department was aided in its action against DuPont by a suit brought in 1906 by Robert Stuart Waddell (1849–1915), who blamed DuPont for the failure of his firm, Buckeye Powder Company. Waddell had worked for DuPont previously and had taken internal corporate documents with him when he departed. He made these documents available to the Justice Department, which exploited them to prove unfair trade practices.

On 21 June 1911, the US Federal Court ruled against DuPont. It took another year for the Justice Department and DuPont to agree on a remedy. That remedy became public on 12 June 1912, when DuPont announced it would divest itself of sufficient assets to create two new firms: Atlas Powder Company and Hercules Powder Company. Assets given over to the new firms included facilities, personnel, patents, and other business information. These assets ensured that Atlas and Hercules would have sufficient resources to produce 50 percent of the demand for black powder in the United States and 42 percent of that for dynamite. The remedy had no effect on DuPont's share of the market for smokeless powder. In fact, DuPont had argued successfully that quality control required by the military demanded a single supplier of this product. Although his documents greatly assisted the Justice Department in its antitrust suit, Waddell lost his personal suit against DuPont. In 1914, a jury found that Waddell had created Buckeye Powder principally to force a purchase offer from DuPont.

PREINITIATION. The sequence and timing of events are critical to the proper functioning of a **nuclear explosive**. Preinitiation occurs when the **fissionable** material of a nuclear explosive initiates a chain reaction prior to attaining maximum compression or complete assembly. When preinitiation occurs, the result is either no nuclear explosion or an explosive yield far less than expected based on theoretical calculations. Concerns over preinitiation caused the United States to cease development of the Thin Man bomb design, which was to be a **gun-type nuclear device** design similar to that used in **Little Boy** but having **plutonium-239** as the **fuel** instead of **uranium-235**. In its place, the United States used the **Fat Man** bomb design. The earliest known example of preinitiation of a nuclear device occurred on 22 October 1951, when the test "Able" failed during Operation Buster-Jangle at the

Nevada Test Site. The most recent example is the 6 October 2006 test of a nuclear device by **North Korea**. While the device functioned, the yield was very low and, thus, the device is reported to have suffered from preinitiation. An alternative term used for preinitiation is *fizzle*. *See also* ABSOLUTE DUD; DWARF DUD.

PRESSED BONDED EXPLOSIVE; PRESSED PLASTIC-BONDED EXPLOSIVE. *See* PBX.

PRIMARY EXPLOSIVE. An **explosive** that **detonates** by simple ignition, such as a spark, flame, or other heat source, or through **shock**; also referred to as "primary high explosive" and "initiator." Examples include **lead azide** and **mercury fulminate**. *See also* SECONDARY EXPLOSIVE.

PRIMER. An **explosive** used to initiate the main explosive. The **percussion cap** is one of the earliest primers. When the percussion cap became incorporated into the munition containing the main explosive, the percussion cap became known as a primer.

PRISMATIC POWDER. Explosive formed as geometric prisms (typically, hexagons) having a central channel, typically for use as a **propellant**. Prismatic powders burn in parallel layers. This burn pattern causes a slower rate of burning than is typical with regular shapes of **black powder** or **smokeless powder**. Such slower burning was found by **Thomas J. Rodman** to be highly desirable, especially as a propellant for large-caliber guns. The slower burning resulted in a longer and more nearly even build-up of pressure within the gun barrel. It had the added benefits that it avoided the higher pressures that often were the cause of failure in gun barrels, and it avoided the drop-off in pressure experienced with conventional shapes of black powder. Rodman, an American military officer who devoted much of his career to the systematic study of military ordnance, is credited with inventing prismatic powders during the **American Civil War**. Toward the end of the nineteenth century, **Hudson Maxim** successfully adapted the prismatic powder concept to smokeless powders, and in 1895, he received four US patents related to multi-perforated smokeless powder. These patents, however, conflicted with a similar one granted in the UK in 1894 to his brother, **Hiram Stevens Maxim**, causing some doubt as to the right of priority on this particular invention. *See also* MAMMOTH POWDER; MULTI-PERFORATED SMOKELESS POWDER.

PROGRESSIVE-BURNING POWDER. *See* RODMAN, THOMAS JACKSON (1816–1871).

PROJECT CHARIOT. A proposed project to use **nuclear explosives** to excavate an artificial harbor on the North Slope of Alaska. The proposal was considered by the US Atomic Energy Commission (AEC), beginning in 1958, as part of the **Plowshare Program** involving peaceful uses of nuclear explosions. The project targeted Cape Thompson and called for the **detonation** of five **thermonuclear devices**. The project was to take place shortly after 1959, when Alaska was to be admitted to statehood in the United States. It was championed by many influential Alaskans, including political leaders, newspaper editors, a university president, and church groups, and by **Edward Teller**, the so-called father of the **hydrogen bomb**, who toured the territory promoting it as an important economic development allowing Alaska's North Slope resources of coal and petroleum to be shipped from Cape Thompson. Opposition arose from environmentalists and, especially, from residents of Point Hope, an Ipiutak village 30 miles from the proposed site of the blasts. In 1962, the AEC announced Project Chariot would be held in abeyance, a step short of outright cancellation, which has never occurred.

PROOF POWDER. Black powder used to demonstrate (or "prove") that an alcoholic beverage has at least a minimum content of ethanol. The process of proving alcohol is believed to have arisen in the eighteenth century, when British sailors received a ration of rum as part of their pay. To prove that the rum or other alcoholic beverage had not been watered down, some of it was poured onto a heap of black powder. A flame was then touched to the mixture. If the mixture failed to burn or burned with difficulty, the beverage was deemed "underproof." If the mixture sustained burning or ignited with a slight **explosion**, the beverage was deemed "proof spirit" or "100 proof." If the mixture ignited with a violent **explosion**, the beverage was deemed "overproof." Eventually, more precise and less destructive test methods were developed for proving alcohol. However, test conditions for determining ethanol content vary internationally. For example, the United States specifies determining content at 60°F (15.5°C), while the European Union (EU) specifies using 68°F (20°C). As a result, the ethanol content equating to 100 proof in the United States differs from that equating to 100 proof in the EU.

1,2,3-PROPANETRIOL TRINITRATE. *See* NITROGLYCERIN.

PROPELLANT. Any material used to move or propel an object. Propellants can use chemical reactions or mechanical forces as energy sources. In the context of **explosives**, a propellant is an explosive used to propel an object, typically out of the barrel of a gun or other firearm or from a mount, as with a missile or rocket. The term is used to distinguish an explosive used for propulsion from one used for **blast effects**. Some explosive substances can be used for both purposes, perhaps even in the same object. An essential aspect of the behavior of a propellant is that it reacts rapidly but in a controlled manner. This rapid, controlled reaction produces gases, and the pressure exerted by these gases creates the thrust that propels the object. Examples include **black powder** and **smokeless powder**, both of which have been used extensively as propellants in munitions.

PYROCOLLODION. Nitrocellulose uniformly soluble in an alcohol-ether mixture. The term was coined by the Russian chemist **Dmitri Mendeleyev**, who invented this form of nitrocellulose in 1891. Because pyrocollodion is uniformly soluble in an alcohol-ether mixture, its characteristics should be uniform both within any manufactured batch and between batches. Therefore, its performance should be predictable when used as a **propellant**. This situation contrasted with earlier formulations of nitrocellulose, such as **Poudre B**. The benefits to such uniformity were recognized in particular in the United States, which learned details of Mendeleyev's invention through **John Baptiste Bernadou**. At the time, Bernadou was assigned to the **Torpedo** Station, Newport, Rhode Island. He used his fluency in Russian and other foreign languages to test various formulations of **smokeless powder**, eventually settling on a formulation based on pyrocollodion. By 1897, a plant had been constructed and was operating at the Torpedo Station, producing about 200–250 pounds of this smokeless powder per day. Military officials recognized that this capacity was too modest for their needs, however, and by the fall of 1897, **DuPont** and the California Powder Works had begun commercial-scale production of a pyrocollodion-based smokeless powder for the military. By the First World War, the United States had become a major manufacturer of such powders, unlike Russia, which failed to exploit Mendeleyev's invention properly.

PYROGLYCERINE. Name given to **nitroglycerin** by its discoverer, **Ascanio Sobrero**.

PYRONITE. *See* TETRYL.

PYROTECHNICS. Explosives whose intended effect is to produce light, noise, or smoke. Familiar examples include fireworks and signaling flares. Fireworks are widely believed to have originated in ancient **China** and migrated to Europe along with knowledge of **black powder**, the basic ingredient of early fireworks. Signaling flares are a more modern development, arising in the nineteenth century when various Europeans found that the addition of elements such as **aluminum** and magnesium brightened firework displays. This finding was exploited by the military, and the combat use of various signaling flares and other pyrotechnics became commonplace by the First World War. *See also* COSTON FLARE; VERY LIGHT.

PYROTOL. Blasting agent comprised of 60 percent finely ground **smokeless powder**, 34 percent **sodium nitrate**, and 6 percent **dynamite** containing 40 percent **nitroglycerin**. When the Armistice was signed in November 1918, the US government and its private industries that manufactured **explosives** jointly confronted the issue of how best to use an apparent abundance of smokeless powder. The quantity on hand and nearing delivery was estimated to be in great excess of what might be consumed by the military during peacetime target practice, and future wars were imagined to be unlikely. Further complicating the issue was that smokeless powder by itself is not an explosive; therefore, this excess could not be diverted readily to commercial use. However, it was shown that smokeless powder can be used as a blasting agent if it is ground to a fine powder, mixed with an **oxidizer** and some dynamite or **TNT**, and with a **blasting cap** added as a **detonator**. Two blasting agent products were created out of this necessity of using the excess smokeless powder: "Pyrotol," which was provided as a US government product but was produced by **DuPont**, and "Dumorite," which was produced and sold commercially by DuPont, the primary US producer of smokeless powder during the First World War.

In the early 1920s, the US government ordered 100 million pounds of smokeless powder from DuPont to be used in making Pyrotol. The order was reduced to 67 million pounds after the 10 August 1924 fire at the Old Hickory, Tennessee, plant consumed nearly 50 million pounds of smokeless powder. Despite this fire, the first supplies of Pyrotol left DuPont on 4 September 1924, less than one month later. After it was produced, Pyrotol was available through the US Department of Agriculture to farmers in quantities up to 1,000 pounds per farmer. The only costs were for preparing the Pyrotol for shipping and for shipping. The notion was that Pyrotol would be used to create ditches and to remove tree stumps and other obstacles, thus clearing land for farming. Pyrotol began to be distributed nationally in 1924. By 1928, the stocks had been exhausted, and no new stocks were created. After these supplies were exhausted, DuPont offered "Agritol" as an alternative, citing it as less flammable and more moisture resistant than Pyrotol. Dumorite experienced a similar fate to that of Pyrotol. It was first sold in 1922 as a blasting agent, especially for use in highway construction, but at a higher cost than Pyrotol. It disappeared from the market, too, around 1928, once excess stores of smokeless powder had been exhausted. The fate of Agritol is uncertain. *See also* BATH SCHOOL BOMBING; LAND CLEARING CAMPAIGN.

PYROXYLIN. Nitrocellulose that contains approximately 11 percent nitrogen and is soluble in an ether-alcohol mix. Pyroxylin is a yellowish-white matted mass of filaments resembling raw cotton and is highly flammable. It is used in making **collodion** and directly as a component in some **smokeless explosives**. French chemist **Théopile-Jules Pelouze** coined the term "pyroxylin" in 1838, when he noted the flammable nature of what had previously been termed by Henri Braconnot as "**xyloidine**," from the Greek πῦρ (*fire*) and ξυλον (*wood*). *See also* GUNCOTTON.

Q

QUATERNARY EXPLOSIVE. Composition explosive comprised of four explosive compounds. An example of a quaternary explosive is **depth bomb explosive (DBX)**.

QUESTION OF THE PACIFIC. *See* WAR OF THE PACIFIC (1879–1884).

R

RACK-A-ROCK. Commercial **Sprengel explosive**, comprised of a paper or cloth container of **potassium chlorate** dipped into **nitrobenzene** immediately prior to use. Nearly 240,000 pounds of Rack-a-Rock were used on 10 October 1885, along with 75,000 pounds of **dynamite**, to demolish **Flood Rock**, a barrier to navigation in New York City harbor. In this case, solid potassium chlorate was the **oxidizing agent**, while liquid nitrobenzene was the **fuel**. The requirement for dipping in a suitable fuel immediately prior to use proved inconvenient, and commercial use in the United States suffered. However, the US Army used Rack-a-rock extensively as a demolition explosive.

Rack-a-rock was covered by various US and British patents, issued to Silas Reynolds Divine (1838–1912) of Loch Sheldrake, New York, and assigned to the Rend Rock Powder Company of New Jersey. These patents, however, became a matter of bitter dispute between **Hermann Sprengel** and Divine. Sprengel was granted patents in the UK on 6 April 1871 (British patent 1,921) and on 2 October 1871 (British patent 2,642) for explosives and **blasting agents** in which either the **fuel** or the **oxidizer** is a liquid or in which both are liquids. Sprengel also openly published the concept for such mixtures in the scientific literature in 1873. Divine, on the other hand, received his first patents a decade later (US patent 243,432, "Explosive Compounds," and US patent 243,433, "Blasting and Blasting Cartridges," both issued 28 June 1881; and British patents 5,584 and 5,596, issued in 1881). Therefore, Sprengel asserted priority of invention, writing that "it seems to me that Mr. Divine's numerous patents are the efforts of a legal mind anxious to surround by a fence a piece of ground which I acquired 9 years, 8 months and 1 day before Mr. Divine appeared on the scene and gave this public property the quaint American name 'Rackarock'." Divine responded in writing that while his patents were issued in 1881, he had filed a caveat with the US Patent Office on 9 January 1871, covering the same concept as that covered in Sprengel's 1871 patents. Noting that he had filed his caveat three months prior to the issuing of the first patent to Sprengel, Divine asserted the right to claim priority of invention. The matter of priority of invention failed to be resolved to the satisfaction of Sprengel, and Divine continued to obtain US patents covering other applications and formulations of Sprengel explosives. However, most scientists credit Sprengel with priority of invention based, in particular, on his 1873 scientific publication.

RADIOLOGICAL DISPERSAL DEVICE (RDD). A **conventional explosive** that has been modified to include radioactive materials. This modification can enhance its destructive and disruptive effects. Such a device is referred to as a radiological dispersal device (RDD) or, more colloquially, as a "dirty bomb." In addition to any destruction of property and harm to humans caused by the explosive component of the RDD, the presence of radioactive materials means that those materials can be scattered around and carried beyond the scene of the blast. Such radioactivity can make the scene uninhabitable until the radioactivity decays or steps are taken to remove the radioactive debris. Any radioactive materials might be used for this purpose, including radioisotopes used for legitimate medical, engineering, and research purposes. By contrast, a **nuclear weapon** requires relatively rare **fissionable** materials that are supposed to be controlled and monitored. Therefore, the construction of an RDD is perceived to be more readily achievable by terrorists than is the construction of a nuclear weapon.

Use of an RDD was considered during the Second World War in the United States. On 25 May 1943, **J. Robert Oppenheimer** sent a classified letter to fellow **Manhattan Project** scientist Enrico Fermi that began, "I wanted to report to you on the question of radioactively poisoned foods." Oppenheimer went on to discuss using radioactive products of fission as weapons to "poison food sufficient to kill a half-million men." He also opined that strontium appeared to offer the highest promise among radioactive materials produced through fission. Another group of Manhattan Project scientists went further than Oppenheimer's proposal for an RDD to poison food. On 30 October 1943, James B. Conant (1893–1978), Arthur H. Compton (1892–1962), and Harold Urey (1893–1981) submitted a report to Leslie Groves on the "Use of Radioactive Materials as a Military Weapon." Their report noted that such a weapon would make evacuated areas uninhabitable, contaminate railroad yards, airports, and other important transportation hubs, promote panic, cause deaths among civilians, remain a potential hazard for a long time, and poison reservoirs, wells, and food. There is no evidence that the United States took any meaningful steps toward producing such weapons during the Second World War.

RAINS, GABRIEL JAMES; RAINS, GEORGE WASHINGTON. *See* AMERICAN CIVIL WAR (1861–1865).

RDX. Colorless, odorless, and tasteless solid crystalline substance used extensively as an **explosive**, especially during the Second World War. The preferred, systematic IUPAC name for RDX is 1,3,5-trinitro-1,3,5-triazinane, and its chemical formula is $C_3H_6N_6O_6$.

The first patent covering RDX was granted on 15 July 1898 to **Georg Friedrich Henning**, a German pharmacist searching for novel drugs. German patent 104,280 fails to note its explosivity. In 1920, Austrian-born chemist **Edmund von Herz** obtained British patent 145,791 covering the use of RDX as an explosive. On 3 January 1922, he obtained US patent 1,402,693 for this same purpose. In the US patent, von Herz observed that the "most surprising feature of this compound, however, is it [*sic*] extraordinary explosive and shattering power." Testing would prove this statement correct, as RDX has a **Relative Effectiveness Factor** of 1.60, one of the highest for any explosive in routine use. Von Herz also noted its relative non-toxicity, commenting that this property should prove useful in manufacturing and in loading shells. Other important features of RDX are its high stability and its low sensitivity to **shock** and **friction**. Most often, RDX is used in combination with other explosives and **phlegmatizing agents**. Use of RDX requires a **detonator**.

The patents granted to von Herz, as well as his continuing work in the explosives industry, stimulated interest in various nations in the use of RDX for military applications. Clarence J. Bain, a military chemist at the US Army Picatinny Arsenal, New Jersey, explored RDX, which was then known by the chemical name "cyclotrimethylenetrinitramine." He dubbed it "cyclonite" because of its cyclic structure, and that name came to be used extensively in the United States, especially. The name "RDX" originated with the British. The substance was studied by the Research Department at the Royal Gunpowder Factory, Waltham Abbey, where it became known as "Research Department Composition X." This name was shortened to "RDX." In Germany, it was designated "hexogen," while in Italy, it was "T4." *See also* BACHMANN PROCESS.

RDX TYPE A. Designation for pure **RDX**. The designation was used to differentiate pure RDX produced by direct nitrolysis of **hexamethylene tetramine** from RDX produced by the **Bachmann Process**. The direct nitrolysis method of making RDX is occasionally referred to as the Woolwich method, a reference to the Royal Gunpowder Factory, Woolwich, where the method was perfected. The designation is also given as "RDX (A)," "Type A RDX," and "Type (I) RDX."

RDX TYPE B. Designation for the mix of **RDX** and **HMX** produced by the **Bachmann Process**. The designation is also given as "RDX (B)," "Type

B RDX," and "Type (II) RDX." The mix typically contained 8–12 percent HMX, with the residual material being exclusively or predominately RDX. In some cases, **BSX** was also present.

REAGAN ASSASSINATION ATTEMPT. *See* EXPLOSIVE BULLET.

RED BEARD. Code name for a British tactical **nuclear weapon**. Developed as a follow-on to **Blue Danube**, Red Beard had a similar composite core of **plutonium** and **uranium**-235 but featured improvements that enhanced operational use and service life. The lead-acid batteries used in the Blue Danube as the electrical means for firing the weapon were replaced by twin ram-air turbine electrical generators. This replacement meant that last-minute installation of batteries was eliminated, as was the chance for stray electric discharges prior to the release of the weapon. Red Beard was produced in two variants: the Mk 1, having a theoretical yield of 15 kiloton **TNT-equivalent**, and the Mk 2, with a 25 kiloton yield. Both variants had a similar weight of slightly under 2 tons. Approximately 110 Red Beards were deployed, beginning in 1962, and all had been withdrawn by 1971. *See also* WE.177A.

REFRACTION OF NITER. *See* INDIA SALTPETER.

RÉGIE DES POUDRES ET SALPETRES. *See* GUNPOWDER SERVICE (*SERVICE DES POUDRES*).

RELATIVE EFFECTIVENESS (RE) FACTOR. Measure of the effectiveness of the **explosive** nature of one substance compared to that of **TNT**. The Relative Effectiveness (RE) Factor is a dimensionless ratio, determined on a weight/weight basis. The numerator is always TNT. It is used to calculate the amount of TNT that would be needed to substitute for a substance with regard to the explosive power. RE Factor is used in particular in blasting operations such as demolition, where equations are often based on the amount of TNT that would be required for a particular effect. For instance, if the RE Factor for Substance X is 1.50, then 1.50 kilograms (kg) of TNT would be needed as a substitute for 1.00 kilograms of Substance X. Similarly, if the RE Factor for Substance Y is 0.85, then 0.85 kilograms of TNT would be needed as a substitute for Substance Y. Therefore, the lower the RE Factor is, the less the relative effectiveness in comparison to that of TNT. Conversely, the higher that the RE Factor is, the greater its relative effectiveness compared to that of TNT. Sample RE Factors include:

- **Ammonium nitrate**—0.42
- **Black powder**—0.55
- **Amatol**—0.91
- **TNT**—1.00
- **Composition C4**—1.34
- **Nitroglycerin**—1.50
- **RDX**—1.60
- **PETN** and **HMX**—1.66

RENDER SAFE. The process of using special knowledge, methods, and tools to separate components of unexploded ordnance (UXO). This process might be as straightforward as disconnecting wires, although even in this example, expertise is required in identifying which wires might be disconnected and in what sequence. Rendering safe typically is performed by specially trained personnel, such as members of an **explosive ordnance disposal (EOD)** team.

REPAUNO CHEMICAL COMPANY. *See* DU PONT, LAMMOT (1831–1884).

RIPPLE ROCK EXPLOSION. The Canadian government conducted a controlled explosion on 5 April 1958 in the Seymour Narrows, British Columbia. For many years, two peaks of a submerged mountain, known as "Ripple Rock," had presented a hazard to navigation within the Discovery Passage. These peaks caused dangerous eddies, especially during low tide. Removing this hazard to navigation had been attempted unsuccessfully prior to 1958. For this ultimately successful attempt, nearly 1,400 tons of explosive were emplaced within Ripple Rock and detonated around 9:31 a.m., local time. A commercial explosive, Nitramex 2H, was selected for this effort. It is comprised of **ammonium nitrate**, dinitrotoluene, **TNT**, and paraffin wax and was developed expressly for **blasting** rock. The explosion was broadcast as it happened, becoming one of the earliest live coast-to-coast television broadcasts of an event in Canada. The explosion removed the mountain tops, enabling ships to transit Seymour Narrows safely and is regarded as among the largest planned non-nuclear explosions ever conducted. *See also* FLOOD ROCK.

ROBERTS, EDWARD A. L. (1829–1881). American inventor who pioneered the use of **explosives** to increase the flow in gas, oil, and water wells. The source of Roberts's inspiration for his invention is uncertain, but it is most likely tied to his service in the military, especially during the **American Civil War**. Born in Moreau, New York, Roberts enlisted in the US Army and served as a private during America's war with Mexico (1846–1848). He pursued some formal education before volunteering for service during the American Civil War. On 15 September 1862, he received a commission as lieutenant colonel, 28th Infantry Regiment, New Jersey Volunteers, serving as its deputy commander. By December, he had assumed command of the regiment, in time to lead it during the Battle of Fredericksburg. In early 1863, he attempted to resign his command owing to his poor performance during that battle. His resignation was rejected and, instead, he was dismissed from the military for "resigning in the face of the enemy." This last part of his military service might have been less than stellar, but his time in the military exposed him both to explosives and to the great need for water supplies for the troops in the field. He also took note during his otherwise undistinguished service at the Battle of Fredericksburg that artillery munitions increased the flow of water when they landed on narrow, water-filled channels. Finally, his service with the 28th Infantry Regiment earned him the title "colonel," which he used for the rest of his life.

His military experience likely accounts for his first patent application, filed 15 November 1864. Entitled "Improvements in exploding **torpedoes** in Artesian wells," the application was granted on 25 April 1865 as US patent 47,458. It was followed by several others, most especially US patent 59,936, "Improvements in methods of increasing capacity of oil wells," granted 20 November 1866. Roberts's concept for using torpedoes to improve flow in a well was straightforward. A metal cylinder was filled with several pounds of explosive (**black powder** was used initially, but was replaced eventually by **nitroglycerin**). The cylinder was sized to fit snugly within the borehole of a well and outfitted with a **percussion cap** on its top. Once filled, the cylinder was lowered to the bottom of the well and then covered with water to a depth of about 50 feet. The water would tamp the explosion such that the force was directed laterally into the earth surrounding the well rather than vertically, merely blowing out of the hole. A metal weight, called a **go-devil**, was dropped down the borehole, striking the cap and **detonating** the main explosive charge. The effect was likened by one writer to that of squeezing a sponge. The oil-bearing rock was compressed violently by the explosion. When the compression was released, reservoirs were created. Oil accumulated in these reservoirs and flowed from there into the borehole. The explosion also helped break up paraffin and other solidified petroleum products that blocked oil flow within the oil-bearing rock.

After filing his initial patent application, Roberts set out for the oil fields around Titusville, Pennsylvania. He carried with him six torpedoes of his design, constructed in New York City, where he worked as a dental technician making artificial teeth for his brother, Poughkeepsie dentist Walter Brooke Roberts (1823–1889). Arriving in early 1865, Col. Roberts attempted to interest oil well operators in his proposal to use an explosive to increase oil flow. Most operators resisted his pitch, fearing the explosive would damage the well. Eventually, Roberts talked the agent for the Ladies' Well in Watson Flats into trying the torpedoes. A large crowd gathered on 21 January 1865 and witnessed two **shots** that succeeded in getting oil to flow in an otherwise unproductive well. That same year, the Roberts brothers formed the **Roberts Petroleum Torpedo Company**.

The Ladies' Well demonstration stirred some interest in Col. Roberts's invention. But his big breakthrough came in December 1866, when he detonated two torpedoes at the Woodin Well on the Blood farm near Titusville. This well had been abandoned as a dry hole. The first detonation caused oil to flow at a rate of 20 barrels per day, and the second detonation upped the flow to 80 barrels per day. This demonstration made clear the financial rewards of the technique of shooting wells, solidified Roberts's reputation, and caused demand for the services of the Roberts Petroleum Torpedo Company to soar. The business prospered, prompting Roberts's brother to close his dental practice in New York and relocate to Titusville. Thereafter, Dr. Roberts devoted himself full-time to the torpedo business and to aiding his brother in pursuing legal action against individuals and firms violating Roberts's patent positions. *See also* GREAT LAW WAR.

ROBERTS PETROLEUM TORPEDO COMPANY. American firm founded in New York City in 1865 by **Edward A. L. Roberts** and his older brother, Walter Brooke Roberts, to exploit a patented process of using **explosives** to improve the flow of gas, oil, and water wells. The process proved successful, and the firm prospered. Initially, the firm used **black powder** as the principal component in its **torpedoes**. In 1871, the firm acquired the rights to manufacture and use **nitroglycerin** under the **Nobel** patents. Thereafter, it established a factory near Titusville, Pennsylvania, where it manufactured nitroglycerin for use in its torpedoes and became the dominant party in **shooting** oil wells. Its dominance was assured by the patents granted to the younger Roberts. But the high fees charged by the firm inspired the practice of **moonlighting**, and the firm's efforts to enforce its patents prompted the **Great Law War**. The firm prevailed in these efforts. These patents expired in 1883, two years after the death of the younger Roberts. The next year, the firm changed ownership and became the Otto Cupler Torpedo Company, which remains active in the business of shooting wells. However, the process of using explosives to increase the flow of gas, oil, and water in wells was eventually surpassed by hydraulic fracturing, which is more commonly called "fracking." Fracking is the process of creating fractures in rocks and rock formations by injecting specialized fluids, rather than using explosives. The result is similar, though. These fluids force cracks to open further, allowing more oil, gas, or, less commonly, water to flow into the well's borehole.

RODMAN, THOMAS JACKSON (1816–1871). American military officer and inventor. Educated at the US Military Academy, West Point, New York, Rodman spent his career in the military, where he became renowned for his contributions to ordnance. In 1844, he started experiments designed to overcome the principal difficulty in casting large iron guns, such as cannons. This difficulty effectively set a maximum size limit for guns in terms of caliber and length. When Rodman began his experiments, large guns were cast solid and were cooled only from the outside. This practice caused the cooling metal to contract toward the outer surface of the barrel. For a large casting, the use of external cooling created internal strains and structural irregularities in the metal as well as "pipes" or "blowholes," actual cavities within the casting. In short, large guns were prone to cracking in cooling, breaking in transport, and finally bursting when fired.

From his studies, Rodman developed a theory to account both for internal strains and imperfections and for variations in the density, hardness, and tensile strength of the metal in cast-iron guns. He perfected a process for casting gun barrels around hollow cores and cooled from the inside, via a jacketed stream of running water, rather than external cooling. The process allowed the manufacture of much larger caliber guns. This casting process proved important to the successful fielding of such guns, which were often referred to as "Rodman guns."

While working on this improved casting process for large guns, Rodman began the systematic study of the **propellants** used with these guns. His studies led him to develop what he termed "the principle of special powders," which held that there is a particular combination of size, shape, and density of a propellant that will give optimal performance for a particular type of gun. His studies led to the development of progressive-burning **black powders**, including **Mammoth Powder**, perforated-cake powder, and **prismatic powders**. At the time Rodman began his studies, many developers of **explosives** assumed that any particular black powder would work equally well as a propellant in all types of guns. This assumption, however, led to disappointing results with regard to large-caliber guns. As Rodman learned, the disappointing results could be traced to the physics underlying muzzle velocity. When a gun is discharged, the force of the propellant pushes a projectile down the barrel toward the muzzle. The projectile travels down the barrel, with the volume of the barrel behind the projectile continually increasing until the projectile exits the muzzle. Normal black powder grain burns from the outside so that its burning surface area continually decreases. Therefore, the pressure produced initially is the highest. While the projectile travels down the barrel, the volume behind it is continually increasing. Simultaneously, the rate at which the powder is burning is dropping. These two factors—increasing volume and decreasing burn rate—result in decreasing internal pressure within the barrel as a function of time following the initial firing of the gun. For small arms and small-caliber guns, the net effect is unnoticeable owing to the relatively short length of the barrel. But for larger caliber guns, the net effect is one of a reduction in the velocity of the projectile at the time it exits the barrel (i.e., the muzzle velocity). Coupled with this situation was one of irregularities in the sizes and shapes of normal black powder, resulting in variable burn rates from batch to batch as well as within batches. The overall result was one of inconsistent performance for black powder when used as a propellant for large-caliber guns.

Rodman, often working in conjunction with **Lammot du Pont**, experimented with various sizes, shapes, and densities of powder, searching for a combination that would perform best in a 15-inch gun. The results showed that a powder having a relatively large grain size (about 15 mm in diameter or approximately the size of a mature grape) but having holes distributed throughout its mass would burn at a more nearly uniform rate. The final design was a powder pressed into hexagonal grains perforated with several longitudinal holes so that each grain burned inside and out simultaneously, thereby increasing the burning surface of each grain. While the initial burn rate was reduced, the overall pressure within the barrel remained nearly constant, rather than dropping off as happened with conventional shapes of black powder. The result was increased range, owing to the higher muzzle velocity, and enhanced accuracy, owing to the reproducibility from batch to batch and within batches of powders. Additionally, the new shape, referred to as "perforated-cake powder," reduced the chances of a catastrophic rupture of the gun barrel on firing because the initial pressure produced on firing was reduced. The final powder produced by Rodman and du Pont resembled small pebbles rather than the fine powders then typically being used as propellants. Its relatively large size caused it to be dubbed "mammoth." By the outbreak of the **American Civil War**, mammoth powder was a standard propellant for 15-inch guns and was used by both sides during that conflict.

ROSS, JAMES HAMILTON. *See* BACHMANN PROCESS.

RPC/12. Smokeless powder credited as the first **solventless powder**. Its properties contributed greatly to the ability of the Central Powers (the German Empire, the Austro-Hungarian Empire, the Ottoman Empire, and the Kingdom of Bulgaria) to prolong the hostilities during the First World War.

"RPC" is an abbreviation for *Röhrenpulver mit Centralit*, which is German for *tubular* ***gunpowder*** *with* ***Centralite***. The "/12" refers to 1912, the year when its manufacture was begun at a factory in Düneberg, Germany. RPC/12 was prepared by mixing soluble **nitrocellulose**, **guncotton**, **nitroglycerin**, and Centralite. The mix was then rolled to promote drying and gelatinizing, and then formed into tubes. RPC/12 offered several advantages over **Cordite** and other smokeless powders then in use. First, it contained no volatile solvents; consequently, RPC/12 could be dried in a few hours, versus the days needed for drying Cordite. Second, its manufacture avoided the use of acetone, which came to be in short supply for the Central Powers during the First World War. Finally, water-soluble salts such as potassium oxalate or potassium tartrate could be incorporated into the mix. These salts came into use during the First World War to reduce or suppress the flash associated with **detonation**, especially of **propellants**. Other nitrocellulose-based propellants then in use required a step in which the mix was soaked in water. That step would remove these water-soluble salts. The composition and manufacturing process for RPC/12 were protected as state secrets throughout the First World War. Following the war, however, its properties became known, and other nations developed their own formulations of solventless powders, such as **Cordite SC**.

RUSSO-JAPANESE WAR OF 1904–1905. Conflict fought from 8 February 1904 to 5 September 1905 between the Empires of Japan and Russia. Japan emerged victorious from this 17-month conflict, which began with a surprise attack by the Imperial Japanese Navy on the Russian Baltic Fleet at Port Arthur. Many factors contributed to Japan's overall victory, among which was Japan's superiority over Russia in their respective use of **explosives**.

This situation came into stark contrast during the Battle of Tsushima on 27–28 May 1905, when the Russian fleet was virtually annihilated, with seven battleships and 14 other ships being sunk and four other battleships being captured, while the Japanese fleet lost merely three torpedo boats. Casualties were similarly one-sided, with more than 5,000 Russians killed but fewer than 120 Japanese killed. One contributing factor to this decisive victory was Japan's use of shells having **smokeless powder** as the **propellant**, a **high explosive** as bursting charge, and an ingenious **fuze** that armed itself in flight and that exploded on contact. **Cordite**, a smokeless powder developed in Europe, had been embraced by the Imperial Japanese Navy and was used as the propellant in their naval munitions. Use of this propellant meant that the naval gun spotters had a relatively clear view as they attempted to adjust gunfire. In 1893, the Imperial Japanese Navy adopted the smokeless powder **Shimose**, an invention of Japanese chemist and naval engineer Shimose Masachika (1860–1911), for the bursting charge. The Ijuin fuze, an invention of Ijuin Goro (1852–1921), an Imperial Japanese Navy officer, meant that shells **detonated** on contact with a ship's superstructure, setting it and nearby stores of ammunition or coal on fire, or, if off-target, on contact with the water, creating a geyser that aided the spotters in correcting aiming.

The Russians, by contrast, were using smokeless powder on the main guns of merely five battleships out of 11 at the Battle of Tsushima. None of the other guns, nor those on any of the other 27 ships in the Russian fleet present during this battle, were using smokeless powder. Instead, they used **black powder** or **brown powder** as the propellant, **guncotton** as the bursting charge, and relatively insensitive delayed-action detonation fuzes. Use of brown or black powder created a large amount of smoke when each gun was fired, hindering observation by spotters aboard the ship who were attempting to adjust the gunfire. The spotters were hampered further by the insensitive fuzes, which routinely failed to detonate when a round fell into the water. Failure to convert its arsenal to smokeless powder by 1904 was surprising in that Russian chemist **Dmitri Mendeleyev** had invented **pyrocollodion**, a smokeless powder, in 1891 while working for the Ministry of the Navy. His invention had been adopted sparingly owing to rivalry between the Ministry of the Navy and the Ministry of War, rather than for technical or operational reasons. This rivalry had tragic consequences for the Russian military and for Russia. *See also* PANPUSHKO, SEMYON VASILY'EVICH (1856–1891).

S

SAFETY BLASTING POWDER. Explosive developed for use in coal mining and similar rock blasting and quarrying applications; also known as "Cahuecit" and "Carboazotine." Raymond Louise Cahuc (1838–?) of Toulouse, France, invented this powder, receiving British patent 3,934 on 14 November 1874. At that time, the powder was comprised of any of various nitrates, **sulfur**, and one or all of the following: **sawdust**, soot, and **tanbark**. Cahuc noted that this powder was useful as an insecticide for *Phylloxera vastatrix*, an insect that harms grapes and, thus, is of concern to French and other winemakers. On 12 December 1877, he received British patent 4,732 for improvements to his earlier invention. The patent covers three formulations, each containing iron (II) sulfate, **potassium nitrate**, sulfur, soot, and sawdust or tanbark, with the proportions varying according to the intended blasting application.

SAFETY FUSE. *See* BICKFORD, WILLIAM (1774–1834).

SAINT PETERSBURG DECLARATION OF 1868. International arms control agreement prohibiting the wartime use of **explosive bullets**. The full name of the agreement is "Declaration Renouncing the Use, in Time of War, of Explosive **Projectiles** Under 400 Grammes Weight."

The Declaration is the first formal agreement among nations prohibiting the use of certain weapons in war. The Lieber Code of 24 April 1863 was an earlier attempt at developing laws governing armed conflict. Among its provisions was Section I, paragraph 16, which provided that "Military necessity does not admit of cruelty, that the infliction of suffering for the sake of suffering or for revenge, nor of maiming or wounding except in fight." The language of this paragraph might be interpreted to include expanding bullets and explosive bullets, both of which can inflict wounds that cause intense suffering and are maiming, with amputation frequently required. However, the language is inexact and open to interpretation. Furthermore, the code, which was adopted by the US military as General Order Number 100, applied solely to the armies of the United States.

While expanding bullets had been known since at least the early part of the nineteenth century, explosive bullets were a more recent development, dating to 1856. One contemporary witness to a demonstration of their effects proclaimed such bullets to be "the most formidable missile ever invented by man." The Russian Empire was among the nations that tracked this invention. By 1863, its military had created an explosive bullet that detonated on contact with a hard substance, consistent with its use against ammunition wagons, which was the preferred use noted by an early developer of such bullets, John Jacobs. By 1867, though, the Russian military was aware of a modified projectile that would explode on contact with a soft substance, including a human body. This projectile used **Pertuiset's Powder**, which had been demonstrated to Tsar Alexander II and others that same year. The Russian government viewed the resultant exploding bullet as an inhumane instrument of war and advocated banning its use.

As a result of this view, Russian Prince Alexander Gorchakov (1798–1883) invited representatives from 18 nations to join with Russia in considering such a ban. They met in Saint Petersburg, Russia, in late 1868. On 11 December (29 November, by the old-style calendar then in use in the Russian Empire), all 19 nations in attendance adopted the declaration. The limit of 400 grams (approximately 14 ounces) was chosen most likely because larger projectiles were imagined to be too large for fitting into and being fired from handheld weapons, such as rifles and handguns. Furthermore, it covered explosive bullets, but it omitted expanding bullets, such as Minié balls, hollow-point bullets, and soft-point bullets. The 19 nations adopting this declaration included most European nations but excluded the United States, which in the view of Russia was not deserving to be considered a "grand power," unlike the 18 other nations invited to the conference. The Saint Petersburg Declaration prompted the adoption of further declarations of a similar nature at the two Hague Peace Conferences of 1899 and 1907. Treaties drafted during those conferences expanded the laws governing armed conflict to include, among other things, bans on launching or dropping **explosives** from balloons.

SAKHAROV, ANDREI DMITRIYEVICH (1921–1989). Russian physicist and peace activist. Sakharov described himself as the father of the Soviet **hydrogen bomb**. His work on controlled **thermonuclear** reactions was essential to the successful development of this bomb. Sakharov is better known, however, for his efforts to promote human rights and world peace. His efforts were recognized internationally, most visibly by his selection for the 1975 Nobel Peace Prize. His outspoken views in regard to human rights and world peace resulted in his arrest by Soviet authorities in 1980, who exiled him from Moscow, requiring that he remain in Gorky (now Nizhny Novgorod), Russia. *See also* JOE-1, JOE-4, JOE-111; TSAR BOMBA.

SALTED WEAPON. A **nuclear weapon** to which elements or isotopes that capture neutrons have been added in order to produce radioactive products that are not produced by a nuclear weapon of normal design. Therefore, these devices pose hazards unlike those seen with weapons lacking this design feature. The term derives from the expressions "to salt the earth" or "sowing with salt," meaning to render an area uninhabitable. The concept of a salted weapon is credited to **Leo Szilárd**, whose intent was to show that nuclear weapon developments could reach a point where they could end human life on earth. *See also* NEUTRON BOMB.

SALTPETER. Archaic term for **potassium nitrate**. The term comes from the Latin *sal* (*salt*) + *petra* (*rock*), reflecting its presence as a white efflorescence appearing on rocks and on the rocky walls of caves and mines. *See also* INDIA SALTPETER.

SALTPETER FARMER. *See DROIT DE FOUILLE.*

SAWDUST. Natural material used in preparing a form of **nitrocellulose**, as well as a **fuel** used in some **explosive** powders. The potential of nitrating sawdust to prepare an explosive was raised as early as 2 November 1846 in a publication by London chemist Edward Frederick Teschemacher (1791–1863), who referred to it as "gun-sawdust," by analogy to **guncotton**. Safety concerns surrounding guncotton, though, likely delayed adoption of nitrated sawdust explosives as well, but this delay had no effect on the adoption of explosives in which sawdust was a fuel. In fact, the use of sawdust as a fuel proved popular. An 1895 edition of "A Dictionary of Explosives" by the British inspector of explosives, Jocelyn Home Thomson (1859–1908), lists 31 explosive powders having sawdust as an ingredient. One such early explosive mixture was haloxyline, which had been invented by 1865 and for which British patent 1,341 was granted on 10 May 1866. It consisted of 75 percent **potassium nitrate**, 15 percent sawdust, 8.5 percent **charcoal**, and 1.5 percent potassium ferrocyanide. Haloxyline was produced in the Austro-Hungarian Empire and used as an explosive filler in hand grenades. The history of use of sawdust in commercial and military explosives and the ease of its acquisition without arousing suspicion suggest sawdust might be used in **homemade explosives**, although no evidence of such use has emerged. *See also* DUALIN; DUALINE; SAFETY BLASTING POWDER; SCHNEBELITE; VIGORITE.

SAWDUST POWDER. *See* SCHULTZE POWDER.

SCHIESSLER, ROBERT WALTER. *See* BACHMANN PROCESS.

SCHLAGWETTERSICHERE SPRENGSTOFFE. German term for **permissibles**; literally, "**firedamp**-safe explosives." *See also* WETTER DYNAMITE.

SCHNEBELITE. Gunpowder for use as a **propellant** in small arms. Schnebelite was invented by a Roman Catholic prelate, Abbé Schnebelin, and his brother of Valence-en-Brie, France, for which they were granted British patent 9,359, on 17 February 1892, followed shortly by a French patent granted on 25 February. The powder consisted of 70.4 percent **potassium chlorate**, 28.2 percent **sawdust** or the pith of elder trees, and 1.4 percent starch. According to their patent filings, the latter two ingredients could be replaced by any farinaceous vegetable, such as beans, potatoes, or rice. Abbé Schnebelin went on a tour in Britain during the 1890s, demonstrating his invention. The *Ann Arbor* (Michigan) *Register* carried a report on 16 November 1893 of one such demonstration before a crowd of spectators. As they watched, the Abbé made some of the powder, showing the ease of its preparation. Then he placed this preparation on an anvil and pounded it with a hammer, which merely made it into a fine dust, thereby demonstrating its resistance to **shock**. He set a lighted match to it, producing a bright light but neither **deflagration** nor **detonation**. Placing some in a cartridge, then sealing it, he fired a gun, propelling a bullet a distance of 100 yards through an 8-inch steel plate and 4 inches of wood behind the plate. Despite such promising demonstrations, samples of Schnebelite sent to the French Ministry of War were returned unopened. Soon thereafter, Abbé Schnebelin abandoned his pursuit of explosives, presenting himself as a master of the occult, capable of exorcising demons from haunted houses. In any event, the composition of Schnebelite testifies to the ease with which ordinary materials might be used to create **homemade explosives**, although no evidence of criminal use of this particular formulation has emerged.

SCHNEIDERITE. Ammonium nitrate-based **explosive** used extensively in the First World War by France, Italy, and Russia. Schneiderite is comprised of approximately 88 percent ammonium nitrate and 12 percent **dinitronaphthalene**. Some sources give naphthalene rather than dinitronaphthalene as the minority constituent of Schneiderite. Its presence, however, seems unlikely, given that naphthalene is generally of little value as an explosive and that its presence would likely contribute to the generation of a gray or black smoke on **detonation**, versus the nearly smoke-free performance desired of **smokeless powders**. Schneiderite was developed in France, where it was also known as *Explosif S* and *Explosif Sc*. The name *Schneiderite* most likely is taken from the firm that marketed it, Schneider et Cie., an international armaments enterprise based in Le Creusot, France, and founded by the brothers Joseph Eugène Schneider (1805–1875) and Adolphe Schneider (1802–1848).

SCHÖNBEIN, CHRISTIAN FRIEDRICH (1799–1868). German-Swiss chemist who recognized the military potential of **guncotton**. Schönbein was born in Germany and learned chemistry by apprenticing in a German pharmaceutical firm. He spent time in England, where he was influenced by lectures given by the great experimentalist Michael Faraday (1791–1867). Despite his lack of a formal education in chemistry, Schönbein was appointed to the faculty at the University of Basel, Switzerland, in 1829, remaining there until his death. On 27 May 1846, Schönbein announced the discovery of "Schiessbaumwolle" (German for *guncotton*) at a meeting of the Society of Scientific Research, Basel. That year marked significant developments regarding guncotton and **nitrocellulose**, with scientists in France, Germany, and Russia publishing results highlighting the unusual nature of these materials. However, the activities of Schönbein and the German chemist **Rudolf Böttger** that year are particularly noteworthy for their scope and influence. In April 1846, Schönbein carried out experiments with guncotton at a German arsenal in Ludwigsburg in the presence of artillery officers and, shortly thereafter, in Stuttgart in the presence of King William I of Württemberg. During May, June, and July, Schönbein conducted experiments in which he used guncotton as a **propellant** in small arms and mortars. These experiments culminated on 28 July 1846 when he successfully used guncotton to propel a projectile from a cannon, marking the first such use of guncotton with a larger caliber gun. During this time, he also demonstrated the use of guncotton as a **blasting agent**, removing rock at Istein in the Grand Duchy of Baden and demolishing old walls in Basel. In August, he learned of similar work by Böttger. Soon thereafter, they appeared jointly in England, demonstrating the nature of guncotton at various military arsenals. These presentations influenced the firm **John Hall & Son** to acquire exclusive rights to manufacture guncotton. Also in August, Schönbein traveled to Cornwall, where he and Richard Taylor of Falmouth conducted experiments using guncotton as a blasting agent in the local tin mines.

Based on these activities and on his 1846 announcement in Basel, Schönbein claimed priority of discovery for guncotton. In truth, establishing such a priority is problematic. The work of **Henri Braconnot** in 1833 and **Théophile-Jules Pelouze** in 1838 had already established the nature of certain nitrocellulose materials. While Braconnot failed to highlight the flammable and potentially explosive nature of these materials, Pelouze had gone so far as to suggest their use in **fireworks**. In addition to Böttger, other scientists were active during 1846 and earlier in researching guncotton. Most notably, the German scientist Friedrich Julius Otto (1809–1870), professor of applied chemistry at the Collegium Carolinum, Braunschweig, had published a scientific paper describing guncotton. This publication occurred after the presentation by Schönbein to the Society of Scientific Research, Basel, but before Schönbein had published his results in the scientific literature, which occurred in 1847. But Schönbein clearly promoted the prospects for using guncotton as an **explosive** and gave the material the name by which it has become best known. In addition to his work on guncotton and other nitrocellulose materials, he is credited with the discovery of ozone and with naming this substance.

SCHULTZE, JOHANN FRIEDRICH EDUARD. Prussian military officer and inventor of **explosives**. Biographical data on Schultze is meager. Various accounts, including patent records, show that he was an officer in the Prussian Royal Artillery. He is credited with inventing **Schultze powder**, which is one of the earliest **smokeless powders**. On 2 June 1863, he was granted US patent 38,789, "Improved Gunpowder." That invention identifies him as a captain in the Prussian Royal Artillery residing in Potsdam. Around the time this patent was granted, Schultze appears to have relocated to Eyeworth Lodge, Hampshire, England, where he took over management of a gunpowder factory that later bore his name. The Schultze Gunpowder Factory was later purchased by **Nobel Industries** and continued operating until the early 1920s. It is uncertain, however, how long Schultze remained in England with this factory. For example, when he was granted US patent 359,289, on 15 March 1887, he is identified as residing in Darmstadt, Germany.

SCHULTZE POWDER. Any of various formulations of **smokeless powder** in which **nitrocellulose** is mixed with **potassium nitrate** and barium nitrate. **Johann Friedrich Eduard Schultze** is credited with developing this basic formulation, although there were eventually several imitations. Some form of Schultze powder may have been the first smokeless powder to be produced and used on a commercial scale. The nitrocellulose used in these formulations was crude by comparison with those that came to be used in standard military **propellants**, such as **Ballistite**, **Cordite**, and **Poudre B**.

The basic process developed by Schultze in the late 1850s or early 1860s involved extensive preparation of wood to remove sap and resinous materials, followed by nitration using nitric acid and sulfuric acid. This crude nitrocellulose was impregnated with a mix of barium nitrate and potassium nitrate. Initially, Schultze used wood cut into strips and then punched into grains. He referred to this product as "granulated-wood gunpowder," for which he obtained US patent 38,789, "Improved Gunpowder," on 2 June 1863. Later, he found he could use **sawdust** directly, and this product was referred to as "sawdust powder." He described his invention in a pamphlet in London in 1865, "The new chemical gunpowder (sawdust-powder) and its advantages over black **gunpowder** and its substitutes." Rights to the invention were purchased by an Austrian firm, Collodin-Fabriks-Action-Gesellschaft in Marchegg, which marketed the product under the names "collodin" and "nitroxylin." Testing of Schultze powder demonstrated that it **detonated** with no smoke, no residue, and greater force than did **black powder**. However, it burned too quickly for use in large-caliber guns then in use by the military. Later, mineral oil, paraffin, turpentine, and wax were used to coat the product in the hopes that the coated granules would burn more slowly.

Schultze powder failed to find much use in military munitions, with the notable exception of a cartridge used successfully with the **Dreyse needle-gun**. Schultze powder and imitations based on the basic concept of nitrated wood plus a nitrate salt saw extensive use as propellants in cartridges for guns and rifles used for sport and hunting game. Because such powders were often white, they came to be called "white gunpowder," distinguishing them from conventional black powder. American folk heroine Annie Oakley (1860–1924), famed for her skill as a sharpshooter, proclaimed she only used cartridges filled with Schultze powder for her guns.

SCHÜPPHAUS, ROBERT CARL (1861–1932). German-born chemist and inventor of **explosives**. After completing graduate studies in chemistry at the University of Göttingen, Schüpphaus immigrated to the United States in 1884, where he took charge of **nitrocellulose** manufacturing for the American Zylonite Company, near North Adams, Massachusetts. In 1893, he joined **Hudson Maxim** and his Maxim Powder & Torpedo Company, near Farmingdale, New Jersey. The association with Maxim was fruitful because it shortly led to the development of the **Maxim-Schüpphaus Smokeless Powder**, the first successful **smokeless powder** to be invented in the United States.

SCHWARZ, BERTHOLD. Franciscan monk credited in some traditions with discovering **black powder** during the thirteenth or fourteenth century. According to these traditions, Konstantin Anklitzen entered a monastery in Freiberg, Germany, taking the name "Berthold." His interest in **alchemistry** and the black arts led some to dub him "Berthold der Schwarze" (Berthold the Black). In some legends, Schwarz became insane and, during his insanity, discovered black powder. This discovery is supposed to have led directly to its use as **gunpowder** and the development of firearms. Stories surrounding Schwarz are difficult to verify, as the records of the Freiberg monastery were destroyed during the Reformation. In any event, a monument honoring Schwarz and his alleged role in discovering gunpowder was erected in Freiberg in 1853. However, this honor is undeserved, because alchemists in China had discovered the **explosive** nature of a mixture of **sulfur**, **saltpeter**, and a carbonaceous material as early as 142, predating Schwarz by more than a millennium. *See also* ALCHEMISTS AND EXPLOSIVES; BACON, ROGER (circa 1214–1292).

SECOND BOER WAR. Conflict fought from 11 October 1899 to 31 May 1902 in southern Africa. The major combatants were the British Empire and the two independent Boer republics of the Orange Free State and the South African Republic, which had been created as an outcome of the defeat of the British in the First Boer War (1880–1881). The British prevailed in the Second Boer War, which is also referred to as the South African War. As a result, both Boer republics lost their independence and became part of the British Empire. The war is notable for extensive use both of **Lyddite** as a **high explosive** in artillery shells and of **smokeless powder** as a **propellant** for artillery and firearms. Combat use of Lyddite first took place in September 1898 when the British used Lyddite-filled shells at the Battle of Omdurman during the Anglo-Egyptian conquest of Sudan. But that battle was brief, and the use of Lyddite was confined to two days. Consequently, the full potential of Lyddite as well as its defects had to wait until it could be tried over a longer period and in various roles. The Second Boer War provided such an opportunity.

Initial British use of Lyddite during the Second Boer War was confined to a 50-pound shell fired by the Ordnance Breech-Loading (BL) 5-inch howitzer. Later, Lyddite shells, having twice the payload of high explosive, were developed for the Ordnance BL 6-inch 30-hundredweight (cwt) howitzer. These shells proved effective at injuring or killing troops in the open and mildly effective at demolishing fortified structures. However, the war highlighted two deficiencies associated with Lyddite. First, Lyddite requires a powerful **detonator**. Contemporary British accounts noted that many Lyddite-filled shells failed to detonate, and the cause was traced to a faulty detonator. Second, the British observed that Lyddite tended to detonate on contact with a dense surface, such as an armor plate. This surface detonation dissipated the impact of the explosive, limiting its effectiveness in penetrating armor and other defenses. Although the British noted both deficiencies of Lyddite during the Second Boer War, they failed to implement solutions prior to the outbreak of the First World War in August 1914.

Combat use of smokeless powder dates at least to the **Franco-Prussian War**. It was also used to advantage by the Spanish during the **Spanish–American War**. Both conflicts demonstrated the tactical advantages afforded by smokeless powder in terms of maintaining a clear line-of-sight and avoiding the puffs of smoke characteristic of **black powder** that betray one's position. In neither conflict, however, did the use of smokeless powder prove decisive in terms of the outcome of the conflict. These advantages were demonstrated again during the Second Boer War. In the case of the Boers, the advantages may have helped the Boers in waging their guerrilla campaign that prolonged the conflict but ultimately failed to defeat the British. In particular, the Boers adopted the tactic of firing from concealed positions, and their use of smokeless powder enabled them to remain concealed. Although the Boers used various firearms, including some that used black powder, they had a significant inventory of German-made Mauser model 1894 and model 1895 rifles as well as various models of the Danish-made Krag-Jørgensen rifles. All of these rifles fired smokeless powder cartridges. The British had developed and fielded **Cordite**, a smokeless powder, as a propellant for their artillery and the .303 Lee Enfield Mk I rifle, which fired a smokeless cartridge. Unlike the Boers, the British failed to modify their tactics to exploit the advantages of smokeless powder, especially with regard to the infantry. This failure persisted into the First World War.

SECONDARY EXPLOSIVE. An **explosive** that typically requires the use of a **detonator** to function fully; also referred to as "secondary high explosive." A secondary explosive can be used as the main charge in an overall explosive mixture or as a **booster**. Examples include **ammonium nitrate**, **dynamite**, **guncotton**, and **TNT**. *See also* PRIMARY EXPLOSIVE; TERNARY EXPLOSIVE.

SEMTEX. Any of various formulations of a **plastic-bonded explosive** (PBX). Most formulations contain **PETN**, **RDX**, a **plasticizer**, a **binder**, and a dye. Formulations presently manufactured and available for purchase are deliberately adulterated with **ethylene glycol dinitrate** (EGDN) as a **detection agent**. Stanislav Brebera (1925–2012) is credited as the primary developer of Semtex. According to an interview in 2002, Brebera claimed the development took place in 1952. The product was introduced commercially over a decade later. At the time of its development, Brebera worked for the firm VCHZ Synthesia, located in Semtin, then part of Czechoslovakia and now part of the Czech Republic. This town inspired the name for this explosive. VCHZ Synthesia later changed its name to Explosia, which continues in operation and continues to market Semtex, mostly for domestic consumption.

Previously, Semtex was widely exported, especially to Libya, which imported large quantities from 1975 to 1981. These imports may account for its use by various terrorist groups, especially those persons responsible for the destruction of **Pan Am Flight 103**.

SENATE SPECIAL COMMITTEE ON INVESTIGATION OF MUNITIONS. *See* NYE COMMITTEE.

SERRAU, JACQUES ROSE FERDINAND ÉMILE (1837–1904). French engineer who devoted much of his professional career to practical studies of **explosives**. Émile Sarrau was closely associated with the **Gunpowder Service** of France, where he worked closely with **Marcellin Berthelot** and **Paul Vieille** of the Laboratoire Centrale des Poudres et Salpêtres (Central Laboratory of Gunpowder and **Saltpeter**). Sarrau is credited with correctly deducing the equation of motion that describes the behavior of a projectile in the bore of a gun. This equation, which he developed around 1875, allowed the behavior of projectiles and **propellants** to be measured and modeled, thereby aiding in the development of improved projectiles, propellants, and guns. In 1878, he was appointed Inspector General of Gunpowder and Saltpeter for France, and in 1883, he became a professor at Ecole Polytechnique, Paris. His scientific contributions were recognized by his election in 1886 to both the Institute of France and the French Academy of Sciences. His research into the shock waves generated by explosives prompted some French scientists and engineers to honor him by using the term *Numbre de Sarrau* (*Sarrau number*) for the dimensionless ratio of the flow velocity of a wave to that of the speed of sound. This ratio is more familiarly known as the Mach number, after Ernst Mach (1838–1916).

SERVICE DES POUDRES; SERVICE DES POUDRES ET SALPÊTRES; SERVICE NATIONAL DES POUDRES ET SALPÊTRES. *See* GUNPOWDER SERVICE (*SERVICE DES POUDRES*).

7/7 TERRORIST BOMBINGS. Four attacks involving **homemade explosives** (HME) that took place in London on 7 July 2005. The attacks targeted the London public transit system during the morning rush hour. Three devices exploded on the London underground, while the fourth exploded on a double-decker bus. All four attackers committed suicide by **detonating** their devices, which they had placed in their backpacks. These four were labeled Islamic extremists based on videotaped testimonies that they left. The four explosions claimed 52 lives beyond those of the four attackers and injured more than 800. These attacks were followed on 21 July by more attacks, but none caused any injuries or significant damage. Investigation into the attackers found that they had created their HME in a dwelling in Leeds, England. **Concentrated hydrogen peroxide (CHP)** was prepared by boiling down commercially available aqueous hydrogen peroxide, such as might be used as an antiseptic or for bleaching hair. This CHP served as the **oxidizer** and was mixed with the ordinary condiment, black pepper, which was the **fuel**. *See also* PERSON-BORNE IMPROVISED EXPLOSIVE DEVICE (PBIED).

SHAPED CHARGE. A cylinder of **explosive** having a symmetrical hollow cavity at its base. If the cavity is unlined, then the explosive is referred to as a hollow charge or an unlined charge. If the cavity is lined, then the explosive is referred to as a shaped charge or a cumulative charge. If the liner is bow-shaped, then the explosive is referred to as an explosively formed penetrator or explosively formed projectile (EFP).

A column of explosives with a hollow at its base causes more damage than does the same mass of explosive in any other configuration when placed directly against a target. The amount of damage is increased further by lining the hollow cavity with a high-density material, such as metal. In this case, the **detonation** pressure collapses the surface of the liner. The liner material then behaves as a liquid and is projected from the front of the explosive charge at high velocity.

Discovery. The shaped charge effect apparently has been discovered and developed independently on at least four occasions. The earliest discovery of the principle of the shaped charge might be credited to Franz Xaver von Baader (1765–1841), remembered most often as a Roman Catholic philosopher and theologian. At an early point in his career, the German-born von Baader studied mining engineering in England prior to working as a mining engineer in Bavaria. He presented his suggestion of what is now known as the shaped charge in the article "Versuch einer Theorie der Sprengbarkeit" ("Investigation of a theory of blasting"), published in the March 1792 issue of *Bergmannisches Journal von Kohler und Hoffman*. That article inspired other mining engineers to adopt the use of explosives having a dome-shaped air space. Such explosives were used for blasting in mines in Norway and Germany in the early part of the nineteenth century, according to an 1811 manual by L. Hausmann, *Leitfaden zur Bergbaukunde* (*Manual on mountain mining knowledge*). Results from using the explosive configuration suggested by von Baader apparently were disappointing, as use of such shaped charges appears to have been abandoned by the second decade of the nineteenth century. In any event, use of an explosive with a hollow cavity would have failed to achieve results consistent with those of the modern shaped charges for the simple reason that **black powder** was the primary explosive then in use for blasting. Black powder is a **low velocity explosive**. It **deflagrates** rather than detonates, and detonation is now known to be required to achieve the penetration associated with the modern shaped charge.

Re-discovery of the shaped charge effect is credited to Karl Hermann Waldemar Maximillian (Max) von Foerster (1845–1905). He began his career by serving as an officer in the Westphalian Pioneer Battalion No. 7, where he dealt extensively with explosives. Leaving military service in 1872, he eventually became technical director of the Wolff & Co. nitrocellulose factory, Walsrode. During testing of explosives in 1883, von Foerster recognized the effect of a cavity in the explosive charge and published an article in 1884 in *Nostrand's Engineering Magazine* about what came to be known as the hollow charge effect. The third and fourth independent discoveries of the shaped charge effect are credited to Charles Edward Munroe (1849–1938) in the United States in 1888 and Max Neumann in Germany in the early twentieth century. American chemist Munroe, while working at the Naval Torpedo Station, Newport, Rhode Island, publicized his discovery in both the scientific literature and the popular press. Neumann was the earliest party to obtain a patent relating to the effect, doing so in Germany in 1910 and in the UK in 1911 and assigning patent rights to the Wittenberg, Germany, firm for which he worked, Westfälische-Anhaltische-Sprengstoff-AG (WASAG). Despite the earlier discoveries by von Baader and von Foerster, the shaped charge effect or hollow charge effect came to be known as the "Munroe Effect" in the United States and the UK and the "Neumann Effect" in Germany.

Applications. The earliest applications for shaped charges were in mining, where the effects of the explosives aided in breaking up rock during quarrying and tunneling. As the technology matured, it proved useful in demolishing buildings and other structures, with the shaped charge effect being useful in cutting through metal piles, columns, and beams. Other uses include felling trees, breaking up ice jams and logjams, and perforators in the gas, oil, and water well industries. Applications of shaped charges for military purposes emerged in the decades after the First World War. The potential of shaped charges was demonstrated dramatically by the capture of the Eben-Emael fortress, a seemingly impregnable Belgian military citadel situated along the Albert Canal between Liège and Maastricht. During the early morning of 10 May 1940, a force of 78 German paratroopers landed by gliders atop the fortress and used shaped charges to destroy or disable gun cupolas. The defending Belgian forces were surprised by both the arrival of the paratroopers and their success in neutralizing the firepower of the fortress. With the arrival of German reinforcements on 11 May, the fortress was captured, and 1,200 defending Belgian troops surrendered.

Shaped charges would continue to play a significant role throughout the Second World War. In November 1940, the British fielded what was likely the first successful high-explosive, anti-tank (HEAT) round. Designated the No. 68 grenade, it could be fired from a specially adapted rifle, giving a lone infantryman a means of destroying a tank or other armored vehicle. Various fills were used in this grenade, including **Lyddite**, **Pentolite**, and a mix of **RDX** and beeswax. German variants of the HEAT round entered service in 1943 and included the Panzerfaust series of man-portable, single-shot weapons, which used black powder as the **propellant** and a mix of **TNT** and **hexogen** as the explosive fill. Shortly after its entry into the war in December 1941, the United States fielded a shoulder-fired rocket launcher that fired a 2.36-inch HEAT round having **PETN** as the explosive fill. The launcher was given M1 as its official designation, but it became popularly known as the bazooka, because of its resemblance to a musical instrument of that name popularized in the 1930s by American comedian Bob Burns (1890–1956).

In addition to the HEAT rounds intended for use by the infantryman, the Germans developed the Schwere Hohlladung (SHL) or heavy shaped charge weapons. For example, the SHL 500 had a diameter of 65 cm (25.6 inches) and was intended for use against lightly armored ships. The SHL 1000, which was given the code name "Beethoven," was even larger, having a diameter of 180 centimeters (70.9 inches) with 5,000 kilograms (11,000 pounds) of explosives. These weapons used a 60/40 mix of RDX and TNT. Development of the SHL weapons led to the "Mistel" (German for "mistletoe") warhead, containing 1,720 kilograms (3,800 pounds) of explosives and fitted into the nose of an aircraft. This aircraft was then placed atop piloted aircraft, then cast loose to drop onto Allied ships or into dams. It emerged late in the war and was used sparingly. But its design was shared with the Japanese, who converted it into a smaller version, called the Sakura bomb. Unlike the pilotless Mistel, Sakura was designed to be delivered in a piloted aircraft on a suicidal mission, and whose pilots were known as kamikazes. While their use proved effective, such use emerged late in the war and failed to be decisive. Perhaps the most devastating single use of the shaped charge effect during the Second World War was with **Fat Man**, the **implosion nuclear explosive** detonated on 9 August 1945 over Nagasaki.

SHAW, JOSHUA (1776–1860). Artist and inventor of **explosives**. Shaw was born in England, where he became a sign painter and eventually an artist noted for his landscapes. He immigrated to the United States in 1817, settling in Philadelphia, where he remained until his death. Shaw became a key figure in the development of landscape painting in the United States, but he also took an interest in explosives and firearms. In his writings, Shaw claimed to have invented the **percussion cap** in 1814, while living in England. His first cap was made of steel and intended to be reused. A year later, he was making disposable caps out of pewter. By 1816, he had perfected the copper percussion cap. The explosive in these caps was a mixture of **charcoal**, **potassium chlorate**, and **sulfur**. Later, **mercury fulminate** would come into common use as the explosive. When Shaw sought to patent his invention, he was advised that his application would prove unenforceable because it infringed on the British patent granted to **Alexander Forsyth** in 1807 for the "**detonating** principle for exploding **gun powder** in firearms." Upon receiving this legal advice, Shaw immigrated, perhaps in hopes of obtaining a US patent for the percussion cap. He made an application soon after settling in the United States, but his application was rejected on procedural grounds. At the time, US patent law required that an applicant either be a US citizen or be an alien who had resided in the United States for two years. Shaw failed to meet this requirement. However, the US patent office did recognize his claim of the right of priority to the invention of the percussion cap.

In 1822 and 1828, Shaw succeeded in receiving US patents covering various improvements to explosives and firearms, including the copper percussion cap as well as a pasteboard percussion cap. This latter invention was a piece of pasteboard into which holes were punched, the holes filled with a suitable explosive, and then the hole was covered in wax to hold the explosive in place. Such pasteboard caps came to be called wafer **primers**. Shaw continued to press his claims with the US government, in the hope of being compensated for his original invention of the percussion cap and the wafer primer, both of which had come into common use. On 20 February 1847, the US Congress passed legislation directing the Secretary of War to pay Shaw $25,000 for "his invention of percussion caps and locks for small arms, and percussion locks and wafer primers to be applied to the firing of cannon." This payment was subsequently reduced to $17,000. *See also* EGG, JEAN JOSEPH (1775–1837).

SHEEHAN, JOHN C. *See* BACHMANN PROCESS.

SHEET EXPLOSIVE. Plastic explosive manufactured in the form of sheets. The resultant sheet is rubber-like in its flexibility and can be cut into a desired shape and size. Examples include Primasheet, which is comprised of 65 percent **PETN**, 27 percent **plasticizer**, and 8 percent **nitrocellulose**, and the various types of **Detasheet**.

SHELL CRISIS OF 1915. Acute shortage of **high explosive** munitions experienced by the British military was caused largely by a change in tactics, and whose resolution brought about governmental, political, and social changes of note. An anonymous military correspondent for *The Times* (London) brought this shortage to the attention of the public. He reported from France on 14 May 1915 that the "want of unlimited supply of high explosives was a total bar to our success" during the Battle of Aubers Ridge, 9–10 May. Other British newspapers carried similar reports. Details soon emerged showing that the preponderance of British artillery was in poor condition owing to overuse and faulty maintenance, with much of this artillery being of designs deemed obsolete. By the end of the fighting at Aubers Ridge, the British military had merely 3,000 artillery shells remaining for its 18-pounder guns; the majority of its 4.7-inch guns were defective; and the **fuzes** for the munitions for its 15-inch guns failed to detonate when the shells hit wet soil. These details contradicted a speech by Prime Minister H. H. Asquith on 20 April 1915 that the military had sufficient munitions. By May 1915, the daily rate of domestic production of munitions by the Central Powers (the Austro-Hungarian, German, Ottoman Empires, and Bulgaria) was 250,000. By contrast, the daily production rate in the UK amounted to slightly over 15,000—a mere 6 percent of the rate of the Central Powers. One factor contributing to this shortage was a change in artillery tactics. As the battlelines on the western front settled into trench warfare, the effectiveness of **Shrapnel**-producing munitions was more limited than had been the case when targets were troops in the open. Such munitions were still effective in cutting the barbed wire emplacements in front of the trenches as well as the communication wire within and between these trenches. But high explosive munitions were needed to demolish breastworks, gun emplacements, and dugouts.

The shell crisis prompted Prime Minister H. H. Asquith to dismiss his cabinet on 17 May, replacing it with a coalition government. The munitions department was separated from the Ministry of Supply and elevated to cabinet level as the Ministry of Munitions, with David Lloyd George as its director, effective 2 July 1915. The Ministry of Munitions would remain throughout the war, being abolished on 21 March 1921. Additionally, to meet the needs for munitions production, the UK government adopted policies limiting profits by those private parties engaged in producing munitions, suspending the right of Trade Unions to strike or take other work actions, requiring all skilled men either to fight by joining the military or to work in war-related trades, and promoting the hiring and training of women as munitions factory workers. The results of these efforts were impressive. By the end of 1915, daily production of munitions had increased by twelvefold to nearly 200,000. This level was still below that of the Central Powers, but one that was being augmented by deliveries from other nations, especially the US. Production increases were seen in other areas, with output of machine guns having increased fivefold and that of hand grenades by 40-fold. Pledges by Trade Unions to withhold work actions for the duration of the war had reduced the number of strikes from June through December to three. As part of its social contract with munitions workers, the Ministry of Munitions had built housing on or near factories to accommodate 60,000 workers. In an effort to ease working conditions for women, many munitions factories had established day-care facilities and were limiting overtime. The Ministry set up a medical department to handle issues brought about by handling potentially harmful chemicals used in munitions such as TNT. By July 1916, 90 arsenals had been built in the UK, with weekly output of rifle cartridges exceeding pre-war annual output. By mid-1916, the UK had transitioned from being a munitions-deficient nation to being one that was supplying both France and Russia. *See also* CANARY GIRLS.

SHELLITE. Picric acid-based **explosive** adopted by the British during the First World War. The Royal Navy learned during the Battle of Jutland, 31 May to 1 June 1916, that **Lyddite** performed inadequately when used as the filling in an armor-piercing shell. In particular, Lyddite tended to **detonate** on contact with the surface of the armor, limiting penetration. A solution was devised by combining Lyddite with a less **shock-sensitive** but still powerful explosive. The British selected **dinitrophenol** (DNP) for this purpose. DNP is less shock-sensitive, but it is also a less powerful explosive than pure Lyddite. A mix of 60 percent Lyddite and 40 percent DNP was found satisfactory and used in British projectiles as of 1918. After the war, the amount of DNP was reduced to 30 percent, and testing showed that this combination was sufficient to guard against premature detonation. The combination of Lyddite and DNP became known as Shellite.

SHIMOSE. Smokeless powder comprised of **picric acid** and **trinitrocresol** and adopted by the Japanese navy in 1893 for use as a **high explosive**. Its name, which is occasionally spelled "schimose," honors Japanese chemist and naval engineer Shimose Masachika (1860–1911), who is credited with its development. As with other smokeless powders, the use of shimose enabled greater precision in gunnery than had been possible with **black powder**. Its use in naval guns contributed to the Japanese victory in the **Russo-Japanese War of 1904–1905**. The development and manufacturing of smokeless powder in Russia lagged that in Japan. During the war with Japan, the Russian arsenal for its land and naval artillery relied almost exclusively on **Shrapnel** projectiles, having a mere 15 percent of its munitions loaded with **smokeless powder**. *See also* BALLISTITE; CORDITE; LYDDITE.

SHOCK-SENSITIVE. The relative propensity of a substance to **deflagrate** or **detonate** as a result of jarring, dropping, shaking, or other sudden movement. The greater the sensitivity to shock, the greater the likelihood of an unintentional or deliberate **explosion**. For example, **acetone peroxide** and **nitroglycerin** are notoriously shock-sensitive, while **TNT** and **RDX** are significantly less shock-sensitive. *See also* FIGURE OF INSENSITIVITY; FRICTION SENSITIVITY.

SHOE BOMBER. London-born Richard Reid boarded American Airlines flight 63 on 22 December 2001. The flight was bound from Paris to Miami. During the flight, he was seen with a lighted match that he was using in a futile attempt to light a **fuse** that would **detonate explosives** within his running shoes. Once this attempt was recognized, Reid was restrained by the crew and passengers. The flight was diverted to Boston, where he was taken into custody. His shoes were analyzed later and found to contain **PETN** and **TATP**. It was also determined that one reason these explosives failed to detonate was the perspiration from his feet, which caused the explosives to be too damp to detonate. Because the explosives were in his shoes, Reid became known as the "Shoe Bomber." Reid ultimately pled guilty to eight terrorism-related charges. On 4 October 2002, his plea was accepted, and he was sentenced to life imprisonment. In confessing his role in this bombing attempt, he stated that he had acted alone. However, in 2005, Saajid Badat, a British citizen, admitted to authorities that he had conspired with Reid and that both had been given explosive-laden shoes to use in blowing up an airliner. Badat, though, never went through with his part in the conspiracy, although he saved the shoes. He confessed to his role and was sentenced to a term in prison. Forensic analysis of the detonator cord in his shoe-bomb was determined to match that from Reid's shoes. In 2006, the US Transportation Security Administration (TSA) enacted a requirement that all passengers must remove shoes for security screening prior to boarding an aircraft. This requirement was relaxed in 2011 to permit those travelers ages 12 and under, as well as those aged 75 and older, to keep their shoes on during security screening. The requirement for removing shoes was due in large measure to the experience gained from the episode with the Shoe Bomber. *See also* PERSON-BORNE IMPROVISED EXPLOSIVE DEVICE (PBIED).

SHOOTING. Use of **explosives** to improve flow of gas, oil, and water wells. Persons using explosives in this manner are called "shooters," and the explosive devices used for this purpose are known as **torpedoes**. *See also* ROBERTS, EDWARD A. L. (1829–1881).

SHRAPNEL. Fragments released when an **explosive** is **detonated**. The term was originally applied to lead balls used as filling in a munition and released when it exploded. These balls proved effective against personnel in the open. The term honors British Royal Artillery officer Henry Shrapnel (1761–1842), credited with inventing the ball-filled munition around 1783. The British adopted Shrapnel munitions in 1803 and used them effectively against the Dutch on 5 May 1804 at the Battle of Surinam. This use inspired other nations to adopt similar ball-filled munitions. Eventually, the term came to be applied to any shell fragments or to other items bundled within an explosive device and released when it explodes. These other items can include ball bearings, nails, and slugs.

SIGNALING FLARES. *See* PYROTECHNICS.

SIGNIFICANT QUANTITY (SQ). Approximate amount of **nuclear material** for which the possibility of making a **nuclear explosive** cannot be excluded. Two categories of SQ exist: "direct use nuclear material," meaning materials that can be used directly in making a nuclear device without transmutation or further enrichment, and "indirect use nuclear material," meaning all other nuclear materials. The IAEA gives the SQ for direct use of nuclear material as 8 kilograms (kg) (17.6 pounds) for **plutonium**, 8 kilograms for **uranium**-233, and 25 kilograms (55 pounds) for **highly enriched uranium** (HEU). Similarly, the IAEA gives the SQ for indirect use of nuclear material as 75 kilograms (165 pounds) for **low-enriched uranium** (LEU), 10 tons for natural uranium (i.e., uranium as it occurs in nature), 20 tons for uranium that is depleted in uranium-235, and 20 tons for **thorium**.

SIMS-DUDLEY DYNAMITE GUN. Breech-loading, 2.5-inch **pneumatic gun** capable of firing a **dynamite**-laden munition and designed to be portable. The gun used a **hybrid explosive** to propel a munition containing 5 pounds of either **blasting gelatin** or gelatin dynamite. The gun featured two chambers, one atop the other and with the upper chamber being the gun barrel. A charge of 7–9 pounds of **black powder** was placed in the lower chamber; when the charge was **detonated**, it drove a piston, compressing air in the chamber. This compressed air could then be released and directed into the upper chamber, expelling the munition.

The Sims-Dudley dynamite gun appears to have been the earliest pneumatic gun to have been used in combat. Credit for inventing this artillery piece is shared by Winfield Scott Sims (1844–1918) and Dana Dudley (1853–1938). On 23 July 1889, Lynn, Massachusetts, machinist Dudley received US patent 407,476 for an early version of a pneumatic gun. Sims enhanced this version, receiving US patent 619,025 on 7 February 1899, and undertook its manufacture at his firm, New York Ordnance Co. He demonstrated the new design, dubbed the Sims-Dudley dynamite gun, to a group of Cuban insurgents around 1896 or 1897. The demonstration was successful, prompting the insurgents to purchase three prior to the outbreak of the **Spanish–American War**. One of these guns was used in combat, reportedly to good effect. On 28 August 1897, Cuban insurgents used a Sims-Dudley dynamite gun to destroy Spanish fortifications around Victoria de las Tunas. The remaining two guns appear to have played no direct role in the fighting in Cuba. Around the same time, the US military purchased 16 of these guns. Some of these guns were used in combat in Cuba during the Spanish–American War. The Sims-Dudley dynamite gun produced little noise and released minimal smoke on firing. Therefore, its firing had the advantage of being difficult or impossible to detect. This advantage, though, was insufficient to overcome its disadvantages in terms of limited range, inaccurate fire, and a relatively small explosive projectile. The US military abandoned use of this gun shortly after the Spanish–American War, and there is no record that it or an artillery piece of similar design was adopted for use by any other military. *See also* SPANISH–AMERICAN WAR (APRIL–AUGUST 1898); ZALINSKI, EDMUND LOUIS GRAY (1840–1909).

SINGLE-BASE POWDER. Smokeless powder comprised solely of some form of **nitrocellulose** as the explosive constituent. Other substances might be present, such as **binders**, **stabilizers**, and residual solvents from the manufacturing process. An example of a single-base powder is **Poudre B**, the original smokeless powder. *See also* DOUBLE-BASE POWDER; TRIPLE-BASE POWDER.

SITES W, X, & Y. Code names used during the **Manhattan Project** for sites where **nuclear weapons**-related work occurred. Site W was the Richland, Washington, site of the Hanford Engineer Works, where **plutonium** was produced. This site eventually became the Pacific Northwest National Laboratory. Site X was the Oak Ridge, Tennessee, site of the Clinton Engineering Works, where **uranium** was enriched. This site eventually became the Oak Ridge National Laboratory. Site Y was the Los Alamos, New Mexico, site where much of the research and design of nuclear weapons took place. This site eventually became the Los Alamos National Laboratory.

SMILING BUDDHA. Code name used by **India** for its first **nuclear explosive** device. The detonation occurred on 18 May 1974 at the Pokhran Test Range, Rajasthan. The device was an **implosion-type design** having a

plutonium core and a yield of 8–10 kilotons **TNT-equivalent**. When news of this detonation became known, India characterized it as a **peaceful nuclear explosion**. But the event likely prompted its neighbor, **Pakistan**, to hasten its own efforts to develop **nuclear weapons**. Smiling Buddha exploited technology and materials supplied by Canada and the United States, prompting a consortium of nations to create the Nuclear Suppliers Group (NSG), which coordinates, through non-binding guidelines, their national export controls on dual-use materials, technology, and equipment, such as nuclear fuel, that could be used to support a **nuclear weapon** program. *See also* BARATOL; COMPOSITION B; KHAN, ABDUL QADEER (1935–2021); PEACEFUL NUCLEAR EXPLOSIONS (PNE) TREATY.

SMITH, HENRY JULIUS (1844–1901). American businessman and inventor, known especially for his work on electric **detonators**. Smith garnered his first patent, US patent 79,268, on 23 June 1868 for an electric detonator having two wires embedded in a **mercury fulminate blasting cap**. He was 24 years old at the time and working as a mechanic for the Oriental Powder Company, Boston. Over the next several years, he made more than 20,000 of these detonators for **George Mowbray** to use in constructing the **Hoosac tunnel**. In 1871, Moses G. Farmer (1820–1893) received a patent for what was called a dynamo-electric machine, which saw use by the US Navy as an electric detonator. However, this machine weighed 120 pounds and proved unwieldy for routine commercial or military use. Smith improved on Farmer's design, reducing the weight to 77 pounds and receiving US patent 173,682 in 1876 for the improvement. His greatest success, however, came two years later when he received US patent 201,296 on 12 March 1878 for a plunger-operated electric detonator. Marketed by Laflin & Rand Powder Company as the "Magneto No. 3," it was a wooden box standing 16 inches by 8 inches by 5 inches and weighing a mere 18.5 pounds. This box, with its plunger affixed to the top, became a well-recognized symbol of the **explosives** industry and would be copied by cartoonists and artists intending to depict an explosion or an attempted explosion. A notable example would be the Looney Tunes cartoons featuring Wile E. Coyote in his attempts to capture the Road Runner.

SMOKELESS POWDER. Explosive that burns with few visible combustion products. The lack of visible combustion products for smokeless powder contrasts with **black powder**, which typically produces a gray or black smoke on ignition. The explanation for this difference rests with both the extent of combustion and the products of combustion for each type of powder. Smokeless powders undergo nearly total combustion, and their combustion products are mostly gaseous substances invisible to human sight. By contrast, black powders undergo incomplete combustion, and their combustion products plus the residues from this incomplete combustion include solids that are readily visible.

The term "smokeless powder" can be misleading. Combustion of a smokeless powder is accompanied by some smoke. Also, the individual grains of smokeless powders are often much larger than one might associate with the term "powder," with sizes ranging from a fraction of an inch in diameter for those used as **propellants** for small arms, such as handguns and rifles, to an inch in diameter for those used as propellants for artillery munitions, and finally to several feet in length for those used as propellants in rockets.

Discovery and Development. The 1833 discovery of **nitrocellulose** by **Henri Braconnot** marked the advance in technology that enabled the development of smokeless powder. This discovery prompted various studies of nitrocellulose, including the 1838 observation by **Théophile-Jules Pelouze** that nitrocellulose left little residue when it burned. In 1846, **Rudolf Böttger** and **Christian Schönbein** discovered **guncotton** and popularized their discovery, causing intense research into the prospect that this material might be used as a propellant. Nearly four more decades would pass, however, before a practical smokeless powder was in routine use by the military. The experience in the United States is representative of the issues encountered with attempts to exploit nitrocellulose in general and guncotton in particular. After Schönbein made public his information on guncotton in 1846, he was contacted by William Robertson, the US Consul at Bremen. Schönbein provided Robertson a small quantity of guncotton as well as directions on its preparation. Robertson then sent these to the United States for testing. That testing was overseen by **Alfred Mordecai** at the Washington, D.C., Arsenal in 1846–1847. Based on these tests, Mordecai noted he observed "no smoke from the discharge [of the guncotton] and only a slight acid odor." He wrote, "I conjectured that the explosive or bursting effect of the gun-cotton [*sic*] is much greater in proportion to its propelling force than that of **gunpowder**." He failed to pursue further development of guncotton, however, concluding that "in consequence of the quickness and intensity of action of the gun-cotton [*sic*] when ignited, it cannot be used with safety in our present firearms."

Mordecai's conclusion speaks to one of the two technical challenges facing the use of guncotton and other nitrocellulosic materials for military purposes. First, nitrocellulose burns very rapidly in a confined space, much more rapidly than does an equivalent mass of black powder. This rapid rate of combustion resulted in pressure increases above those that could be handled safely by the military firearms then in use, especially by large artillery pieces. The second technical challenge facing nitrocellulose lay in finding a safe method for large-scale manufacturing.

The first practical smokeless powder appears to have been the one developed by **Johann Friedrich Eduard Schultze**, a Prussian artillery officer. In 1863, he was granted US patent 38,789 for an explosive that used nitrated wood products (a crude form of nitrocellulose) mixed with **potassium nitrate** and barium nitrate. This mixture left little residue and produced little smoke, as is characteristic of any smokeless powder. However, its **detonation** proved too powerful for the artillery then in use by the military. Consequently, the use of the mixture was restricted to firearms. A form of the mixture was marketed as **Schultze powder** and proved popular as the propellant in firearms used for sport and hunting. Schultze powder failed to find much use by the military in firearms, however, with the notable exception of a cartridge used successfully with the **Dreyse needle-gun**. Use of that cartridge during the 1870–1871 **Franco-Prussian War** likely marks the first combat use of smokeless powder. More importantly, its use demonstrated the effectiveness of smokeless powder in allowing users to keep an unclouded line-of-sight while avoiding the telltale puffs of smoke that might give away their positions. However, during the 1860s and 1870s, several massive explosions, often accompanied by injuries and fatalities, occurred in guncotton factories. These explosions were likely owing to a lack of appreciation for the sensitivity of guncotton and to inadequate means of stabilizing it. Manufacturing and use of guncotton declined significantly as a result of the explosions. For example, the Austrian government banned all use and manufacturing of guncotton as of 11 October 1865. Progress regarding the manufacturing and use of any smokeless powder lagged.

This situation changed dramatically in 1884. French chemist **Paul-Marie-Eugène Vieille** announced his invention of a novel explosive that he called **Poudre B**. It was made from a mixture of ethanol, ether, and gelatinized guncotton, which was then extruded through rollers, forming thin sheets. These sheets could be cut into flakes of almost any desired size. Poudre B alleviated or eliminated the safety issues associated with guncotton itself, sparking interest in its use for military and civilian purposes. Poudre B itself soon found use by the French military in the **Lebel Model 1886 Rifle**, and its invention helped spark a brisk period of research into other possible formulations of smokeless powders. Results of this research included the 1887 invention of **Ballistite** by **Alfred Nobel**, followed shortly by the 1889 invention of **Cordite** by **Frederick Abel** and **James Dewar**, as well as the invention that same year of **Maxim's Powder** by **Hiram Stevens Maxim**.

Military use of smokeless powders played prominent roles in the **Spanish–American War** (April–August 1898) and the **Second Boer War** (1899–1902), and the results of their use in combat demonstrated the operational effectiveness of smokeless powders and encouraged many other nations to adopt their use, especially for their military forces. By the First World War, smokeless powder had largely replaced gunpowder for use in firearms, from sidearms to large-caliber artillery pieces. *See also* LENK VON WOLFSBERG, WILHELM, FREIHERR (1809–1894); MAXIM, HUDSON; MAXIM-SCHÜPPHAUS SMOKELESS POWDER.

SMOLIANINOFF, SERGEY. *See* AMERICANITE.

SNPE. *See* GUNPOWDER SERVICE (*SERVICE DES POUDRES*).

SOBRERO, ASCANIO (1812–1888). Italian chemist credited with the original synthesis and discovery of **nitroglycerin**. Sobrero studied medicine and chemistry, spending 1840–1842 in Paris, working in the laboratory of **Théophile-Jules Pelouze**, who had discovered **nitrostarch** in 1838, and 1842 in the Giessen, Germany, laboratory of Julius von Liebig (1803–1873). In 1845, Sobrero was appointed professor of industrial chemistry at the University of Turin, where he remained until his retirement in 1883. In 1847, at a meeting of the Royal Academy of Science in Turin, Sobrero announced his discovery. Calling this substance "pyroglycerine," he noted its incredible **explosive** properties. In letters to Pelouze and in an article in a scientific journal, Sobrero warned of the difficulties of handling nitroglycerin. Later that same decade, his face was badly scarred by a laboratory explosion, most likely caused by nitroglycerin. He considered nitroglycerin to be too dangerous to be of any practical use. Sobrero is quoted to have said, "When I think of all the victims killed during nitroglycerin explosions, and the terrible havoc that has been wreaked, which in all probability will continue to occur in the future, I am almost ashamed to admit to be its discoverer." *See also* NOBEL, ALFRED BERNHARD (1833–1896).

SOCIÉTÉ NATIONALE DES POUDRES ET EXPLOSIFS. *See* GUNPOWDER SERVICE (*SERVICE DES POUDRES*).

SODA. Term for **sodium nitrate**, especially in the sense of **Chile Saltpeter**. The use of this term can cause confusion because "soda" is also a synonym for sodium carbonate, sodium bicarbonate, and sodium hydroxide. *See also* DU PONT, LAMMOT (1831–1884); INDIA SALTPETER; SALTPETER.

SODA BLASTING POWDER. *See* "B" BLASTING POWDER.

SODA NITER; SODA NITRE. Term for **sodium nitrate**. *See also* AMERICAN CIVIL WAR (1861–1865); DU PONT, LAMMOT (1831–1884); NITER AND MINING BUREAU.

SODIUM NITRATE. Strong **oxidizing agent** used as an ingredient in **black powder**. Historically, **potassium nitrate** rather than sodium nitrate was the oxidizer in black powder. This situation changed, however, with the 1857 invention by **Lammot du Pont** of a process for making black powder using sodium nitrate. It is a colorless crystalline solid and has the chemical formula $NaNO_3$. Another name for sodium nitrate is soda niter. *See also* AMERICAN CIVIL WAR (1861–1865); CHILE SALTPETER; CHILEAN NITRATES; WAR OF THE PACIFIC (1879–1884).

SOLUBLE NITROCELLULOSE. *See* INSOLUBLE NITROCELLULOSE.

SOLVENTLESS POWDER. Term for a **smokeless powder** having low **nitroglycerin** content. The term has been used to distinguish certain formulations of smokeless powder from **Ballistite**. The term is inexact and problematic, similar to the situation that applies to the term "**insoluble nitrocellulose**." A solventless powder often does require a solvent during its manufacture. Therefore, it is not truly solventless, although the quantity of solvent is less than that associated with the manufacturing of Ballistite and other high nitroglycerin content **explosives**. Additionally, most solventless powders are produced as tubes or sticks rather than as powders. The first solventless powder was **RPC/12**; in this case, the solvent used was **Centralite**, and the final product was tubular, not powdery, in shape. Its properties contributed significantly to the ability of the Central Powers to continue producing explosives during the First World War. *See also* CORDITE; POUDRE SD.

SOUTH AFRICA. On 24 March 1993, F. W. de Klerk, President of the Republic of South Africa, announced in a speech to the South African Parliament that his nation had pursued a **nuclear weapons** program. The program began in 1974 but had its roots in work decades earlier in "Atoms for Peace" initiatives. Beginning in 1948, South Africa pursued scientific and engineering work related to the use of nuclear power to generate electricity and for **peaceful nuclear explosions**, in which such explosions might be used for non-military purposes. According to de Klerk, the possession of nuclear weapons was viewed as a deterrent to Soviet ambitions in sub-Saharan Africa. Rather than acknowledge this program, though, the government of South Africa pursued a policy of deliberate ambiguity regarding its status as a nuclear weapons state.

The first operational nuclear weapon was completed in 1982 and given the code name "Hobo," which was later changed to "Cabot." This device was eventually disassembled, with its warhead components used elsewhere. Shortly after 1982, work began on a newer model nuclear weapon, with the code name "Hamerkop." Several of these models were built, all being TV-guided glide bombs. The weapons were tested in the Kalahari Desert at the Vastrap weapons range. Preparations were underway for a full-scale nuclear weapon test at Vastrap, but were detected by intelligence methods and sources. Bowing to pressure from foreign governments, the test was cancelled, and the test site was abandoned. The available evidence suggests that South Africa never conducted a test of a nuclear weapon, despite speculation that the 1979 **Vela Incident** can be attributed to such a test in the waters off the South African coast. The nuclear weapons program was abandoned by South Africa in 1989 following intense pressure from the governments of the Soviet Union and various Western nations, and South Africa acceded to the **Treaty on the Non-Proliferation of Nuclear Weapons** in 1991. In total, South Africa made six **gun-type nuclear devices**, each containing 55 kilograms (121 pounds) of highly enriched **uranium** (HEU) as the **nuclear explosive**. On 19 August 1994, the IAEA confirmed that six nuclear weapons plus one partially assembled weapon had been dismantled, making South Africa the only nation thus far to have eliminated its nuclear weapon stockpile voluntarily. South Africa would later take a leading role in declaring the African continent a **nuclear weapon-free zone** (NWFZ). It signed the **Treaty on the Prohibition of Nuclear Weapons** in 2017 and ratified it on 25 February 2018.

SPANISH–AMERICAN WAR (APRIL–AUGUST 1898). On 15 February 1898, the US battleship *Maine* exploded in the harbor off Havana, Cuba, and sank with the loss of about 260 lives. The cause of the **explosion** remains uncertain, but the sinking of this ship helped propel the United States into declaring war against Spain in April. The war was brief, with an armistice being declared on 12 August. From a US perspective, the war was successful. The Treaty of Paris, signed on 10 December, forced Spain to grant independence to Cuba and to cede Guam, the Philippines, and Puerto Rico to the United States.

The conflict witnessed one novelty in the use of explosives in combat. The ***Vesuvius***, equipped with **pneumatic guns** that launched **dynamite** projectiles, entered the conflict and saw action in Cuba. The results were disappointing, and the US Navy abandoned plans for building a second such ship. A second pneumatic gun, the **Sims-Dudley dynamite gun**, was also used, in this instance as an artillery piece with land forces. This same gun had been used by Cuban insurgents in 1897, prior to the US declaration of war against Spain. Some 16 Sims-Dudley dynamite guns were purchased by the US Army, and many were used in combat during the conflict. The United States did make extensive use of **brown powder**, but the experience was mixed. While this type of **black powder** proved superior for use in rifled guns of long range, it proved very sensitive to **friction**, and stocks of this powder would ignite from the simple act of handling or shaking the canvas bag in which it was stored. Further use of brown powder was abandoned.

The sole use of **smokeless powder** by the United States during the Spanish–American War took place aboard the ***New Orleans***, a "Protected Cruiser" purchased from Brazil in March 1898. That use appears to have been limited to engagements against the coastal defenses at Santiago de Cuba on 6 and 16 June 1898. However, the experience of the US Navy with the smokeless powder used by the 6-inch guns of the *New Orleans* helped spur development of its own smokeless powder, which was clearly superior to the black powder and brown powder otherwise in use. The Spanish military, however, did use smokeless powder, with dramatic results that impressed the Americans. The Spanish infantry used the Mauser model 93 rifle, which fired a smokeless cartridge, and the Spanish artillery fired smokeless powder projectiles. American artist and war correspondent Fredric Remington wrote of the American experience during one battle that "no one's glass could locate the fire of the Spanish guns, and we could see [the American artillery's] smoke miles away on our right." Smokeless powder probably contributed to the ability of a Spanish force of some 750 riflemen to hold off 15,000 US troops during the Battle of San Juan Hill in Cuba. The United States eventually won the battle, but at much greater cost than had been anticipated. Firing smokeless powder helped the Spanish conceal their location, while the black and brown powder used by the United States both gave away their locations and hampered their ability to see reliably on the battlefield beyond the initial volley. The United States had learned a lesson, though, and the Spanish–American War helped stimulate the United States' move to weapons using smokeless powder cartridges and projectiles. One concrete, near-term result was the US adoption of the Springfield M1903 rifle.

SPECIAL ATOMIC DEMOLITION MUNITION (SADM). Common name given to a **nuclear weapon** deployed by the United States. Its official designation was the Mk-54. The Special Atomic Demolition Munition (SADM) was among the smallest nuclear weapons ever deployed by the United States, estimated to have been 3 feet long and 2 feet in diameter, with an overall weight of around 160 pounds in its shipping container. It used the W54 nuclear warhead, which was a compact **implosion device** that weighed around 50 pounds, was slightly under 1 foot in diameter, and was slightly over 1 foot in length. The modest size of a SADM allowed it to be deployed by US military Special Forces units and was intended for both defensive and offensive use, including being carried behind enemy lines to destroy key facilities. It was estimated to have a yield of 10 tons to 1 kiloton **TNT-equivalent**. SADMs were stockpiled in Europe, but all had been removed and returned to the United States by 1985. The United States began dismantling its SADMs shortly thereafter, and by 1992, none remained in the US nuclear weapon inventory. *See also* LULU; WE.177A.

SPECIAL NUCLEAR MATERIAL (SNM). The US Atomic Energy Act of 1954 defines Special Nuclear Material (SNM) as any isotope of **plutonium** or **uranium** that has been enriched in either uranium-233 or uranium-235. *See also* WEAPONS-GRADE MATERIAL.

SPRENGEL EXPLOSIVE. An **explosive** in which the **fuel** or the **oxidizer** is a liquid, or possibly both are. Development of this class of explosive is credited to **Hermann Sprengel**, who obtained British patents in 1871 for such mixtures (British patent 921, issued 6 April 1871, and British patent 2,642, issued 5 October 1871). Sprengel also described such explosives in a scientific report, published in *The Journal of the Chemical Society*, August 1873. Sprengel wrote that "to avoid the danger of spontaneous explosion of these compounds during their manufacture, storage, and transport, we can keep apart the oxidizing from the combustible agent [that is, the fuel], until their chemical combination is to take place in obedience to the will of men." But mixing of the fuel and oxidizer immediately prior to use proved to be inconvenient for many applications, and the use of Sprengel explosives failed to gain wide acceptance.

More recently, concerns have surfaced over potential and actual terrorist use of **ammonium nitrate-fuel oil mixture** (ANFO) and **peroxide-based explosives**, which might be correctly termed "Sprengel explosives." "Sprengel explosive" is sometimes considered to be synonymous with "**liquid explosive**"; however, many consider "liquid explosive" to refer to an explosive comprised entirely of liquids. Because a Sprengel explosive can contain a solid as the fuel or oxidizer (as in the case of **Rack-a-Rock**), the terms "Sprengel explosive" and "liquid explosive" have different connotations and are not truly synonymous. *See also* HELLHOFFITE; PANCLASTITE; TURPIN, FRANÇOIS EUGÈNE (1848–1927).

SPRENGEL, HERMANN JOHANN PHILIPP (1834–1906). German-born chemist and inventor. Sprengel received his doctorate in chemistry in 1858 from Heidelberg University and then moved to England, where he spent the rest of his life. He took up various positions as a research assistant at Oxford University and spent five years as a manufacturing chemist. In 1870, he became an independent consultant, where he flourished. In 1871, he received two patents for a novel type of **explosive** in which the **fuel**, **oxidizer**, or both were liquid. Such **liquid explosives** came to be called **Sprengel explosives** in his honor. In 1873, Sprengel discovered the explosive nature of **picric acid**, spurring intense research into picric acid-based explosives, such as **maximite** and **mélinite**. He was elected a fellow of the Royal Society of London in 1878 in recognition of his various contributions to chemistry.

SQUIB. Small **explosive** device. The earliest squibs used a tiny amount of **black powder**, wrapped in parchment or stuffed inside the shaft of a feather. The Chinese likely invented these devices. By the eighteenth century, similar devices had found use worldwide in mining, where a squib could be used as a safe means of loosening rock. In the mid-twentieth century, squibs found novel use for special effects in the film industry. These squibs typically were flat-shaped discs containing 0.5–1.0 grains (30–60 mg) of a **smokeless powder** that would burst a container of fake blood on **detonation**, thereby simulating the effects of a gunshot wound. Alternatively, the container might hold such materials as dust, feathers, or water, again to simulate penetration by a gunshot. Detonation might be done by the actor wearing the squib or might be controlled remotely.

SRI LANKA EASTER BOMBINGS. Eight bombs were **detonated** in locations around Sri Lanka on Easter Sunday, 21 April 2019. The bombings left at least 269 dead, including the eight bombers. The first detonation occurred around 8:25 a.m., and the eighth and final one took place around 2:15 p.m. The Catholic Shrine of St. Anthony in Colombo was the first to be hit. A Catholic church in Batticaloa was the second target, and a Protestant congregation in Negombo was the third. In all three cases, Easter services were underway, and the churches were packed with worshippers. Soon thereafter, suicide bombers struck three hotels in Colombo. After a pause of five hours, a seventh suicide bomber struck a hotel in Dehiwala. The eighth and final bomb hit a housing complex in Dematagoda. Later that same day, an unexploded device was discovered at the international airport in Colombo and **rendered safe**. Three days later, a van was discovered near the site of one of the bombings and found to be loaded with **explosives**. Efforts by **explosive ordnance disposal** (EOD) experts to render these explosives safe failed, but no one was injured in the blast.

Analysis of debris collected on the scene and at locations associated with the suicide bombers identified the **primary explosive** as **acetone peroxides**, with each bomb being packed with nails and ball bearings, which served as **Shrapnel** to maximize casualties in the ensuing explosion. Sticks of **blasting gelatin** were recovered from a location associated with one of the suspected bombers. Subsequent investigation determined that the bombers were associated with National Thowheeth Jama'ath, a local militant radical Islamist group believed to have links to the Islamic State of Iraq and the Levant (ISIL). The same tactics and materials had been used in both the 13 November 2015 bombings in Paris that killed 130 and the 22 March 2016 bombings in Brussels that killed 32. Both explosions were carried out by suicide bombers, who were later found to have links to ISIL.

STABILIZER. Any substance added to a compound or mixture of compounds to keep it from changing its form or chemical nature. In the context of **explosives**, a stabilizer is often used to retard degradation that, if left unchecked, would diminish the desired effects of the explosive or would cause an unplanned **detonation** of the explosive.

The need for stabilizers was demonstrated dramatically and with tragic consequences by the explosion of **nitrocellulose**-based **smokeless powders** during storage. For example, on 11 September 1905, the Japanese battleship *Mikasa* exploded while in Sasebo harbor, Japan, with the loss of some 600 lives. *Aquidaban*, a Brazilian battleship, exploded near Rio de Janeiro on 21 January 1906, claiming at least 212. The French battleship *Iéna* exploded on 12 March 1907 near Toulon, with the loss of at least 117. On 30 April 1908, the Japanese cruiser *Matsushima* exploded near the Pescadores Islands off Taiwan, killing more than 220. And on 25 September 1911, another French battleship, *Liberté*, exploded in Toulon harbor, killing perhaps 300. In all cases, the explosion was blamed on munitions containing unstable smokeless powder—**Cordite** in the case of the Brazilian and Japanese ships, **Poudre B** in the case of the French ships. In each case, as the powder deteriorated, it produced acidic byproducts that, in turn, produced further deterioration. As the deterioration progressed, it released heat. Eventually, the build-up of heat was sufficient to detonate the powder. That initial detonation prompted other munitions to detonate, adding to the fire and destruction.

A solution to stabilizing nitrocellulose had been identified shortly after its introduction as an explosive. In the late 1880s, **Alfred Nobel** discovered that polymeric amines, such as **diphenylamine**, would scavenge the nitrates liberated by the deterioration of nitrocellulose. In 1889, he obtained German patent 51,471 for his discovery. Although the German government attempted to keep the discovery secret, it became widely known over the next two decades.

By 1908, most nations had taken to the aggressive use of diphenylamine owing to its superior performance as a stabilizer for nitrocellulose. France, however, resisted the use of diphenylamine, preferring to stick with amyl alcohol, its original choice as a stabilizer for Poudre B. This choice proved ineffective, as demonstrated dramatically by the 1911 explosion of the French battleship *Liberté*. As a result of that tragedy, France was forced to abandon the use of amyl alcohol and convert to the use of diphenylamine.

STABILLITE. Smokeless powder consisting of **nitrocellulose** and **trinitroanisole**. This formulation was developed by **Hudson Maxim** for **DuPont**, where an executive dubbed it "Stabillite" owing to expectations that it would have exceptional stability during storage. Furthermore, the finished product was ready for use directly, requiring no solvent, as was the case for other smokeless powders. However, the inclusion of trinitroanisole proved unfortunate. Despite the name "Stabillite," trinitroanisole proved unstable in storage, decomposing to **picric acid** and methanol. Furthermore, trinitroanisole is poisonous and a potent skin irritant. Therefore, Stabillite proved a commercial failure.

STEAM GUN. Weapon design in which steam is used as a **mechanical explosive**. The steam is used either directly as a **propellant** or indirectly to turn a wheel that imparts centrifugal force to propel a projectile. **Jacob Perkins**, an American-born engineer working in England, may have invented the first practical steam gun in the early part of the nineteenth century. On 15 May 1824, he received British patent 4,592 for a steam gun that used steam directly to propel shot at a rate of fire of 500–1,000 per minute. He exhibited his invention for several years at the Adelaide Gallery, London. He also demonstrated it to engineers, military officials, and various dignitaries, including the Duke of Wellington, when they visited the gallery. Although the design of his steam gun was deemed ingenious, most military officials felt it was impractical for use in combat. In September 1861, US senator Henry Wilson recommended that the Union military adopt "Perkins' Steam Gun," as it was called, for use against the Confederate forces in the **American Civil War**. The US Army rejected the senator's recommendation, but the US Navy eventually followed up by designing and constructing an experimental steam-driven gun. The gun successfully propelled 15 12-pound projectiles in 16 seconds and had a range of 1.5 miles. But its accuracy was poor, and no further development was attempted.

Prior to the American Civil War, American inventor **Charles S. Dickinson** had designed a gun in which steam rotated metal spheres within the gun. At the proper time, the shot was released and flung from the gun, using the centrifugal force imparted by rotation. Various names were given to this gun, including "Baltimore Steam Battery," "Centrifugal Steam Gun," "Dickinson's Patent Centrifugal Gun," and "Winans Steam Gun." The last name reflected the role of Maryland inventor and industrialist Ross Winans in promoting the gun. When the war began, at least one gun of Dickinson's design had been built, perhaps in Boston or in Maryland. It was seized by Union forces while it was being transported from Baltimore to Harper's Ferry, Virginia (now West Virginia), to support the Confederate forces. The merits of Dickinson's design are a matter of controversy, despite his claims that the gun could fire shots ranging in size from 1 ounce up to 24 pounds and propelling the shot with range and force similar to those achieved with **black powder**.

While the use of steam to propel projectiles might have a certain appeal, it suffers from assorted shortcomings that would limit its usefulness to the military. These shortcomings include the need for extensive logistical support similar to that needed for any other steam engine, such as those in locomotives; scaling from the trace minerals in the water heated to create steam; and the size, mass, and relative immobility of the steam-generating machinery.

STERLING HALL BOMBING. Early on the morning of 24 August 1970, a **vehicle-borne improvised explosive device** (VBIED) **detonated** outside Sterling Hall on the campus of the University of Wisconsin-Madison. Subsequent investigation showed that the intended target was the Army Mathematics Research Center, housed in Sterling Hall. The explosion killed one and left three injured. Four perpetrators were identified as responsible for the bombing, and three were eventually arrested and convicted for their roles. The bombing is notable, according to the Board of Chemical Science and Technology, US National Academies of Science, Engineering, and Medicine, because it was the first major attack in the United States using precursor chemicals associated with **homemade explosives**. In this case, the explosive was approximately 2,000 pounds of a mix of **ammonium nitrate** and **fuel oil** (ANFO). That same mix would be used 25 years later in the highly destructive and much deadlier **Oklahoma City Bombing**.

STOWMARKET GUN-COTTON EXPLOSION OF 1871. Around 2:00 p.m. on 11 August 1871, an explosion rocked the town of Stowmarket, Suffolk, England. The proximate cause was the **detonation** of three magazines containing approximately 3 tons of **guncotton** on the site of the **Patent Safety Gun-cotton Factory**. A fire ensued, and employees worked briskly to remove guncotton from packing sheds to safer locations. Despite their efforts, a second explosion took place around 3:00 p.m., followed almost immediately by a third and final explosion. These explosions left a crater 10 feet by 40 feet on what had been the site of the magazines and caused more than 180 cases of deafness. Of the 130 employees working at the time of the explosions, 28 were killed, including eight children under 15 years old.

In an effort to understand what happened, the newly appointed British chief inspector of explosives, **Vivian Majendie**, convened an inquest. Its findings established that the explosions resulted from adding sulfuric acid subsequent to the guncotton having passed required tests. The addition of sulfuric acid at this late stage was known to make the guncotton unstable and liable to detonation. The person or persons doing this addition and the reason for doing so remain unknown. The inquest also established that the factory personnel in general practiced poor standards of hygiene and exhibited a disregard for their own health and safety. The inquest was notable as an early—perhaps the earliest—emergence of the discipline of **forensic analysis of explosives**. The factory was located within Stowmarket, with the River Gipping running through its grounds. In the aftermath of this event, communities and local governments took an increased interest in seeing that similar factories were sited more carefully, moving them outside towns and cities and into more remote locations where future explosions might cause less damage to infrastructure and less harm to residents.

STRAIGHT DYNAMITE. *See* DYNAMITE.

SULFUR. Yellow solid element used as a **fuel** in **explosives**, especially **black powder**. An ancient name for sulfur is brimstone. This name likely derives from Old English meaning "burning stone," reflecting the familiar way in which natural deposits of sulfur could be burned.

SULFUR-FREE MEAL POWDER (SMP). Explosive containing 79.5 percent **potassium nitrate** and 20.5 percent **charcoal**. Sulfur-free meal powder is a type of **black powder**; as the name implies, it contains no sulfur. Meal powder is the fine dust remaining when black powder has been screened into different grain sizes, a process referred to as mealing. Sulfur-free meal powder (SMP) was developed following the introduction of **smokeless powders**, which displaced black powder from many of its traditional uses. However, most smokeless powders used as a propellant require a **primer**, and black powder continued to see use for this purpose. Use of certain smokeless powders was found to be associated with enhanced corrosion of gun barrels. This situation was especially acute for **Cordite MD**. The corrosion was traced to the sulfur in the black powder primer. When SMP was used in place of traditional black powder, the corrosion returned to normal levels.

SULTANA**.** Steamboat destroyed 27 April 1865 in a **mechanical explosion** with the loss of over 1,500 lives. Some sources credit the explosion and subsequent sinking of the *Sultana* as the greatest non-military, maritime disaster in the history of the US. Commissioned in 1863, the *Sultana* was a paddle-wheeler that regularly traveled the Mississippi River between New Orleans, Louisiana, and St. Louis, Missouri, delivering cotton and transporting Union troops. On 21 April 1865, she departed New Orleans, traveling north toward St. Louis and carrying over 100 passengers plus livestock and other cargo. Two days later, the *Sultana* stopped in Vicksburg, Mississippi, for repairs to her steam boilers. While in port, an unknown number of passengers boarded, mostly Union soldiers recently released from Confederate prisons and anxious to get home. The government offered a bounty to boat operators of 5 dollars for each enlisted man and 10 dollars per officer for transporting Union soldiers home. Anxious for the revenue, the operators of the *Sultana* crammed perhaps as many as 2,100 soldiers onto a ship designed to carry 375 passengers. Shortly after passing Memphis, Tennessee, one or more of her coal-fired boilers exploded. The explosion threw metal shards and hot coals from the boilers onto the wooden superstructure, causing it to catch on fire. It also tossed passengers into the Mississippi River, which was swollen and frigid from spring rains.

A US government investigation into the accident estimated the loss of life from the explosion, fire, and drowning at 1,547. However, that estimate is likely to have been low owing to the uncertainty over the numbers and identities of the soldiers who embarked in Vicksburg. News coverage of the *Sultana* disaster was meager, perhaps owing to competition from other events, especially the ongoing surrender of the Confederate forces, which began with those under General Robert E. Lee at Appomattox, Virginia, on 9 April, and developments related to the 14 April assassination of President Abraham Lincoln in Washington, D.C.

SWEAT. On prolonged storage, **dynamite** can degrade. Specifically, **nitroglycerin** tends to migrate out of the inert matrix making up the bulk of a stick of dynamite and collect on the exterior of the stick or pool under the stick. This process is commonly referred to as sweating or weeping, and the crystals of the nitroglycerin are known as "sweat." The phenomenon presents a substantial hazard, owing to the **shock-sensitive** nature of nitroglycerin when it has migrated out of the inert matrix of the stick.

SZILÁRD, LEO (1898–1964). Hungarian-born physicist. Szilárd made important contributions to nuclear physics, including discovering a novel means of isotope separation, solving the classical thought experiment known as "Maxwell's demon," and conceiving the idea of a nuclear chain reaction. Finding his prospects for education stymied owing to his Jewish heritage in his native Hungary, Szilárd emigrated to Germany but left following Hitler's coming to power in 1933, eventually settling in the United States in 1938. By this time, he and two other Hungarian-born physicists were working in the United States, **Edward Teller** and Eugene Wigner (1902–1995), had grown concerned that Germany might be developing a **nuclear weapon**. They wrote a letter in October 1939 to US president Franklin Roosevelt, warning of their concerns, explaining the possibility of nuclear weapons, and urging the creation of a national program to develop such weapons. The letter prompted the formation of the **Manhattan Project**, and Szilárd became part of it. After the war, Szilárd gained some notoriety for his 26 February 1950 radio broadcast in which he described how a **salted weapon** might be developed and how its use might annihilate mankind. His views were discounted in the popular press, with *Time* magazine labeling him "chicken little" and the US Atomic Energy Commission dismissing his ideas as impossible. Despite this reaction, his concept of a salted weapon was eventually accepted as feasible.

T

T4. Synonym (Italian code name) for **RDX**.

TAGGANT. *See* DETECTION AGENT.

TAIWAN. Taiwan possesses no nuclear weapons and has never been a nuclear weapon state (NWS). However, during the Taiwan Strait Crisis of 1954–1955, the United States placed nuclear weapons in Taiwan. These weapons remained in Taiwan until 1974, when they were removed on orders from US president Richard Nixon.

T-AMMONAL. *See* AMMONAL.

TAN; TANBARK. Tree bark used in tanning hides. Spent tanbark was used in certain formulations of **explosives**, most likely as a **fuel**. For example, De Tret's Powder, also known as pyronome, was a slow-burning powder consisting of 52.5 percent **sodium nitrate**, 27.5 percent tanbark, and 20 percent **sulfur**. It received British patent 1,226 on 17 May 1859, but appears to have disappeared from use by the twentieth century. Another example was Brain's Powders, which were various explosives covered by British patent 2,984, issued on 11 September 1874. These explosives contained 40–50 **nitroglycerin** and varying proportions of **potassium chlorate**, **potassium nitrate**, **sawdust**, and tanner's sumach, which is the bark obtained from the Rhus coriaria and is used in tanning Moroccan leather.

TANNERITE. Brand name of a **binary explosive** available in kit form, comprising separate packages of **ammonium nitrate**, which serves as the **oxidizer**, and **aluminum**, which serves as **fuel**. Because the individual packages are not explosive by themselves, the kit falls outside US regulations regarding explosives for purposes of sale, transport, and storage, as given in 27 CFR 555.11. The kits are popular for making targets to be used in firearms practice. The brand name is taken from the surname of Daniel Jeremy Tanner, who patented the original formulation in 1995.

TATP. Abbreviation for **triacetone triperoxide**.

TELLER, EDWARD (1908–2003). Hungarian-born physicist. Teller made important contributions to theoretical nuclear physics but may be remembered best for his role during the Second World War effort to develop a **nuclear weapon** and for his work after the war in the development of the **hydrogen bomb**. Born in Budapest, Teller left Hungary in 1926 and emigrated to Germany. After Hitler came to power, Teller relocated to the UK, Denmark, and ultimately the United States, arriving in 1935 to take up an academic position. At the outset of the Second World War in Europe, Teller and his fellow Hungarian expatriate scientists **Leo Szilárd** and Eugene Wigner (1902–1995) grew concerned over the prospect that Germany might develop a nuclear weapon. Their concerns gave rise to an October 1939 letter to US president Franklin Roosevelt. The letter prompted the formation of the **Manhattan Project**. At the invitation of **J. Robert Oppenheimer**, Teller joined the project in Los Alamos, New Mexico, in March 1943. His work was particularly important to the **implosion design** used for **Gadget** and **Fat Man**.

After the Second World War, Teller returned briefly to an academic position but returned to Los Alamos in 1950, where he and Stanislaw Ulam collaborated to produce a new weapon design that made possible the **hydrogen bomb**. In 1954, he testified during an investigation into the character of Oppenheimer. Teller's testimony cast doubt on the reliability and suitability of Oppenheimer for such a clearance, contributing to its permanent revocation and earning Teller the enmity of many of his fellow scientists. When proposals for banning tests of nuclear weapons emerged, Teller mounted a strong opposition, testifying before Congress against ratification of the Limited Test Ban Treaty. His opposition failed to prevent ratification, however. He became a vocal advocate for the peaceful uses of nuclear explosives, especially Project Oilsands, a plan to use such devices to extract petroleum from the tar sands of northern Alberta, Canada, and **Project Chariot**. Neither project was adopted. *See also* ALARM CLOCK; GEORGE; NEUTRON BOMB.

TERNARY EXPLOSIVE. Composition explosive comprised of three explosive compounds. Several ternary explosives were in use prior to the Second World War, including the **ammonals** and **torpex**. Later developments included Amatex 20 (40 percent RDX, 40 percent **TNT**, and 20 percent **ammonium nitrate**); the **High Blast Explosives**; and HTA-3, which is a mix of **HMX**, TNT, and **aluminum**.

TETRYL. Yellow crystalline solid used principally as a **booster** in **high explosive** munitions and in **detonators**. Tetryl was first reported in the scientific literature in 1879 by Wilhelm Michler (1849–1889) and Victor Meyer (1848–1897), German chemists then working at the Polytechnic Institute, Zurich, Switzerland. The structure of tetryl was correctly deduced in 1886. The preferred, systematic IUPAC name for tetryl is 2,4,6-trinitrophenyl-N-methylnitramine, and its chemical formula is $C_7H_5N_5O_8$. Other common names for tetryl include composition exploding (CE), nitramine, pyronite, and tetralite.

Testing found tetryl to be **shock-sensitive** and **friction-sensitive** and to burn readily. These properties worked against its use by itself as an explosive. However, around the First World War, it was found that mixing tetryl with other explosives, especially **TNT**, proved useful. Thereafter, tetryl remained in common use, primarily for military applications. Tetryl is extremely irritating to the skin, giving rise to a condition called "tetryl dermatitis" that is common among those personnel who are exposed to it during manufacturing or handling. It also causes yellow staining of the skin and hair, similar to but less dramatic than that seen in **canary girls**. *See also* TETRYTOL.

TETRYTOL. **Binary explosive** comprised of **tetryl** and **TNT**. A typical composition is 70 percent tetryl and 30 percent TNT, but the proportions can vary. Tetrytol came into use when it was found that tetryl, which is highly sensitive to **shock** and **friction**, could be melted and mixed with TNT, forming a liquid that could be cast into munitions and was comparatively safe to handle, store, and use. This mix was called "tetrytol" and was used primarily for military applications.

TEXAS CITY DISASTER. The SS *Grandcamp*, a French-registered cargo vessel, was moored in the Port of Texas City, Galveston Bay, Texas, on the morning of 16 April 1947. The ship was loaded with assorted goods destined for Europe, with the majority of its load being 2,300 tons of fertilizer-grade **ammonium nitrate**. Around 8:00 a.m., smoke was spotted in the cargo hold, signaling a fire. Efforts to put out the fire using a bucket of water and fire extinguishers failed. The captain ordered all hatches over the hold to be secured, in hopes of snuffing out the fire. These hopes proved misplaced. Around 8:30 a.m., the pressure from accumulating steam grew so great that the hatches blew off. At 9:12 a.m., the combination of heat and pressure in the hold caused the ammonium nitrate to **detonate**. This detonation touched off a cascade of events, leading the Texas City disaster to be labeled the largest industrial accident in US history, in terms of human deaths and injuries as well as destruction of property. It has also been considered one of the largest non-nuclear explosions experienced to date.

While efforts were underway throughout that day and the next to fight the fires from the initial explosion and to contain the damage, a second detonation occurred. Like the *Grandcamp*, the SS *High Flyer* carried a mixed cargo, including ammonium nitrate. This cargo caught fire, and attempts were made to tow the ship safely away from other vessels. These attempts failed, and around 1:10 a.m. on 17 April, its cargo exploded. Some witnesses would later describe this second explosion as greater than that of the *Grandcamp*.

The death toll from the Texas City disaster was placed at 581, and more than 5,000 were injured. Damages from the explosions and fires were immense. The port of Texas City was obliterated. More than 500 homes were destroyed, and their occupants were left homeless. Other losses included property damage estimated at $100 million and another $500 million in fuel oil products consumed or otherwise ruined. These losses were followed by more than a thousand legal claims, holding the US government liable and seeking compensation. Many of these claims were combined into a single case that ultimately made its way to the US Supreme Court in 1953. By the narrow margin of 4–3, the court found that the government was exempt from liability. This finding prompted the US Congress to enact legislation providing relief to many claimants. The final claim was settled in 1957, 10 years after the disaster, and capped off 1,394 awards totaling more than $17 million. *See also* HALIFAX EXPLOSION; OPPAU EXPLOSION.

THERMOBARIC WEAPON. *See* FUEL-AIR EXPLOSIVE (FAE).

THERMONUCLEAR. The process in which rapid **fusion** reactions are made possible by an increase in temperature. A thermonuclear reaction is the basis of the **hydrogen bomb**.

THORIUM. A naturally occurring element, with atomic number 90 and symbol "Th." In the context of **nuclear explosives**, the **isotope** thorium-232 is significant. It is a fertile isotope that can be converted to **uranium**-233 when irradiated with neutrons. The **Significant Quantity (SQ)** of thorium associated with indirect use for making a nuclear explosive is 20 tons.

TNT. Commonly used abbreviation for trinitrotoluene, an important military and industrial **explosive**. Six positional isomers of trinitrotoluene exist. The primary isomer of interest for use in explosives is 2,4,6-trinitrotoluene. The 2,4,6-positional isomer is the most stable form of trinitrotoluene, and the abbreviation "TNT" normally refers to this isomer. The other five isomeric forms of trinitrotoluene are 2,3,4-trinitrotoluene; 2,3,5-trinitrotoluene; 2,3,6-trinitrotoluene; 2,4,5-trinitrotoluene; and 3,4,5-trinitrotoluene. The preferred, systematic IUPAC name for TNT is 2-methyl-1,3,5-trinitrobenzene, and its chemical formula is $C_7H_5N_3O_6$.

Discovery. TNT was discovered by **Julius Wilbrand** (1839–1906), a German chemist. He prepared TNT by mixing nitric acid with toluene or with toluol, which is a coal-tar product having a high toluene content. At the time, Wilbrand was searching for yellow dyes. He reported his discovery in the scientific literature in 1863 but failed to note the explosive nature of TNT. Most likely, Wilbrand prepared a mixture of the isomeric forms of trinitrotoluene. His work inspired others to prepare and characterize the trinitrotoluenes. In 1883, Adolf Karl Ludwig Claus (1838–1900) and Heinrich Becker of the University of Freiburg established the structure of TNT. But it was not until 1889—26 years after its discovery—that the explosive nature of TNT was reported in the scientific literature. That report, by **Carl Häussermann** (1853–1918) of the Technical College of Stuttgart, apparently prompted an intense period of research into the use of TNT as a commercial and military explosive. Within two years, bulk production of TNT had begun in Germany, and by 1902, the German military had replaced **picric acid** with TNT as a bursting charge in munitions. Other nations soon followed suit. By the 1914 start of the First World War, TNT was being used by military forces of all of the combatant nations and had become the dominant **high explosive** for use as a bursting charge in shells and bombs. TNT remains an important commercial and military explosive. However, its use has diminished owing to more recent discoveries, such as **EDNA**, **PETN**, and **RDX**, which possess better **brisance** and are less **sensitive** to **shock** and **friction**.

Properties. Certain properties of TNT favor its use as an explosive. Its sensitivity to **friction** and **shock** is low, reducing the risk of accidental **detonation**. Its melting point of 80°C is far below the temperature at which it detonates, allowing it to be melted safely and either poured into munitions or mixed thoroughly with other explosive substances. Finally, TNT neither absorbs water nor dissolves in water, allowing it to be used underwater and to be stored without concern that exposure to moisture will ruin its effectiveness.

Production. TNT is routinely produced on an industrial scale by nitration of toluene (methylbenzene, C_7H_8) using nitric acid or, more commonly, a combination of nitric acid and sulfuric acid. The process involves three steps, with a nitro group ($-NO_2$) being added in each step. In the first step, toluene is nitrated to form mononitrotoluene. In the second step, mononitrotoluene is nitrated to form dinitrotoluene. In the final step, dinitrotoluene is nitrated to form trinitrotoluene. Various chemicals and procedures are used at each step to remove undesired reaction products and to concentrate the desired isomeric form of trinitrotoluene.

When industrial-scale production of TNT began in the early twentieth century, nitric acid and sulfuric acid were readily available, but toluene was less available. At that time, toluene was produced from coal as a by-product of coke ovens and, to a lesser extent, by extraction from illuminating gas (a coal-distillation product). Toluene was also found in toluol, a coal-tar derivative. Early in the First World War, all combatant nations were faced with competing needs for toluene, including its use as an additive to boost the octane of aviation fuel. These competing needs and the high rate at which explosives were being consumed in combat led to shortages of toluene and TNT, perhaps its most important product. The shortages were sufficiently severe that various schemes were attempted either to develop substitutes or to increase toluene availability. For example, the British determined that petroleum from the Pacific region of Borneo and used by Royal Dutch Shell could be distilled to give a fraction high in toluol. The government arranged to dismantle a Shell distillation plant in Rotterdam, the Netherlands, and brought this plant under naval escort to London. From there, the dismantled plant was transported inland to Portishead and re-erected. A duplicate plant was erected by Shell at Barrow-in-Furness. These two plants would furnish about half of the toluene used by the British during the First World War. The British also developed **amatol**, a blend of **ammonium nitrate** and TNT that essentially diluted the TNT content in munitions and thereby conserved existing stocks of TNT.

The United States, which was supplying TNT to various combatants prior to its entry into the war in April 1917, hit upon the scheme of mixing TNT with trinitro-*m*-xylene (TNX, $C_8H_7N_3O_6$). This scheme seems reasonable, given the structural similarity between TNT and TNX and given the availability of *m*-xylene (1,3-dimethylbenzene, C_8H_{10}), an essential precursor in manufacturing TNX but a chemical not otherwise considered vital to the war efforts. TNX has a significantly higher melting point than TNT and is insoluble in TNT. Consequently, TNX needed to be blended thoroughly into molten TNT to ensure a homogeneous mixture. Anticipating the use of TNX as a means of extending its existing stocks of TNT, the United States set about building plants to produce TNX. However, construction of facilities to produce TNX was unfinished when the Armistice was announced on 11 November 1918, and the TNT-TNX mix failed to come into common use as a military explosive. The French, faced with similar shortages of TNT, developed a mix of 77 percent **ammonium nitrate** and 23 percent TNX. This mix, designated *Explosif NX*, was used as a bursting charge in shells during the First World War.

Following the First World War, the German chemical cartel I.G. Farben invented a process for making toluene from petroleum. Standard Oil of

New Jersey, a US firm, licensed this process in 1927, allowing it to produce toluene from domestic supplies of petroleum. When the Second World War began in 1939, the United States was in an improved position relative to the supply of toluene. However, US producers of TNT were reluctant to use petroleum-derived toluene versus coal tar-derived toluene, imagining that there was some difference. Eventually, the US military influenced the producers to experiment with the petroleum-derived toluene. Results were favorable, and the switch took place.

By the time the United States entered the war in December 1941, US producers still suffered from insufficient stocks of TNT to meet projected wartime consumption rates, despite the switch to using petroleum-derived toluene. Two solutions were possible: increase the production rate for the existing production lines or build more production lines. **John P. Harris**, an officer in the US Army Ordnance Corps, solved this problem when he visited a TNT production facility in Beloeil, Quebec, Canada. His visit was unplanned, having been arranged at the last moment to fill time while he awaited transport for his return to the United States. While there, Harris noticed that the Beloeil plant added toluene to a mix of nitric acid and sulfuric acid. This addition step was the reverse of the process used in the United States. On his return to the United States, he urged US producers to try what came to be referred to as "reverse nitration." Producers were reluctant to make this change, which seemed counterintuitive, and which would require substantial re-engineering of existing production lines. Eventually, the Canadian process was tried on a trial basis at the Keystone Plant, Meadville, Pennsylvania. The results of reverse nitration convinced the producers of its merits. TNT output could be tripled, and production lines designed to turn out 33,000 pounds of TNT per day were suddenly able to produce more than 100,000 pounds of TNT per day. This increase eliminated the need to use substitutes for TNT, as had occurred during the First World War, and the price of TNT dropped to half the then prevailing price. Reverse nitration remains the standard practice for industrial-scale production of TNT. *See also* CANARY GIRL.

TNT-EQUIVALENT. Amount of **TNT** that would be required to release an equivalent amount of energy. The energy released by an **explosive** is commonly expressed in terms of "TNT-equivalent."

TNX PROJECT. *See* DUPONT (E. I. du PONT de NEMOURS & COMPANY).

TOLITE. French designation for **TNT**.

TOLLENS, BERNHARD CHRISTIAN GOTTFRIED (1841–1918). German chemist credited with discovering **PETN**. Tollens spent most of his professional career at the University of Göttingen, Germany, where he received his doctorate in chemistry. His primary area of research was fundamental studies into the chemistry of aldehydes and ketones. During the course of this research, he examined the preparation of polyhydric alcohols from formaldehyde and acetaldehyde. One such alcohol was pentraerythritol (or, more formally, 2,2-*bis*(hydroxymethyl)-1,3-propanediol). Tollens and his associate, Peter Wigand, found that when this alcohol was treated with nitric acid, a reaction ensued, forming pentaerythritol tetranitrate (PETN). They reported their results in the scientific literature in 1891; however, the **explosive** nature of PETN was outside the main area of interest for Tollens and his colleagues in Göttingen, and they appear to have taken no further interest in the substance. Instead, Tollens continued his fundamental studies on aldehydes, including his discovery of diamminesilver (I) complexes as an oxidizing agent to test for aldehydes and ketones. This latter discovery was widely recognized, and such complexes became known as "Tollens' reagent."

TOLUOL-AMMONAL. *See* AMMONAL.

TONITE. Smokeless powder containing 52.5 percent **guncotton** and 47.5 percent barium nitrate (an **oxidizer** similar to **potassium nitrate** and **sodium nitrate**). Tonite (pronounced "toe-nite") was developed at the **Cotton Powder Company, Ltd.**, Faversham, England, by George Trench (1838–1912), Camille Faure (1840–1898), and Samuel Joseph Mackie (1823–1902), for which they received a British patent in 1874. One drawback associated with guncotton as an **explosive** was that its combustion was incomplete, resulting in significant quantities of carbon monoxide being produced. While the presence of carbon monoxide might prove no barrier to the use of guncotton for most military applications, it was potentially harmful when used in mining and other commercial applications in enclosed environments. The addition of barium nitrate to guncotton was intended to reduce the carbon monoxide production by driving the combustion reaction to complete oxidation.

This property of reduced carbon monoxide fumes helped Tonite to enjoy some use as a **blasting agent**, especially for mining and civil construction projects such as building tunnels. It was advertised as superior to dynamite because it was less expensive, produced reduced carbon monoxide fumes on detonation, and was very stable toward extreme heat and cold. Tonite initially gained wide acceptance in Britain, but its use was prohibited by all other European countries owing to the perceived instability of guncotton. By 1880, Trench of the Cotton Powder Company had relocated to Stege, California, to aid in the startup of the Tonite Powder Company. This company became the first US producer of a **nitrocellulose**-based explosive in commercial quantities. However, success for this product was short-lived. While Tonite initially gained wide acceptance in Britain, its use was prohibited by all other European countries owing to the perceived instability of guncotton. An 1881 explosion in Mexico that claimed 21 lives was attributed to the Tonite, fueling rumors that Tonite could explode spontaneously. More harmful to its commercial success, though, was the combination of the rise in prices for cotton, which made guncotton more expensive, and the drop in prices for glycerin, which made **dynamite** and other **nitroglycerin**-containing explosives less expensive. By 1885, Tonite Powder Company had ceased operations. *See also* POTENTITE.

TORPEDO. (1) Device, typically cigar-shaped, containing an **explosive** and designed to be launched underwater, where it is self-propelled and often capable of being directed to the intended target. In the nineteenth and early twentieth centuries, any explosive device used underwater was referred to as a torpedo, including those floating and stationary devices now commonly known as "mines." (2) Device containing an explosive and designed to be **detonated** in gas, oil, and water wells to improve flow. *See also* ROBERTS, EDWARD A. L. (1829–1881); ROBERTS PETROLEUM TORPEDO COMPANY.

TORPEX. Cast **ternary explosive** comprised of 41–45 percent **RDX**, 37–41 percent **TNT**, and 18 percent powdered **aluminum**. Torpex was developed by the British at the Royal Gunpowder Factory, Waltham Abbey, around 1942. "Torpex" is a shortening of "**torpedo** explosive," and the original applications were for use as a blasting charge in torpedoes and undersea mines. Eventually, beeswax was added to Torpex as a **phlegmatizing agent**, and calcium carbonate was added to absorb water that might mix with the explosive to produce hydrogen. The presence of water was particularly concerning in high-humidity environments, such as those aboard submarines. *See also* H-6; HIGH BLAST EXPLOSIVE (HBX); PBX.

TRAUZL LEAD BLOCK TEST. Test to measure the strength of **explosives**. In its original form, the test involves placing a 10-g foil-wrapped sample of the material to be tested into a hole drilled into a lead block. The block is a cylinder, having a length of 200 mm and a diameter of 200 mm. The hole measures 115 mm deep and 25 mm in diameter. After placing the sample in the hole, sand is poured into the hole, then tapped down to cover the sample thoroughly. The sample is then **detonated** electrically. The resultant explosion deforms the lead block by expanding the hole. The volume of this hole is measured, the measurement is corrected for the original volume as well as the expansion due to the detonator, and the final corrected volume is taken as a crude indication of the strength of the explosion. Results are more likely to be meaningful if they are taken from a single test report conducted by an individual or a group rather than from various sources, where slight differences in materials and methods could confound comparisons.

Isidor Trauzl (1840–1929) is credited with developing the test in 1885, when he worked as the manager of the **Alfred Nobel** & Company **dynamite** factory in Zámky, near Prague, which was then part of the Austro-Hungarian Empire. Prior to joining Nobel, Trauzl had served with the Imperial Austrian Engineers, where his research on **high explosives** is credited with leading to the introduction of dynamite in Austria. The Trauzl lead block test, or more simply, the Trauzl test, was widely adopted, becoming an international standard in 1904. Variants on the Trauzl test include the use of aluminum in place of lead and the use of a smaller block. The test suffers from several drawbacks, including the mass of the lead block (approximately 75 kg), cost (the block must be melted down and recast following each use), and the qualitative nature of the results. Eventually, it was supplanted by other tests, such as the pendulum test. In this latter test, the strength of an explosive is measured by the extent to which a piston is pushed out of a horizontal cylinder, with the movement of the piston being measured by the arc of motion of a pendulum affixed to it.

TREATY ON NONPROLIFERATION OF NUCLEAR WEAPONS. International arms control agreement having as its goals preventing the spread of **nuclear weapons** and technologies usable in making such weapons, promoting peaceful uses of nuclear energy, and, ultimately, achieving nuclear disarmament. The 1968 Treaty on the Nonproliferation of Nuclear Weapons, commonly referred to as the Nonproliferation Treaty (NPT), forms the legal foundation for the global nuclear nonproliferation regime. The treaty opened for signature on 1 July 1968 and entered into force on 5 March 1970. As of 2024, 191 states are party to the NPT. Notable states that have not become party to it are **India** and **Pakistan**, both of which have tested nuclear weapons, and **Israel**, which is suspected to possess nuclear weapons. **North Korea** was a party to the NPT but withdrew on 10 April 1993, becoming the only party to have done so thus far. The NPT obliges its members not to "assist, encourage or induce" non-nuclear weapon states (NNWS) from manufacturing or acquiring nuclear weapons or nuclear explosive devices, or from exercising control over such weapons and explosive devices. The treaty also obliges each NNWS to place IAEA safeguards on various nuclear activities carried out under its control. *See also* PEACEFUL NUCLEAR EXPLOSIONS (PNE) TREATY; TREATY ON THE PROHIBITION OF NUCLEAR WEAPONS.

TREATY ON THE PROHIBITION OF NUCLEAR WEAPONS. Comprehensive, legally binding international arms control agreement prohibiting **nuclear weapons**, having the total elimination of nuclear weapons as its goal. Parties to the treaty agree to refrain from acquiring, developing, possessing, producing, stockpiling, testing, using, or threatening to use nuclear weapons. These same parties are prohibited from deploying nuclear weapons on their national territory and from aiding any other State in the conduct of prohibited activities. The Treaty on the Prohibition of Nuclear Weapons, also known as the Nuclear Weapon Ban Treaty, was adopted by a Conference of UN Member States on 7 July 2017. It opened for signature that same year on 20 September. The 50th instrument of ratification or accession was deposited on 24 October 2020, and the treaty entered into force on 22 January 2021. As of September 2024, 73 nations are party to the treaty. Those nine nations that are known or are suspected to possess nuclear weapons—**China**, France, **India**, **Israel**, **North Korea**, **Pakistan**, Russia, the UK, and the United States—have neither signed nor become party to the treaty. *See also* NONPROLIFERATION TREATY; NUCLEAR-WEAPON-FREE ZONE.

TREATY ON UNDERGROUND NUCLEAR EXPLOSIONS FOR PEACEFUL PURPOSES. *See* PEACEFUL NUCLEAR EXPLOSIONS (PNE) TREATY.

TRIACETONE TRIPEROXIDE (TATP). An **acetone peroxide explosive**. Triacetone triperoxide (TATP) is a white crystalline powder. The preferred, systematic IUPAC name for TATP is 3,3,6,6,9,9-hexamethyl-1,2,4,5,7,8-hexaoxonane, and its chemical formula is $C_9H_{18}O_6$. TATP is highly sensitive to **friction**, heat, **shock**, and vibration. It and the similar substance, **diacetone diperoxide** (DADP), are of concern owing to their use or feared illicit use in **homemade explosives** because of the relative ease with which they might be produced from commercially available chemicals. *See also* SHOE BOMBER.

TRIAMINOTRINITROBENZENE (TATB). Brown or yellow powdered crystalline solid that is a powerful **explosive**. Triaminotrinitrobenzene (TATB) is extremely insensitive to **fire**, impact, **shock**, or vibration and, thus, is difficult to **detonate** by accident. These characteristics make TATB a prime candidate for use in an **insensitive high explosive (IHE)** where extreme safety is required, such as in **nuclear weapons**. In such cases, it is often used as a **plastic-bonded explosive** (PBX). The preferred, systematic IUPAC name for TATB is 2,4,6-trinitrobenzene-1,3,5-triamine, and its chemical formula is $C_6H_6N_6O_6$.

TRINITROANISOL (TNA). Shock-sensitive crystalline solid used in **picric acid**-based **explosives**, such as **Stabillite**. Trinitroanisol (TNA) is both highly toxic by ingestion and a potent skin irritant. TNA was adopted by the Japanese and used in various munitions either directly or in a mixture with other explosives. For example, "Type 91" was the Japanese designation for 100 percent TNA, and "Type 94" was the mix of 60 percent TNA and 40 percent **RDX**. "Trinol" is another name used for TNA in the context of explosives. The preferred, systematic IUPAC name for TNA is 2-methoxy-1,3,5-trinitrobenzene, and its chemical formula is $C_7H_5N_3O_7$.

TRINITROCRESOL. Shock-sensitive yellow crystalline solid used in **picric acid**-based **explosives**, such as mélinite/O and **shimose**. The preferred, systematic IUPAC name for trinitrocresol is 3-methyl-2,4,6-trinitrophenol, and its empirical formula is $C_7H_5N_3O_7$.

TRINITROGLYCERIN. *See* NITROGLYCERIN.

2,4,6-TRINITROPHENOL. Formal chemical name for **picric acid**.

TRINITROTOLUENE. Chemical name for **TNT**. Six isomers of trinitrotoluene exist. The primary isomer of interest for use in **explosives** is 2,4,6-trinitrotoluene. The other five isomeric forms of trinitrotoluene are 2,3,4-trinitrotoluene; 2,3,5-trinitrotoluene; 2,3,6-trinitrotoluene; 2,4,5-trinitrotoluene; and 3,4,5-trinitrotoluene.

TRINITROXYLENE (TNX). *See* TNT.

TRINITY TEST SITE. Site of the first known **nuclear explosion**. **J. Robert Oppenheimer** selected "Trinity" as the code name for the first test of the **plutonium** nuclear device known as **Gadget**. According to Oppenheimer, the name was inspired by the writings of the seventeenth-century poet John Donne. The 51,500-acre site where this test occurred on 16 July 1945 became known as the Trinity Test Site. The site itself is in an isolated and barren region of southern New Mexico known as the Jornada del Muerto ("Journey of Death"), approximately 210 miles south of Los Alamos. It was selected because it was already part of the Alamogordo Bombing and Gunnery Range, established in 1942, and because it was close to Los Alamos, where much of the scientific and engineering work on Gadget was being done. On 21 December 1965, Trinity Test Site was designated a National Historic Landmark, and on 15 October 1966, it was placed on the National Register of Historic Places. The heat of the Trinity blast vaporized the steel tower that had supported Gadget and melted the surrounding desert sand, turning it into a green, glassy substance. Dubbed "trinitite," this material littered the site for several years and served as a reminder of the historical test.

TRINOL. *See* TRINITROANISOL.

TRIPLE-BASE POWDER. Smokeless powder comprised of a mix of **nitrocellulose**, **nitroglycerin**, and **nitroguanidine**. Other substances might be present, such as **binders**, **stabilizers**, and residual solvents from the manufacturing process. *See also* DOUBLE-BASE POWDER; SINGLE-BASE POWDER.

TRITONAL. *See* BINARY EXPLOSIVE.

TROJAN EXPLOSIVE. *See* NITROSTARCH.

TROTYL. Synonym for **TNT**.

TSAR BOMBA. On 30 October 1961, the Soviet Union **detonated** a **hydrogen bomb**. It exploded with a yield of 50 megatons (mt) **TNT-equivalent**, making it the most powerful **nuclear weapon** detonated to that time, a record that still stands. Designated the **RDS**-220 by the Soviet Union, the device was given various nicknames but became known in Western countries as "Tsar Bomba." The US government dubbed this device "**Joe-111**," consistent with designations used for prior Soviet nuclear devices. Tsar Bomba was a three-stage device in which a **fission** bomb (first stage) compresses a **thermonuclear** secondary device, and the energy from the resulting explosion compresses a larger, additional thermonuclear device. Tsar Bomba weighed nearly 30 tons, was delivered by an aircraft, and exploded over the Mityushikha Bay test range north of the Arctic Circle. It had a theoretical yield thought to be as large as 100 megatons, but measures apparently were taken to lessen the yield both to limit fallout and to provide adequate time for the aircraft to escape the explosion. By contrast with the 50 megatons yield of Tsar Bomba, the largest detonation by the United States—Castle Bravo—was 15 megatons. *See also* SAKHAROV, ANDREI DMITRIYEVICH (1921–1989).

TURPIN, FRANÇOIS EUGÈNE (1848–1927). French chemist and inventor. Turpin began his career as a manufacturer of rubber toys and then branched out to manufacturing rubber topographic maps for the French Ministry of War. In 1884, he invented **mélinite**, a **smokeless explosive** based on **picric acid**. Around this time, Turpin also claimed to have invented a novel class of explosives, which he dubbed "**panclastites**." Both claims proved controversial. Although the French government would eventually adopt mélinite and panclastites for military use, the initial mild reaction from the government caused Turpin to feel free to market his inventions to other governments. For example, in 1886, Turpin obtained German patent 38,734 for the use of picric acid as a bursting charge. His marketing activities and especially the German patent were viewed as treasonous acts by the French government, and Turpin was arrested, tried, and convicted. Sentenced to four years imprisonment, he served two years in Étampes before being released in 1893. Shortly thereafter, Turpin sued Jules Verne, the famed French novelist, for libel. His claim alleged that Verne had used Turpin as the basis for "Thomas Roch," a mad-scientist character in Verne's novel *Facing the Flag*. The suit was unsuccessful, although Verne privately admitted that he had based Roch on Turpin.

Turpin was eventually awarded compensation by the French government for his invention of mélinite. The award was small, however, because the government asserted that the use of picric acid as a base for explosives was already common knowledge. This position seems strangely at odds with the government's position in its successful prosecution of Turpin for treason. Turpin surfaced again during the First World War with his claims regarding **Turpinite**, although no basis for those claims has ever surfaced. *See also* ANILITE.

TURPINITE. The 28 September 1914 issue of the Italian newspaper *La Tribuna* carried a report of French use of a mysterious new **explosive** that invariably caused instantaneous and absolutely painless death to every living thing within 50 or so meters of its **detonation**. Similar reports appeared in an October 1914 issue of the *Daily Mail* (London) and the 6 January 1915 *Sydney Mail*. The Germans, too, took the reports seriously, having been alarmed by what appeared to be a spike in deaths from asphyxiation among their soldiers. On 6 April 1915, the *New York Times* reported that the new explosive had been given a name: "Turpinite," named in honor of its inventor, French chemist **Eugène Turpin**. Reportedly, Turpinite doubled the explosive effects of the 3-inch artillery shell, killing victims "by a paralysis of the heart." Such reports caused the Germans to consider the prospect that the French had resorted to chemical warfare. These prospects were reinforced by contemporary French postcards depicting German soldiers with the bodies of cockroaches but wearing their Pickelhaube (spiked helmets) fleeing the battlefield as Turpinite shells burst overhead. The postcards were labeled "La Turpinite: Poudre Insecticide Français" ("Turpinite: the French insecticidal powder").

In an interview given to the *New York Times* on 11 July 1915, American chemist **Hudson Maxim**, a well-known inventor of explosives, comments on these reports.

> It is reported that soldiers have been found stone dead in sitting postures and without a wound upon them after an explosion of Turpinite in their vicinity. It is the shock of the explosion which kills. This is the explanation of what has been denominated the great mystery of Turpinite. Turpinite is a poem, a creation of the imagination. It does not actually exist.

Maxim's position is likely to have been accurate. Technical data on Turpinite are lacking. German medical specialists who investigated the asphyxiations determined the soldiers died from carbon monoxide poisoning in their poorly ventilated dugouts. Finally, the *Oxford English Dictionary*, second edition, attributes the first use of "Turpinite" to a 6 December 1895 issue of the *Daily Chronicle* (London) and defines it as "a comprehensive word coined in honor of a string of episodes bearing upon the new methods of putting a stop to war by the prospective destruction of all cities." In all likelihood, publicity over the use of Turpinite was a part of a propaganda campaign aimed at frightening the Germans into imagining French possession of a weapon lacking an equivalent in the German arsenal. As it was, the publicity may have added to German interest in the use of chemical weapons. *See also FACING THE FLAG.*

U

UNDERWEAR BOMBER. On 25 December 2009, Umar Farouk Abdulmutallab attempted to **detonate explosives** hidden in his underwear. At the time, the 24-year-old native of Nigeria, was aboard Northwest Airlines flight 253, en route from Amsterdam to Detroit, Michigan. His attempted suicide bombing failed. Shortly after he exited the lavatory, flames were seen on his trouser leg and were extinguished by flight attendants. After they had extinguished the flames, he informed the attendants that he had an explosive device in his pocket. Examination of his person uncovered a packet sewn into his underwear. The packet was analyzed later and found to contain **PETN**. The location of this packet caused him to be referred to in the news as the "Underwear Bomber." On landing in Detroit, Abdulmutallab was taken into custody and charged with various offenses. He told authorities he was acting on behalf of Al Qaeda in his attempt to destroy this Christmas Day international flight. He ultimately pled guilty to eight terrorism-related charges. His plea was accepted, and he was sentenced to life imprisonment. *See also* PERSON-BORNE IMPROVISED EXPLOSIVE DEVICE (PBIED).

UNEXPLODED ORDNANCE (UXO). *See* EXPLOSIVE ORDNANCE DISPOSAL (EOD).

URANIUM. A naturally occurring element, with atomic number 92 and symbol "U." Isotopes of uranium find use as fuel in **nuclear weapons** as well as in nuclear power generation. The most prominent isotopes with regard to nuclear weapons are uranium-233, uranium-235, and uranium-238. Uranium-233 is a man-made radioisotope and is significant because it is fissile. Uranium-235 is the only naturally occurring fissile isotope of uranium. Natural uranium contains about 0.7110 percent uranium-235. It was the fuel in **Little Boy**. During the Second World War, uranium-235 was referred to by the code name "Oralloy." The "OR" referred to Oak Ridge, the Tennessee facility where much of this isotope was produced, while "alloy" imitated the British use of "Tube Alloys" to refer to their early efforts at developing a nuclear weapon. **Low-enriched uranium (LEU)** and **highly enriched**

uranium (HEU) refer to uranium enriched in uranium-235. Uranium-238 is the most abundant isotope of uranium, constituting approximately 99.2830 percent of natural uranium. It is not fissionable, but it is a fertile material, meaning that it forms the fissile isotope **plutonium-239** as a result of neutron bombardment. *See also* CRITICAL MASS; SIGNIFICANT QUANTITY (SQ); SPECIAL NUCLEAR MATERIAL (SNM); WEAPONS-GRADE MATERIAL.

V

VACUUM BOMB. *See* FUEL-AIR EXPLOSIVE (FAE).

VEHICLE-BORNE IMPROVISED EXPLOSIVE DEVICE (VBIED). An **explosive** carried on or within, loaded into, or otherwise attached to a vehicle. Well-known examples of a vehicle-borne improvised explosive device (VBIED) include the **Oklahoma City Bombing**, which involved a truck, and the **Bali Bombings**, where a passenger van was used. The **attempted assassination of Napoléon** in 1800 as well as the 1920 **Wall Street Bombing** were caused by a horse-drawn cart loaded with explosives. Even **mules** and donkeys, which can carry people or goods, have been cited as VBIEDs. Two bombings in British Mandated Palestine—one each in 1938 and 1939—have been attributed to donkey-borne bombs used in terrorist attacks in the Haifa marketplace. More recently, donkeys laden with explosives were used in 2004 in Ramadi, Iraq; by the Taliban in Helmand Province, Afghanistan, in 2009; and, lastly, in an attack in April 2013 in Kabul, Afghanistan that killed a law enforcement officer. The 17 December 2024 assassination of Russian lieutenant general Igor Anatolyevich Kirillov and his aide, Ilya Polikarpov, in Moscow was attributed to explosives planted in an unmanned electric scooter. At the time of his death, Kirillov commanded the Russian Army's chemical, biological, and nuclear defense troops.

VELA INCIDENT. In the late 1950s, the United States began deploying Vela-series satellites to detect the detonation of a **nuclear weapon** or other nuclear device. On 22 September 1979, one of these satellites, Vela 6911, detected a double flash of light. This type of flash is known to occur during a nuclear explosion. The flash occurred in an area approximately 3,000 miles in diameter, encompassing the southern tip of Africa, the Indian Ocean, the South Atlantic Ocean, and part of Antarctica. Shortly thereafter, US president Jimmy Carter convened a panel to investigate what became known as the Vela Incident. In May 1980, this panel issued a report, announcing its conclusion that the flash was likely to have been caused by a meteorite hitting the satellite, with sunlight then reflecting off particles as they spread around the satellite following the collision. This conclusion was not universally accepted within the US government. Other groups, particularly from the intelligence community, believed that the data from the satellite plus signals picked up by hydroacoustic monitoring stations were sufficient to indicate that a nuclear weapon had been detonated. The incident led to speculation at the time that **Israel** and **South Africa** had jointly tested a nuclear weapon as part of suspected cooperation between the nations in the nuclear field. However, no information has emerged that South Africa ever conducted a test of its nuclear weapons, leaving Israel as the leading suspect if this event was actually the detection of a nuclear explosion. The actual cause of the September 1979 detection remains a matter of speculation.

VERNE, JULES (1828–1905). *See FACING THE FLAG.*

VERY LIGHT. A **pyrotechnic** device used as a signaling flare. Edward Wilson Very (1847–1910), an 1867 graduate of the US Naval Academy, is credited with its invention and is its eponym. In 1877, Very received US patent 190,263 for a signal cartridge having an **explosive** charge at its base with two pyrotechnic mixtures above. As designed, the cartridge was shot from a gun, also of his design and called a Very pistol, propelling the cartridge several 100 meters into the air and releasing the pyrotechnic mixtures, which burst into flames. Use of pyrotechnic mixtures producing differing colors enabled a coded system of signaling. Flares following Very's design and from others saw extensive use from the latter half of the nineteenth century and during the First World War, and these flares came to be called *Very lights* regardless of design or maker. *See also* COSTON FLARE.

VESUVIUS. American warship distinguished by its **pneumatic guns**. Authorized on 3 August 1886 during a period of intense interest in pneumatic guns, the *Vesuvius* entered the fleet of the US Navy on 1 October 1890. Her armament included three 15-inch pneumatic guns, built by the Pneumatic Dynamite-Gun Company, New York. The guns fired a shell containing approximately 500 pounds of **dynamite**. The ship was used during the **Spanish-American War** and was credited with eight shore bombardments in action off Cuba. The bombardments apparently had little effect, with **Hudson Maxim** noting that the shells "succeeded in damaging somewhat the green grassy effect" of the lawn in front of Morro Castle.

The gun design proved problematic, as the air compressor arrangement needed for its pneumatic guns left the guns immobile, incapable of being turned to engage a target except by movement of the ship. The range of the guns was judged to be less than that of enemy ships against which it might operate, leaving the *Vesuvius* vulnerable to an attack from which it could not retaliate. The pneumatic gun design had one advantage over other guns using chemical explosives as **propellants**. The use of compressed air to launch the dynamite projectiles meant its operation was virtually noiseless, smokeless, and free of any muzzle flash, robbing an adversary of sound or visual clues that a bombardment was underway. Despite this advantage, the design was judged a failure. Plans for a second ship similarly equipped were dropped, and the *Vesuvius* was taken out of service on 16 September 1898. She would eventually return to the fleet but outfitted with conventional guns using smokeless explosives as propellants in place of her pneumatic guns. The US experience with the *Vesuvius* was followed closely by other nations, which were considering building ships of a similar design. The failure of the *Vesuvius* may have influenced their decisions to abandon any such plans.

VIEILLE, PAUL-MARIE-EUGÈNE (1854–1934). French physicist and inventor of the first successful **smokeless powder**. Vieille made many fundamental and practical contributions to the field of **explosives**. He collaborated with renowned French chemist **Pierre-Eugène Marcellin Berthelot** in the early 1880s on developing the first method for measuring the velocity of **detonation** of explosives, and this method and the results it produced proved important in placing the development of explosives on a scientific rather than trial-and-error basis. Vieille also suggested using amyl alcohol as a **stabilizer** for **nitrocellulose**, and this suggestion would be widely adopted and used in France until 1911. He is best remembered for his development of the material he designated "**Poudre B**." As originally prepared by Vieille, Poudre B was a dense colloidal mass produced from treating both soluble and **insoluble nitrocellulose** with an ether-alcohol mixture, then kneading this material to form a stiff jelly. This material was rolled into thin sheets, cut into squares, allowed to dry, and used as a **propellant**. The French military subjected this product to testing, using a 65-mm cannon. The testing proved successful, and the results were communicated to the Minister of Armaments on 23 December 1884. Thereafter, the French army and navy adopted Poudre B, and it remained in use with various modifications and improvements over the next several decades.

In 1889, Vieille became the first recipient of the Leconte Prize, which recognized his various contributions to the study of explosives. The prize was accompanied by an award of 50,000 Francs. He was also made a member of the French Academy of Sciences. Later, Vieille was promoted to Director, Laboratoire Centrale des Poudres et Salpêtres (Central Laboratory of Gunpowder and Saltpeter), Paris, succeeding **Émile Sarrau**. *See also* NOBEL, ALFRED BERNHARD (1833–1896).

VIGORITE. Trade name for assorted **explosives** developed for use in **blasting**. The original formula for Vigorite was developed by Norwegian chemist Carl Gustav Bjorkmann, for which he received British patent 2,459 in 1875 and US patent 177,988 on 30 May 1876. This formula involved a mixture Bjorkmann named "nitroline" that proved complex and time-consuming to prepare. In 1877, the Vigorite Powder Company, later renamed the California Vigorite Powder Company, licensed his process but refined it, resulting in two variants of Vigorite that were marketed. One consisted of 43.75 percent **nitroglycerin**, 18.75 percent **potassium nitrate**, 17.50 percent **potassium chlorate**, 11.25 percent **sawdust**, and 8.75 percent calcium or magnesium carbonate. The second consisted of 49 percent potassium chlorate, 30 percent nitroglycerin, 9 percent sawdust, 7 percent potassium nitrate, and 5 percent calcium or magnesium carbonate. Patent disputes, though, kept the company from prospering, and it had folded in the 1880s. However, a Canadian firm, Hamilton Powder Company, also licensed the Vigorite process, creating a variant that was 60 percent **sodium nitrate**, 30 percent **nitroglycerin**, 5 percent **charcoal**, and 5 percent nitrated **sawdust**. The presence of such a high percentage of sodium nitrate poses a potential hazard with regard to this mixture. Sodium nitrate is deliquescent; that is, it takes up water from the atmosphere. As it dissolves owing to the presence of atmospheric water vapor or from absorption on exposure to rain or snow, the nitroglycerin becomes unstable and liable to spontaneous explosion. This situation played out, for example, on 5 May 1879, when a consignment of Vigorite exploded accidentally on the Grand Trunk Railroad near Stratford, Ontario, Canada. Two men were killed, many were injured, 24 railway cars were destroyed, and others were seriously damaged. Commercial demand for this Canadian variant of Vigorite waned in the wake of this accident.

VOLKMANN, FRIEDRICH. Austrian chemist and inventor of **explosives**. Biographical data on Volkmann is meager. Various accounts credit him with inventing an early form of **smokeless powder**. The invention likely took place around 1870, as he was granted Austrian patents 21–208 and 21–257 in 1871. These patents describe materials prepared in a manner similar to that used at least a decade earlier by **Eduard Schultze**, the Prussian artillery officer who nitrated various wood products to make a crude form of **nitrocellulose**. The final product prepared by Volkmann, however, omitted **potassium nitrate** and barium nitrate, which were incorporated by Schultze into **Schultze powder**. Therefore, Volkmann appears to have invented the first smokeless powder based solely on nitrocellulose. Around the time he received his patents, he became associated with the firm Collodin-Fabriks-Action-Gesellshaft in Marchegg, Austria. This firm was licensed to manufacture both the smokeless powder invented by Volkmann and the Schultze powder. The products were marketed for use as **propellants** in firearms used for sport and hunting, rather than for military firearms. At the time, the Austrian government enjoyed a monopoly on the manufacture of **black powder**, and any manufacturer of explosives had to be licensed by the government. Volkmann's firm lacked such a license. In 1875, the government forced the firm to cease production, perhaps fearing loss of income from Volkmann's invention. Thereafter, Volkmann and his invention apparently faded into obscurity, as his work appears to have escaped notice among those scientists working on explosives.

W

W60, W70, W77. *See* ENHANCED RADIATION WEAPON.

WAFER PRIMER. *See* SHAW, JOSHUA (1776–1860).

WALL STREET BOMBING. Around noon on 16 September 1920, a horse-drawn cart came to a stop on New York City's Wall Street. It was parked outside the entrance to the US Assay Office, an arm of the Treasury Department, and opposite the headquarters of the J.P. Morgan & Co. bank. The driver departed the cart and disappeared into the lunchtime crowds thronging the streets and sidewalk. At 12:01 p.m., the cart vanished in a violent explosion. Thirty people died nearly instantaneously, and hundreds were injured. Ten more would die in the days and weeks ahead, making the total killed 40 by what became known as the Wall Street Bombing. Reaction was swift, focusing on life-saving efforts to aid the injured. City police soon arrived and would be joined by agents from the US Secret Service and the Justice Department Bureau of Investigation, forerunner of the FBI. The initial investigation examined the possibility that a legitimate load of commercial **explosives** had **detonated** accidentally.

That possibility proved unlikely, reinforced by the finding the next day of four flyers in the vicinity. The spelling on each was crude, but the message was clear. Each flyer demanded that political prisoners be released and was signed "American Anarchists Fighters." The flyers were thought to be similar to those associated with the **Italian-American Galleanisti**, responsible for a nationwide campaign of bombings in April and June 1919. The finding of these flyers confirmed in the minds of investigators that the bombing was a deliberate act by terrorists. The location of the bombing fueled speculation that it was the work of a radical group opposed to capitalism. Suspects included anarchists, Bolsheviks, communists, and militant socialists. Investigative leads were followed in an attempt to locate the driver. A hoof from the horse was recovered among the debris, its horseshoe intact. Visits to blacksmiths failed to provide useful information as to the horse or its owner. Examination of the scene and of the victims showed that the explosive had been packed with metal slugs, which acted as **Shrapnel** to increase the destructive power of the bomb, especially in terms of human injuries. This tactic was consistent with past bombings by the **Italian-American Galleanisti**. Those bombings, though, were minor in scale and appeared to be more precisely targeted, with each bomb sent to a named individual. The explosive itself was widely reported to have been **dynamite**, the explosive of choice of the Galleanisti. However, recovery efforts and cleanup of Wall Street and environs had compromised the crime scene, with what might have proven to be useful physical evidence being tossed as trash or otherwise lost to investigators. Despite exhaustive efforts by local and federal law enforcement agencies, the identity of the driver, the source of the explosives, and the motive were never learned. No one came forward to claim responsibility. Intense speculation then and for decades afterward blamed Galleanisti or other anarchists as responsible. Definitive proof of their responsibility is lacking, and the perpetrator or perpetrators remain unknown. The Wall Street Bombing showed the destructive potential of a **vehicle-borne improvised explosive device**. This lesson would be repeated throughout the next century, as demonstrated by the acts of both domestic terrorists in the **Oklahoma City Bombing** and international groups in the **Bali Bombing**. It also showed the need to balance life-saving and recovery efforts with crime scene management such that evidence was identified and preserved.

WAR OF THE PACIFIC (1879–1884). Conflict involving Bolivia, Chile, and Peru, fought largely over control of **Chilean nitrates** and other valuable mineral resources, especially copper. The conflict arose from a dispute between Bolivia and Chile over control of the parts of the Atacama Desert that lies between the 23rd and 26th parallels on the Pacific coast of South America. This region is rich in naturally occurring inorganic compounds found in abundance in the Atacama Desert but scarce elsewhere across the globe. The 1857 invention by **Lammot du Pont** of a process for the direct use of **sodium nitrate** in making **black powder** vastly enhanced the economic importance of Chilean nitrates as a replacement for the more costly **India Saltpeter**. The national borders in the northernmost portions of the Atacama Desert were poorly defined among Bolivia, Chile, and Peru. Prior to the outbreak of hostilities, Chile had respected Bolivian claims to sovereignty over the Antofagasta Province along the Pacific coast in return for favorable tax treatment for the Chilean Antofagasta Nitrate Company. In 1878, Bolivia levied tax increases on this company, abrogating an 1874 agreement to take no such action for 25 years. Angered by this step and by Bolivian threats to confiscate the company's property, Chile occupied the port city of Antofagasta by force on 14 February 1879. Bolivia retaliated by declaring war on Chile and calling on Peru to come to its aid in fulfillment of a mutual-guarantee pact signed in 1873. On 5 April 1879, Chile formally declared war on both Bolivia and Peru.

The conflict between Chile and Peru lasted until 20 October 1883, when the two nations signed the Treaty of Ancón, by which Peru ceded its southernmost territory, Tarapacá Province, to Chile. However, diplomatic disputes between Chile and Peru continued, chiefly over the occupation of the provinces of Tacna and Arica. These disputes, known as the Question of the Pacific, were finally settled in 1929 through the good offices of the United States. The settlement allowed Chile to keep Arica, while Peru regained Tacna and received other considerations. The conflict between Bolivia and Chile continued until 1884, when a truce was declared. The terms of the truce were costly to Bolivia. It gave up its claims to the Antofagasta Province, thereby losing direct access to the Pacific Ocean and all the mineral industries in the coastal regions. A 1904 treaty made the terms of the truce permanent. By the terms of its treaties with Peru and Bolivia, Chile clearly achieved strategic victories through the War of the Pacific. The economic importance of its newly acquired territories in the Atacama Desert continued into the early decades of the twentieth century, fueled by demands for Chilean nitrates for use in explosives and fertilizers. The invention and industrial development of **nitrogen fixation** in the decade after the First World War ultimately would diminish the importance of Chilean nitrates.

WE.177A. Designation for a tactical **nuclear weapon** fielded by the UK. The WE.177A had a **boosted fission design** and was intended for use as an aircraft-delivered depth charge. The weapon weighed 600 pounds and was slightly over 9 feet long and 1 foot in diameter. It featured a variable yield of either 0.5 or 10 kiloton **TNT-equivalent**. The lower-yield weapon was to be used in shallow waters or in deep ocean water where other shipping might be at risk, while the higher-yield was for deep ocean waters deemed free of risk to other shipping. First delivery of the WE.177A was in 1969. The weapon was retired from the Royal Navy in 1992 and the Royal Air Force in 1998. Two similar weapons were fielded: WE.177B and WE.177C, although neither was intended to be used as a depth charge. *See also* LULU.

WEAPONS-GRADE MATERIAL. Those forms of **special nuclear material (SNM)** are most suitable for direct use in a **nuclear weapon**. Weapons-grade material usually connotes either high-purity **plutonium** whose isotopic content is greater than 90 percent plutonium-239 and less than 7 percent plutonium-240 or high-purity **uranium** enriched to at least 90 percent uranium-235.

WEINDL, CASPAR or GASPAR or KASPAR. Caspar Weindl is widely credited as being the first person to use **gunpowder** as a **blasting agent** for a successful underground blast. Documents attribute a blast on 8 February 1627 at the Oberbieberstollen mine, Schemnitz, Hungary (now Banská Štiavnica, Slovakia) to Weindl. This event was followed eight days later by a public demonstration of the use of gunpowder as a blasting agent. News of this event apparently was publicized widely, as the use of gunpowder in mining operations became commonplace during the seventeenth century. Biographical information on Weindl is lacking.

WETTER DYNAMITE. A group of **permissibles** incorporating metal salts having waters of hydration. The term derives from *Schlagwettersichere*, a German term meaning "safety from **firedamp**." The search for explosives that would be safe to use in firedamp-laden atmospheres led to the development in the 1880s of **nitroglycerin**-based explosives having salts that naturally contain "waters of hydration," such as sodium carbonate decahydrate ($Na_2CO_3 \cdot 10H_2O$) and magnesium sulfate heptahydrate ($MgSO_4 \cdot 7H_2O$). The addition of the hydrated salts lowered the temperature of the explosion, reducing the possibility of an unplanned, secondary explosion.

WHITE GUNPOWDER. An **explosive** used as **gunpowder**. On 30 March 1860, J. Davies received British patent 824 covering a gunpowder composed of 47 percent **potassium chlorate**, 23.5 percent potassium ferrocyanide, 12 percent sugarloaf, 12 percent sugar crystals, and 5.5 percent **sulfur**. As its common name suggests, the mixture was white. Sugarloaf and sugar crystals were **fuel**. The presence of sugar in this **explosive** from 1860 provides an early example of the use of readily available materials that would be exploited in the next century in the form of **icing sugar** in preparing **homemade explosives**. *See also* SCHULTZE POWDER.

WHITE POWDER. *See* POUDRE B.

WILBRAND, JULIUS BERNHARD FRIEDRICH ADOLPH (1839–1906). German chemist credited with the original synthesis of **TNT**. Wilbrand was searching for novel synthetic dyes when he prepared TNT by treating toluene with nitric acid. His search appears to have been inspired by the fame and fortune achieved by William Henry Perkins (1838–1907) from his accidental discovery in 1858 of mauve, the first synthetic dye. That discovery touched off keen interest by the public in mauve-colored clothing and furnishings as well as a frenzied search by chemists for other dyes. Wilbrand most likely produced a crude mixture of isomers of **trinitrotoluene**.

He reported his synthesis in the scientific literature in 1863, noting its yellow color. His discovery saw limited use as a yellow dye owing to its limited stability. Wilbrand, too, failed in this report to note the **explosive** nature of TNT. That property was noted by another German scientist, **Carl Häussermann**, and was reported in the scientific literature in 1889.

WINANS STEAM GUN. *See* DICKINSON, CHARLES S. (1809–1874); STEAM GUN.

WOLFSBERG, WILHELM LENK VON. *See* LENK VON WOLFSBERG, WILHELM, FREIHERR (1809–1894).

WOOLWICH METHOD. *See* RDX TYPE A.

WORLDWIDE NUCLEAR DETONATIONS. The number of **detonations** of commercial **explosives** and conventional military explosives is beyond reckoning. The number of detonations of **nuclear weapons**, however, is sufficiently small that it can be estimated. That number is on the order of 2,050–2,100 detonations that have taken place worldwide. A precise figure is lacking owing to, among other things, uncertainty over detonations that may have occurred prior to modern methods of detecting such detonations. The issue is especially challenging for detonations where **preinitiation** dramatically reduces the yield and, hence, its seismic signature. There are also debates over what qualifies as an individual test. Various sources give a figure of 2,075 detonations, which are distributed as follows:

- United States—first detonation on 16 July 1945, with 1,032 detonations in total, inclusive of the two weapons exploded in combat, **Little Boy** and **Fat Man**. The first successful test of a hydrogen bomb took place on 11 November 1952, when the United States conducted a test "**Ivy Mike**." The last detonation conducted by the United States was on 23 September 1992.
- Soviet Union and Russia—first detonation on 29 August 1949 (**Joe-1**), with 715 detonations in total. Among these detonations is **Tsar Bomba**, the largest nuclear device ever detonated. It was exploded by the Soviet Union on 30 October 1961, with a yield estimated at more than 50 megatons **TNT-equivalent**.
- UK—first detonation on 3 October 1952, with 57 detonations in total, including 24 joint UK-US detonations. Its first detonation was codenamed Operation **Hurricane**.

- France—first detonation on 13 February 1960 (**Gerboise Bleue**), with 210 detonations in total.
- **China**—first detonation on 16 October 1964, with 45 detonations in total. China conducted the last known atmospheric detonation by any nation on 16 October 1980.
- **India**—first detonation on 18 May 1974 (**Smiling Buddha**), with six detonations in total.
- **Pakistan**—first detonation on 28 May 1998, with six detonations in total.
- **North Korea**—first detonation on 9 October 2006, with six detonations in total. On 3 September 2017, North Korea conducted the most recent detonation of any nation.

Of these 2,075 detonations, approximately two-thirds have been conducted underground, with the remainder either in the atmosphere or underwater.

While some sources attribute **Israel** with having a nuclear weapons program, it is not known to have conducted any test detonations. Furthermore, the existence of such a program is unverified. **South Africa** admitted to having possessed nuclear weapons but never conducted any test detonations. *See also* GADGET; KHAN, ABDUL QADEER (1935–2021).

WRIGHT, EDWIN GOODE. *See* GOODE WRIGHT, EDWIN (1783–1859).

WRIGHT, GEORGE F. *See* BACHMANN PROCESS; HMX.

X

XYLOIDINE. Name given by **Henri Braconnot** to the sticky mass obtained from boiling the product resulting from the mixing of nitric acid and either potato starch or a cellulosic material such as sawdust. Braconnot reported that he coined the name from the Greek ξυλον (*wood*). The name came to be used as a general term for both **nitrocellulose** and **nitrostarch**, although nitrocellulose was more often the material being studied.

Z

ZALINSKI, EDMUND LOUIS GRAY (1840–1909). Polish-born American military officer and inventor closely associated with the development of the **pneumatic gun**. In 1883, Zalinski, then a captain in the US Army, witnessed a successful demonstration of a small version of a pneumatic gun. Thereafter, he became a major force in promoting the development and adoption of larger-scale pneumatic guns, primarily for use as coastal defense artillery against ships. His advocacy for this type of gun caused it to be dubbed "Zalinski's dynamite gun," and he was often credited incorrectly as its inventor. However, Zalinski was an inventor in his own right, with his inventions including an entrenching tool, a ramrod bayonet, and a telescopic sight for use with artillery.

Bibliography

CONTENTS

I. INTRODUCTION

The information contained in this work is taken from unclassified primary sources, other authoritative published sources, and Internet resources, such as websites. The sensitivity of the subject area demands that the information presented in this work as well as the sources cited fall short of giving stepwise instructions for acquiring, handling, making, or using explosives. Otherwise, the information or resources might be used by individuals, groups, and other nonstate actors whose aims may be to cause terror or whose interests may be for illegal or illicit purposes. Tragic events such as the 1995 bombing in Oklahoma City and the 7 July 2005 bombings in London, England, point to the relative ease with which common chemicals have been used, either directly or as precursors. The search for less expensive and more effective explosive powders during the mid-nineteenth century illustrates the extent to which readily accessible materials might be used, including flour, moss, sawdust, and sugar.

This section presents an overview of resources regarding conventional explosives and nuclear explosives. The resources are divided further into historical works, technical literature, and biographies and memoirs. The division among these three groups can appear arbitrary. Many historical works present technical information regarding conventional and nuclear explosives. Some technical literature captures historical events, episodes, and personalities. Various biographies and memoirs recount the historical context as well as the technical accomplishments associated with an individual or group.

II. CONVENTIONAL EXPLOSIVES

Conventional explosives can be divided into three groups for purposes of considering the available literature and accessible online resources. First, there are explosives developed expressively for use by the military, such as Cordite, HMX, Lyddite, and RDX. Second, there are explosives originally developed and used for legitimate commercial purposes, although they might have been used later by the military. Examples include dynamite, rack-a-rock, and TNT. The third group is that of chemicals of commerce that have been diverted for illicit or illegal uses. Examples include ammonium nitrate, concentrated hydrogen peroxide, and icing sugar as well as the mining explosives exploited for use in the 2004 Madrid train bombings. Excluded from this section are nuclear explosives, which are treated in Section III.

Many historical works on military explosives and commercial explosives are accessible in book-form online. The US Army, through its Army Center of Military History (www.history.army.mil), has an extensive collection of historical information as well as the magazine *Army History* available at no cost. The US Navy, through its Naval History and Heritage Command (www.history.navy.mil), is similar in concept, albeit oriented toward naval matters. Topics covered by these centers are broader than military explosives, but the centers provide a useable starting point for further reading and research. Other online resources that cover military explosives occasionally are subscription- or fee-based. Historical accounts of events involving military, commercial, and homemade explosives are also presented in book-form as biographies and memoirs (see Section IV).

Some books from the late nineteenth and early twentieth centuries on the explosives industry are available freely online. The information they provide might be dated and, thus, is less useful from a technical perspective. But such books can prove useful in learning historical details overlooked in more modern treatments. Examples include *The Manufacture of Explosives. A Theoretical and Practical Treatise on the History, the Physical and Chemical Properties, and the Manufacture of Explosives* (1895); *Explosives. Their Manufacture, Properties, Tests and History* (1915); and *History of the Explosives Industry in America* (1927). Despite the title of this last book, its coverage includes Canada and Mexico, although in less depth than that devoted to the United States. It is especially useful regarding the men and, occasionally, women, who oversaw research, development, and production of explosives in North American from the earliest explosives industries in North America through the first quarter of the twentieth century.

A challenge regarding availability of technical information on all three groups of explosives is to balance the information such that it might prove useful to specialists while withholding details that might otherwise prove helpful to those seeking to make explosives for illegal or illicit purposes. An example of this challenge is found with Technical Manual 31-210, *Improvised Munitions Handbook*. This handbook was issued in 1969 to special operations personnel in the US Army, which was then engaged in the conflict in Viet Nam. It became widely available, though, and the information was used to create booby traps and improvised explosive devices (IEDs) that were used against US forces and its allies fighting in Viet Nam. The handbook is reported to have been found in more recent times in locations previously used by terrorists or by other hostile groups.

Reducing the Threat of Improvised Explosive Device Attacks by Restricting Access to Explosive Precursor Chemicals, prepared by the Board on Chemical Sciences and Technology, US National Academies of Science, Engineering, and Medicine, is available freely online (www.nap.nationalacademies.org). It offers both historical and technical information regarding the use of commercially available chemicals to make explosives, while withholding details that might enable the information to be misused.

A. Historical

Andrade, Tonio. *The Gunpowder Age: China, Military Innovation, and the Rise of the West in World History.* Princeton, NJ: Princeton University Press, 2016.

Anon. "Colonel Pasley's Operations at Spithead, Against the Royal George." *Hampshire Telegraph and Naval Chronicle* September 30, 1839, 4

Armstrong, Richard L., and Marion O. Smith. *"They Also Served." The Confederate Niter & Mining Bureau. District 4½—Staunton, Va.* Buena Vista, VA: Mariner Media, 2015.

Baer, F. H. "Lieutenant Very's Pistol." *American Rifleman* 141, no. 6 (June 1993): 40–3, 84.

Barthel, Thomas. *Opening the East River: John Newton and the Blasting of Hell Gate.* [Flood Rock Demolition] Jefferson, NC: McFarland, 2021.

Black, Andrew R. *Buried Dreams: The Hoosac Tunnel and the Demise of the Railroad Age.* Baton Rouge, LA: LSU Press, 2020.

Coulter, Matthew W. *The Senate Munitions Inquiry of the 1930s: Beyond the Merchants of Death.* Westport, CT: Praeger, 1997.

Crossland, James. "Radical Warfare's First 'Superweapon': The Fears, Perceptions, and Realities of the Orsini Bomb, 1858–1896." *Terrorism and Political Violence* 35, no. 2 (2023): 355–69.

Department of the Navy. "The Secretary of the Navy Exonerates 256 Defendants from 1944 Port Chicago General and Summary Courts-martial." July 17, 2024. [Press release].

Destrée, Jules, and Georges Clemenceau. "Britain Democratized under Lloyd George. The Munitions Crisis." In *Source Records of the Great War, Vol. 3*, edited by Charles F. Horn, 204–13. New York: National Alumni, 1923.

Dolmage, Victor. "Removal of Ripple Rock, Seymour Narrows, British Columbia." *Engineering Geology Case Histories* 5 (January 1964): 7–13.

Doss, Erika. "Commemorating the Port Chicago Naval Magazine Disaster of 1944: Remembering the Racial Injustices of the 'Good War' in Contemporary America." *ASJ Special Issue of American Studies Journal* 59, no. 6 (2015): 1–11.

Farcau, Bruce W. *The Ten Cents War: Chile, Peru, and Bolivia in the War of the Pacific, 1879–1884.* Westport, CT: Praeger, 2000.

Gold, Russell. *The Boom. How Fracking Ignited the American Energy Revolution and Changed the World.* New York: Simon & Schuster, 2014.

Hadden, R. Lee. "Saltpeter Mining in Sinnett Cave (Pendleton County, West Virginia) During the American Civil War." Presented at Joint Meeting of Northeastern Section and Southeastern Section, Geological Society of America, Washington, D.C., March 25, 2005.

Hamilton, John D. "Gardiner's Explosive Musket Shell." *American Society of Arms Collectors Bulletin* 120 (September 2019): 21–31.

Harding, Phillip. *The British Shell Shortage of the First World War.* Stroud, Gloucestershire: Fonthill Media, 2015.

Johnson, Jeffrey Allan, and Roy M. MacLeod, eds. *Frontline and Factory: Comparative Perspectives on the Chemical Industry at War, 1914–1924.* Dordrecht: Springer, 2014.

Kates, James. "Booming the Cutover: The Campaign for Explosives in the Wisconsin Forest, 1916–1928." *Historical Geography* 47 (2019): 166–93.

Kelly, Jack. *Gunpowder: Alchemy, Bombards, and Pyrotechnics: The History of the Explosive that Changed the World.* New York: Basic Books, 2005.

Kennedy, Donald R. *History of the Shaped Charge Effect: The First 100 Years.* Los Altos, CA: D. R. Kennedy & Associates, 1990.

Kim, David, Benjamin D. Mosher, Chet A. Morrison, Carol Parker-Lee, Razvan C. Opreanu, Penny Stevens, Sarah Moore, and John P. Kepros. "A Modern Analysis of a Historical Pediatric Disaster: The 1927 Bath School Bombing." *Journal of Surgical Research* 158, no. 2 (February 2010): 420–1.

Kochan, Michael P., and John C. Wideman. *Civil War Torpedoes: The History of Improvised Explosive Devices in the War Between the States.* Paoli, PA: Keystone Press, 2012.

Kurzer, Frederick. "Fulminic Acid in the History of Organic Chemistry." *Journal of Chemical Education* 77, no. 7 (July 2000): 851–7.

Lamb, John. *A Strange Engine of War: The "Winans" Steam Gun and Maryland in the Civil War*. Baltimore, MD: Chesapeake Book Company, 2011.

Leveau, A. "Pertuiset's Explosive Bullet. French Ammunition Patents of the 19th Century." *International Ammunition Association Journal* 444 (July/August 2005): 86–91.

Lieber, Francis. *Instructions for the Government of Armies of the United States in the Field.* General Order No. 100 – Adjutant-General's Office. New York: D. Van Nostrand, 1863.

McClintock, Robert Lyle. "The Destruction of Wire Entanglements—A Suggested Method." *The Royal Engineers Journal* XVII, no. 3 (March 1913): 129–38.

McLachlan, Sean, Gerry Embleton, and Sam Embleton. *Medieval Handgonnes: The First Black Powder Infantry Weapons*. Oxford: Osprey Publishing, 2010.

Merriman, John M. *The Dynamite Club: How a Bombing in fin-de-siècle Paris Ignited the Age of Modern Terror*. New Haven, CT: Yale University Press, 2016.

Multhauf, Robert P. "The French Crash Program for Saltpeter Production, 1776–94." *Technology and Culture* 12, no. 2 (April 1971): 163–81.

Needham, Joseph. *Science and Civilisation in China. Vol. 5, Chemistry and Chemical Technology. Part 7: Military Technology; The Gunpowder Epic*. New York: Cambridge University Press, 1986.

Oklahoma Department of Civil Emergency Management. *After Action Report: Alfred P. Murrah Federal Building Bombing April 19, 1995*. Stillwater, OK: Department of Central Services Printing Division, 1996.

Partington, James Riddick. *A History of Greek Fire and Gunpowder*. Baltimore: Johns Hopkins University Press, 1998.

Reinares, Fernando. *Al-Qaeda's Revenge. The 2004 Madrid Train Bombings*. New York: Columbia University Press, 2022.

Salecker, Gene E. *Destruction of the Steamboat* Sultana*: The Worst Maritime Disaster in American History*. Annapolis, MD: Naval Institute Press, 2022.

Stevens, William Harrison Spring. *The Powder Trust, 1872–1912*. South Yarra, Victoria Leopold Classic Library, 2015.

Taubman, Philip. "Explosive Bullet Struck Reagan, F.B.I. Discovers." *The New York Times*.April 3,1981, A-1.

Van Gelder, Arthur Pine, and Hugo Schlatter. *History of the Explosives Industry in America*. New York: Columbia University Press, 1927.

Walters, William. "A Brief History of Shaped Charges." In *24th International Symposium on Ballistics* 1 (September 22–26, 2008): 3–10.

Waters, William Davis, and Joseph I. Brown. *Gabriel Rains and the Confederate Torpedo Bureau*. El Dorado Hills, CA: Savas Beatie, 2017.

Williams, William B. *History of the Manufacture of Explosives for the World War. 1917–1918*. Philadelphia: U.S. Army Ordnance Department, 1920.

Winston, Rachel. "The Lake Denmark Explosion: An Extraordinary Mishap that Changed Military Safety Standards." *NJS: An Interdisciplinary Journal* 7, no. 1 (Winter 2021): 346–60.

B. Technical

A Guide for Explosion and Bombing Scene Investigation. NCJ 181869. Washington, DC: U.S. Department of Justice, June 2000.

Akhavan, Jacqueline. *The Chemistry of Explosives*, 4th ed. Cambridge: Royal Society of Chemistry, 2022.

Bachmann, Werner E., and John C. Sheehan. "A New Method of Preparing the High Explosive RDX." *Journal of the American Chemical Society* 71 (1949): 1842–5.

Bell, Jack. *Civil War Heavy Explosive Ordnance: A Guide to Large Artillery Projectiles, Torpedoes, and Mines.* Denton: University of North Texas Press, 2003.

Bergman, Yoel. "Union and Confederate Views on Guncotton." *Vulcan* 3 (2015): 93–109.

Beveridge, Alexander. *Forensic Investigation of Explosions.* London: Taylor & Francis Ltd., 1998.

Board on Chemical Sciences and Technology. *Reducing the Threat of Improvised Explosive Device Attacks by Restricting Access to Explosive Precursor Chemicals.* Washington, DC: The National Academies Press, 2018.

Brown, David K., and Iain McCallum. "Ammunition Explosions in World War I: A Re-Examination of the Evidence." *Warship International* 38, no. 1 (2001): 58–69.

Carlson, Bob. "The Jacob Rifle and Its Exploding Projectile." *American Society of Arms Collectors* 120 (September 2019): 1–7.

Cestkowski, Gerald P. *American Dynamite: An Illustrated History.* Menlo Park, CA: Schiffer Publishing, 2024.

Cook, Melvin A. *The Science of High Explosives.* Malabar, FL: Robert E. Krieger Publishing Co., 1985.

Cooper, Paul W. *Explosives Engineering.* New York: Wiley-VCH, 1997.

Cooper, Paul W., and Stanley R. Kurowski. *Introduction to the Technology of Explosives.* New York: Wiley-VCH, 1997.

Dastidar, Ashok Ghose. "This Is Not 'Dust,' So I Don't Have a Problem." *Process Safety News* 19, no. 3 (March 2012): 1, 4–5.

Davis, Tenney L. *The Chemistry of Powder and Explosives.* Hollywood, CA: Angriff Press, 1972.

Delannoy, Yann, Tania Delabarde, Isabelle Pu, Laurence Legrand, Marc Taccoen, Antoine Tracqui, and Bertrand Ludes. "Terrorist Explosive Belt Attacks: Specific Patterns of Bone Traumas." *International Journal of Legal Medicine* 133, no. 2 (March 2019): 565–9.

Department of the Army. *Engineering Design Handbook. Explosives Series. Properties of Explosives of Military Interest.* AMC Pamphlet 706–177. Washington, DC: Army Materiel Command, January 1971.

Department of the Army. *Improvised Munitions Handbook.* TM 31–210. Washington, DC: Department of the Army, 1969.

Department of the Army. *Military Explosives.* U.S. Army and U.S. Air Force Technical Manual TM 9-1300-214. Washington, DC: Department of the Army, September 1984.

E. I. Du Pont de Nemours and Company. *Farmers' Hand Book of Explosives: Instructions in the Use of Explosives for Clearing Land, Planting and Cultivating Trees,*

Drainage, Ditching, Subsoiling, and Other Purposes. Whitefish, MT: Kessinger Publishing, 2010 (originally published in 1922).

Encyclopedia of Explosives and Related Items. Vols. 1–10. Dover, NJ: Picatinny Arsenal, U.S. Army Armament Research and Development Command, 1960–1983.

Field, Simon Quellen. *Boom! The Chemistry and History of Explosives.* Chicago: Chicago Review Press, 2017.

Guttmann, Oscar. *The Manufacture of Explosives. A Theoretical and Practical Treatise on the History, the Physical and Chemical Properties, and the Manufacture of Explosives.* London: Wittaker and Co., 1895.

Koch, Ernst-Christian. *High Explosives, Propellants, Pyrotechnics.* Berlin: De Gruyter, 2021.

Le Conte, Joseph. *Instructions for the Manufacture of Saltpeter.* Columbia, SC: Charles P. Pelham, 1862.

Marshall, Arthur. *Explosives. Their Manufacture, Properties, Tests and History.* Philadelphia: P. Blakiston's Son & Co., 1915.

Martin, Geoffrey, and William Barbour. *Manuals of Chemical Technology – III. Industrial Nitrogen Compounds and Explosives.* London: Crosby Lockwood and Son, 1915.

Matyás, Robert, and Jîrî Pachman. *Primary Explosives.* Heidelberg, Germany: Springer-Verlag, 2013.

Meyer, Rudolf, Josef Köhler, and Axel Homburg. *Explosives.* 6th ed. New York: Wiley-VCH, 2007.

Mocella, Chris, and John A. Conkling. *Chemistry of Pyrotechnics. Basic Principles and Theory.* 3rd ed. Boca Raton, FL: CRC Press, 2018.

Mowbray, George M. *Tri Nitro-Glycerin, as applied in the Hoosac Tunnel, Submarine Blasting, etc., etc., etc.* North Adams, MA: James T. Robinson & Son, Printer, 1872.

Price, David James. *Dust Explosions: Theory and Nature of, Phenomena, Cause, and Methods of Prevention.* Madrid, Spain: Hardpress Publishing, 2012.

Schindler, Dietrich, and Jirí Toman, eds. *The Laws of Armed Conflicts: A Collection of Conventions, Resolutions, and Other Documents.* 4th ed. Leiden, The Netherlands: Brill Publishers, 2004.

Schlit, Alfred A. *Perchloric Acid and Perchlorate.* Columbus, OH: The G. Frederick Smith Chemical Company, 1979.

Smith, G. Carlton. *TNT. Trinitrotoluenes and Mono- and Dinitrotoluenes. Their Manufacture and Properties.* New York: Van Nostrand, 1918.

Thomas, Dean S. *Confederate Arsenals, Laboratories, and Ordnance Depots.* 3 vols. Gettysburg, PA: Thomas Publications, 2014.

Urbanski, Tadeusz. *Chemistry and Technology of Explosives, Vol. 1.* Translated by M. Jurecki. Amsterdam: Elsevier, 1963.

Urbanski, Tadeusz. *Chemistry and Technology of Explosives, Vol. 2.* Translated by Wladyslaw Ornaf and Sylvia Laverton. New York: Pergamon Press, 1965.

Urbanski, Tadeusz. *Chemistry and Technology of Explosives, Vol. 3.* Translated by M. Jurecki. Amsterdam: Elsevier, 1983.

III. NUCLEAR EXPLOSIVES

The historical literature for the efforts during the Second World War to develop nuclear weapons is both extensive and readily available in books, scholarly articles, biographies, and memoirs. An excellent starting point for research into those efforts is the co-called Smyth Report (Smyth, Henry DeWolf. *Atomic Energy for Military Purposes: The Official Report on the Development of the Atomic Bomb Under the Auspices of the United States Government, 1940–1945*). Originally issued in 1945, it has been reprinted and re-issued most recently in 1990. Historical literature on developments following the Second World War is less extensive but impressive in its breath of coverage, even if the reporting is difficult to verify independently. Examples of published profiles of national programs to develop nuclear weapons include ones in English on China (1991), India (1998), Iran (2021), Israel (1998), North Korea (2020), Pakistan (2012), South Africa (2015), and Taiwan (2018). The successful penetration by the Soviet Union of the Manhattan Project has received attention and is recounted in *Mortal Crimes: The Greatest Theft in History: The Soviet Penetration of the Manhattan Project*, as well as in assorted biographies and memoirs of Soviet scientists. Richard Rhodes four-volume series *The Making of the Nuclear Age* is accessible, authoritative, detailed, and entertaining.

Unclassified technical data are available and accessible on early nuclear weapon designs, developments, testing, and stockpiles. For example, John Coster-Mullen's *Atom Bombs: The Top Secret Inside Story of Little Boy and Fat Man* provides extensive details of both of these nuclear weapons. Technical data on more recent nuclear weapon designs are limited owing to their sensitivity and to concerns over misuse of the information. Technical literature on nuclear weapon stockpiles and testing is available. In this regard, the four-volume Nuclear Weapons Databooks provide useful, unclassified information.

The history of the nuclear weapon programs of the United Kingdom and the United States is amply documented and accessible. This documentation includes the formative period when the scientific staff on what became the Manhattan Project was multinational, from Canada, France, and Italy in particular. Information has emerged in English and is accessible regarding the nuclear weapon program during the Soviet era, attributable in measure to a relaxation in secrecy after the collapse of the Soviet Union. Renewed secrecy, however, is limiting accessibility to information on the Russian nuclear weapons program. Historical accounts of the nuclear weapons programs in China, India, Iran, Israel, North Korea, Pakistan, and South Africa have appeared, although in most cases the information lacks full validation.

The IAEA, the "Atoms for Peace" arm of the UN, has a website providing unlimited access to scientific and technical publications, guidance documents, and databases related to nuclear energy (www.iaea.org). Some of this information bears on issues related to nuclear weapons, particularly the safeguards program, whereby nations comply with an inspection process aimed at verifying that their nuclear programs are strictly for peaceful purposes. Additionally, the IAEA has been a leader in the emerging discipline of nuclear forensics. That discipline seeks to enhance the ability to detect nuclear materials out of regulatory control, to identify the pathway taken by materials from the point where control was lost to where it was interdicted, to prosecute parties responsibilities for theft or diversion, and to identify areas where safeguards might be enhanced to prevent further theft or diversion. An example of a nuclear forensics guidance document prepared through the IAEA is the 2015 publication *Nuclear Forensics in Support of Investigations*.

The Comprehensive Nuclear-Test-Ban Treaty bans nuclear detonations anywhere and by anyone. The treaty has yet to enter into force (EIF). However, in anticipation of EIF, the Preparatory Commission for the Comprehensive Nuclear-Test-Ban Treaty Organization has been established. Its website, accessible at www.ctbto.org, provides historical and technical information. Examples include the history of nuclear weapon testing, types of nuclear weapons, and effects of nuclear weapon testing. Coverage includes the nuclear weapon programs of China, France, the Soviet Union, the United Kingdom, and the United States. No information is available from this website on the nuclear testing programs of India, North Korea, or Pakistan, nor is any information regarding the nuclear testing program of Israel, if such a program exists.

The US Department of Defense (DoD) plays an important role in maintaining nuclear forces and protecting against nuclear attack. The primary, cognizant office within the DoD is that of the assistant secretary of defense for nuclear, chemical, and biological defense programs. This office is the DoD focal point for the modernization and sustainment of US stockpile of nuclear weapons. One of its activities is publishing the *Nuclear Matters Handbook*, which provides an overview of the US nuclear deterrence. The handbook is updated periodically and is accessible at www.acq.osd.mil/ncbdp/nm/NMHB.

The US Department of Energy (DoE) is an essential part of the US nuclear weapon program, with much of the responsibility for nuclear matters residing with its National Nuclear Security Administration (NNSA). Historical and technical information might be found on and obtained from the internet library of the NNSA. It is accessible at http://nnsa.energy.gov/library. In addition to the NNSA, the DoE operates assorted national laboratories, most often under government-owned/contractor-operated arrangements. Several of these laboratories grew out of the Manhattan Project and provide assorted historical and technical information on their role via their websites. Argonne National Laboratory began as the Metallurgical Laboratory of the Manhattan Project (www.anl.gov). The history of Oak Ridge National Laboratory includes its work during the Manhattan Project to separate uranium isotopes (www.ornl.gov). Its beginnings are described in *City Behind a Fence: Oak Ridge, Tennessee, 1942–1946*. Los Alamos National Laboratory was established as Site Y and given the task of designing and building nuclear weapons (www.lanl.gov). Pacific Northwest National Laboratory can trace its history to the Hanford Site, where plutonium was prepared for use in Fat Man and other nuclear weapons (www.pnnl.gov). Its story is given in part in *Atomic Frontier Days: Hanford and the American West*.

The U.K. Atomic Weapons Establishment works for the Ministry of Defence via a government-owned/contractor-operated arrangement. As such, it serves as a useful link to the U.K. nuclear weapons program. A four-page document on its website outlines the classes of programmatic areas and historical information associated with this program (www.awe.co.uk/document-category/historical-documents). None of this information is accessible from the website, but the document gives guidance on seeking access through the U.K. National Archives (www.nationalarchives.gov.uk), and this guidance might facilitate inquires.

The websites of nongovernmental organizations afford an accessible means of locating historical and technical information on nuclear explosives. Four such websites devoted primarily or largely to nuclear issues may prove especially useful. The Atomic Heritage Foundation is dedicated to the preservation and interpretation of the Manhattan Project and subsequent nuclear-related events. The Foundation hosts the National Museum of Nuclear Science and History. The website for this museum, ahf.nuclearmuseum.org, provides access to the Foundation's extensive collection of oral histories, interpretive vignettes, and articles about the Manhattan Projects and its legacy. A second website is that of the Bulletin of the Atomic Scientists (www.thebulletin.org). It provides access to its monthly journal, as well as an archive. Both the journal and the archive include information on topic related to nuclear weapons. A potentially useful feature is labeled "Nuclear Notebooks," which gives data on the worldwide arsenals of nuclear weapons. Finally, the Institute for Science and International Security (ISIS) focuses on nuclear security, largely but not solely through the analysis of findings put forward by the IAEA. Two examples of their work are cited here: *Taiwan's Former Nuclear Weapons Program: Nuclear Weapons On-Demand*, published in 2018, and *Iran's Perilous Pursuit of Nuclear Weapons*, from 2021. Similar in-depth studies of other national nuclear programs as well as of other issues related to nuclear proliferation are accessible through the ISIS website (www.isis.org). Finally, the website of the Wisconsin Project on Nuclear Arms Control (www.wisconsinproject.org) provides assorted materials bearing on nuclear weapons and, especially, the export controls that might hamper nuclear weapon development by non-nuclear weapon states and nonstate actors. These online materials include a primer on the steps and materials needed to make a nuclear weapon, a description of weapon designs, and a history of nuclear weapon tests. It hosts a website section devoted to monitoring Iran's nuclear program, noting the entities contributing to that program as well as the international efforts to impede Iran's progress toward acquiring a functional nuclear device.

A. Historical

Abraham, Itty. *The Making of the Indian Atomic Bomb: Science, Secrecy, and the Postcolonial State.* New York: Zed Books, 1998.

Albright, David, and Sarah Burkhard. *Iran's Perilous Pursuit of Nuclear Weapons.* Washington, DC: Institute for Science and International Security, 2021.

Albright, David, and Andrea Stricker. *Taiwan's Former Nuclear Weapons Program: Nuclear Weapons On-Demand.* Washington, DC: Institute for Science and International Security, 2018.

Aylen, Jonathan. "First Waltz: Development and Deployment of Blue Danube, Britain's Post-War Atomic Bomb." *The International Journal for the History of Engineering & Technology* 85, no. 1 (January 2015): 31–59.

"Baby Tooth Survey—First Results." 1961. *Nuclear Information* 4, no. 1 (November 1961): 1–6.

Bickel, Lennard. *The Deadly Element: The Story of Uranium.* New York: Stein and Day Publishers, 1979.

Clark, Ronald W. *The Birth of the Bomb: The Untold Story of Britain's Part in the Weapon that Changed the World.* London: Phoenix House, 1961.

Cohen, Avner. *Israel and the Bomb.* New York: Columbia University Press, 1998.

Corera, Gordon. *Shopping for Bombs: Nuclear Proliferation, Global Insecurity, and the Rise and all of the A.Q. Khan Network.* Oxford: Oxford University Press, 2009.

Coster-Mullen, John. *Atom Bomb. The Top Secret Inside Story of Little Boy and Fat Man.* Waukesha, WI: J. Coster-Mullen, 2002.

Findley, John M., and Bruce Hevly. *Atomic Frontier Days: Hanford and the American West.* Seattle: University of Washington Press, 2011.

Goetz, Peter A. *A Technical History of America's Nuclear Weapons. Volume I: Introduction and Developments to 1960.* 2nd ed. Seattle, Washington, DC: Amazon Digital Services, 2020.

Goetz, Peter A. *A Technical History of America's Nuclear Weapons. Volume II: Introduction and Developments from 1960 through 2020.* 2nd ed. Seattle, WA: Amazon Digital Services, 2020.

Gosling. F. G. *The Manhattan Project: Making the Atomic Bomb.* Washington, DC: United States Department of Energy, 2001.

Graetzer, Hans G., and David L. Anderson. *The Discovery of Nuclear Fission: A Documentary History.* New York: Van Nostrand Reinhold, 1971.

Groves, Leslie R. *Now It Can Be Told: the Story of the Manhattan Project.* New York: Harper, 1962.

Hawkins, David. *Manhattan District History: Project Y, the Los Alamos Project, Vol. 1. Inception until August 1945.* LAMS-2532. Washington, DC: Department of Commerce, 1961.

Jackson, Charles O. *City Behind a Fence: Oak Ridge, Tennessee. 1942–1946.* Knoxville, TN: University of Tennessee Press, 1981.

Kaplan, Fred. *The Bomb: Presidents, Generals, and the Secret History of Nuclear War.* New York: Simon & Schuster, 2021.

Khan, Feroz Hassan. *Eating Grass: The Making of the Pakistani Bomb.* Stanford, CA: Stanford University Press, 2012.

Latypov, Nikolai V., Jan Bergman, Abraham Langlet, Ulf Wellmar, and Ulf Bemm. "Synthesis and Reactions of 1,1-Diamino-2,2-Dinitroethylene." *Tetrahedron* 54, no. 38 (September 17, 1998): 11525–36.

Lewis, Flora. *One of our H-bombs is Missing.* [1966 Palomares Incident] New York: Bantam Books, 1987.

Lewis, John Wilson, and Xue Litai. *China Builds the Bomb.* Stanford, CA: Stanford University Press, 1991.

Panda, Ankit. *Kim Jong Un and the Bomb. Survival and Deterrence in North Korea.* London: Hurst, 2020.

Rhodes, Richard. *The Making of the Nuclear Age: Vol. 1. The Making of the Atomic Bomb: 25th Anniversary Edition.* New York: Simon & Schuster, 2012.

Rhodes, Richard. *The Making of the Nuclear Age: Vol. 2. Dark Sun: The Making of the Hydrogen Bomb.* New York: Simon & Schuster, 1995.

Rhodes, Richard. *The Making of the Nuclear Age: Vol. 3. Arsenals of Folly: The Making of the Nuclear Arms Race.* New York: Alfred A. Knopf, 2007.

Rhodes, Richard. *The Making of the Nuclear Age: Vol. 4. The Twilight of the Bombs.* New York: Alfred A. Knopf, 2010.

Szasz, Ferenc Morton. *British Scientists and the Manhattan Project: The Los Alamos Years.* New York: St. Martin's Press, 1992.

Szasz, Ferenc Morton. *The Day the Sun Rose Twice: The Story of the Trinity Site Nuclear Explosion, July 16, 1945.* Albuquerque: University of New Mexico Press, 1984.

Von Wielligh, Nic, and Lydia Von Wielligh-Steyn. *The Bomb: South Africa's Nuclear Weapons Programme.* Pretoria, South Africa: Litera, 2015.

West, Nigel. *Mortal Crimes: The Greatest Theft in History: The Soviet Penetration of the Manhattan Project.* New York: Enigma Books, 2004.

B. Technical

Cochran, Thomas B., William M. Arkin, and Milton M. Hoenig. *U.S. Nuclear Forces and Capabilities*. Nuclear Weapons Databook, Vol. 1. Cambridge, MA: Ballinger, 1984.

Cochran, Thomas B., William M. Arkin, and Milton M. Hoenig. *U.S. Nuclear Warhead Production*. Nuclear Weapons Databook, Vol. 2. Cambridge, MA: Ballinger, 1987.

Cochran, Thomas B., William M. Arkin, Robert S. Norris, and Jeffrey I. Sands. *Soviet Nuclear Weapons*. Nuclear Weapons Databook, Vol. 4. New York: Harper & Row, 1989.

Cochran, Thomas B., William M. Arkin, Robert S. Norris, and Milton M. Hoenig. *U.S. Nuclear Warhead Facility Profiles*. Nuclear Weapons Databook, Vol. 3. Cambridge, MA: Ballinger, 1987.

Department of Energy. *Explosives Safety*. DOE-STD-1212-2019. Washington, DC: National Nuclear Security Administration, November 29, 2019.

Diehl, Sarah J., and James Clay Moltz. *Nuclear Weapons and Nonproliferation: A Reference Handbook*. Contemporary World Issues. Santa Barbara, CA: ABC-CLIO, 2007.

Gibson, James N. *Nuclear Weapons of the United States: An Illustrated History*. Atglen, PA: Schiffer Publishing, 1996.

Hoddeson, Lillian, Paul W. Henriksen, Roger A. Meade, and Catherine Westfall. *Critical Assembly: A Technical History of Los Alamos during the Oppenheimer Years, 1943–1945*. Cambridge: Cambridge University Press, 1993.

International Atomic Energy Agency. *Nuclear Forensics in Support of Investigations*. IAEA Nuclear Security Series, No. 2-G (Rev. 1). Vienna: International Atomic Energy Agency, 2015.

Jacobsen, Annie. *Nuclear War: A Scenario*. New York: E. P. Dutton, 2024.

Mangano, Joseph J., and Janette D. Sherman. "Elevated in Vivo Strontium-90 from Nuclear Weapons Test Fallout among Cancer Decedents: A Case-Control Study of Deciduous Teeth." *International Journal of Health Sciences* 41, no. 1 (January 2011): 137–58.

Moody, Kenton J., Patrick M. Grant, and Ian D. Hutcheson. *Nuclear Forensics Analysis*, 2nd ed. Boca Raton, FL: CRC Press, 2014.

Norris, Robert S., Andrew S. Burrows, and Richard W. Fieldhouse. *British, French, and Chinese Nuclear Weapons*. Nuclear Weapons Databook, Vol. 5. Boulder, CO: Westview Press, 1994.

Norris, Robert S., and Hans M. Kristensen. "The British Nuclear Stockpile, 1953–2013." *Bulletin of the Atomic Scientists* 69, no. 4 (2013): 69–75.

Podvig, Pavel L., ed. *Russian Strategic Nuclear Forces*. Cambridge, MA: The MIT Press, 2004.

Reed, Bruce Cameron. *The Physics of the Manhattan Project*, 4th ed. New York: Springer, 2021.

Siracusa, Joseph M. *Nuclear Weapons: A Very Short Introduction*, 3rd ed. Oxford: Oxford University Press, 2020.
Smyth, Henry DeWolf. *Atomic Energy for Military Purposes: The Official Report on the Development of the Atomic Bomb Under the Auspices of the United States Government, 1940–1945*. [The Smyth Report.] Stanford, CA: Stanford University Press, 1945. Revised and reprinted in 1990.

IV. BIOGRAPHIES AND MEMOIRS

Biographies and memoirs are a rich source of historical information, often extending beyond the recitation of data on birth, death, education, and career. Such works might reveal the envies, jealousies, rivalries, tensions, misunderstandings, and outright deceptions among colleagues seemingly engaged in the same endeavor. A prime example of the intense competition and bitter disputes is found in the writings of the scientists and engineers who contributed to the Manhattan Project. The title of Bertrand Goldschmidt's memoir speaks bluntly to this situation: *Atomic Rivals. A Candid Memoir of Rivalries Among the Allies over the Bomb*. Similarly, Leo Szilárd sought to ensure his version of events was memorialized through *Leo Szilárd: His Version of the Facts—Recollections and Correspondence*. The decades following the Manhattan Project witnessed many other biographies and memoirs by the participants themselves. Their children have sought to ensure the contributions of their parents receive proper attention, with one example being the 2022 release of *Wilhelm's Way: The Inspiring Story of the Iowa Chemist Who Saved the Manhattan Project*, telling the story of Iowa State College chemistry professor Harley Wilhelm as told by his daughter Teresa.

Similar rivalries and score-settling existed among scientists outside the Manhattan Project. An example can be found in the German-language biography of Fritz Haber by Dietrich Stoltzenberg. Haber, best known in the context of explosives for his Nobel Prize-winning invention of the means to synthesize ammonia directly from the nitrogen in the air, was instrumental to the German chemical weapons program during the First World War. Stoltzenberg relates how Haber's colleague Hermann Staudinger expressed displeasure with Haber's role in promoting chemical weapons. In response, Haber retaliated against Staudinger, dismissing his theory regarding polymers as nonsense and using his influence as a Nobel Laureate to keep Staudinger's work from gaining acceptance.

Some literature has emerged in English providing insights into the nuclear weapon programs in nations that were or remain closed to most outsiders. *Arzamas-16. Soviet Scientists in the Nuclear Age: A Memoir* was written by a husband-and-wife team of nuclear physicists who held senior positions in the closed city of Arzamas-16, hub for the Soviet nuclear weapon program. Their memoir relates problems with obtaining proper equipment for their work, despite the high priority assigned to it by the Soviet government. It also offers glimpses of the interference by the Soviet security apparatus as well as clever ways devised to counter it. In 2023, *Hinge Points. An Inside Look at North Korea's Nuclear Program* appeared in print, offering a rare look into a regime and program where success is rewarded with privileges and acclaim but where failure is dealt with harshly.

In general, there has been less attention paid to the contributions by women engaged in work related to both conventional explosives and nuclear explosives then to the contributions of male scientists and engineers. Memoirs and similar accounts by and about these women are rare. Exceptions exist, however, and can prove enlightening. The 1886 memoir *A Signal Success. The Life and Travels of Mrs. Martha J. Coston* in which Coston relates her struggles to gain acceptance for her work, most notably the refusal of the US government to recognize her contribution to the invention of Coston flares, issuing a patent solely to her by-then deceased husband. This refusal was addressed partly by the 2006 election of Martha Coston to the National Inventors Hall of Fame—an honor her husband has yet to receive.

Attention has been paid, too, to the munition workers whose task of filling shells during the First World War was done largely by women. Contemporary accounts of their work and, especially, their sacrifices are lacking, perhaps owing to wartime secrecy as well as their need to return to their pre-war occupations. But this group did garner scholarly and public attention eventually, beginning at least in the 1970s and continuing to the present day, as witnessed by a 2024 announcement of a memorial being erected to the Canary Girls at the site of a former shell-filling factory. *Their Day in the Sun: Women of the Manhattan Project*, published in 1999, and *The Girls of Atomic City: The Untold Story of the Women Who Helped Win World War II*, published in 2013, detail contributions by women engaged in work on nuclear explosives at the various sites of the Manhattan Project in the United States. Similar book-length literature in English of the work done by women in the explosive industries outside of the United Kingdom and the United States would likely be well-received.

Book-length English-language biographies are scarce for many of the people and programs associated with conventional explosives prior to the mid-twentieth century. A few of the most well-known individuals are exceptions—Fritz Haber, Lammot du Pont, Antoine Lavoisier, and Alfred Nobel—as well as the less well-known but fascinating Maxim brothers, Hiram and Hudson. Consequently, biographical information must be extracted from obituaries and tributes. These sources are useful but fall short of giving a broad picture of the struggles, hopes, professional conflicts, and triumphs of the individuals. Examples of such overlooked or understudied contributors to explosives include William Bickford, George Kistiakowsky, Charles Edward Munroe, Edward A. L. Roberts, and François Eugène Turpin. Book-length biographies of these individuals and others are deserved and would likely prove interesting.

Biographies or memoirs of another sort are those featuring a city or an area subjected to an explosives-related disaster. Halifax, Nova Scotia (*Shattered City: The Halifax Explosion and the Road to Recovery*) and Texas City, Texas (*City on Fire: The Explosion that Devastated a Texas Town and Ignited a Historic Legal Battle*) are two excellent examples. Each book captures events surrounding the explosion, the heroism of some who perished, and the consequences endured by survivors. The devastation and injuries suffered by victims of the nuclear weapons dropped on Hiroshima and Nagasaki have received attention, too, beginning shortly after the end of the Second World War and continuing to the present day. Examples include *Hiroshima*, John Hersey's 1946 classic account of the bombing and its aftermath, and *Hiroshima Nagasaki: The Real Story of the Atomic Bombings and Their Aftermath*, published in 2015. Finally, *Penniman: Virginia's Own Ghost City* is an unusual biography or memoir. It tells the story of a "company town" that owed its existence to the production of explosives to meet the needs of the military immediately prior to US entry into the First World War. When that need dropped off, the town was largely abandoned by both the producer (du Pont) and the US military. By June 1919, fewer than 100 residents remained out of a peak population in excess of 10,000.

A. Conventional Explosives

Adler, Cyrus. "Edmund Louis Gray Zalinski." *American Jewish Historical Society* 19 (1910): 196–7.

Anon. "Émile Sarrau." *Nature* 70, no. 1805 (June 02, 1904): 106–7.

Anon. "He Armed the French but Got Scant Thanks. Eugene Turpin Invented Melinite and Many Other Engines of War." *Star (Toronto, Canada) Weekly*, April 09, 1927, 16.

Anon. "Obituary Notices of Fellows Deceased. Pierre Eugène Marcellin Berthelot, 1827–1907." *Proceedings of the Royal Society of London. Series A, Containing Papers of a Mathematical and Physical Character* 80, no. 542 (1908): iii–x.

Anon. "Obituary Notices of Fellows Deceased. Sir Andrew Nobel, 1831–1915." *Proceedings of the Royal Society, London. Series A* 94, no. 664 (August 1918): i–xvi.

Arnaud, Pascal. "Théophile-Jules Pelouze (1807–1867). Un pharmacien pionnier de la chimie organique." ["Theophile-Jules Pelouze (1807–1867). A Pioneering

Pharmacist and Organic Chemist."] *Revue d'Histoire de la pharmacie* 62, no. 385 (March 2015): 79–100.

Browne, Charles E. "Bernhard Tollens (1841–1918) and Some American Students of his School of Agricultural Chemistry." *Journal of Chemical Education* 19, no. 6 (June 1942): 253–9.

Browne, Charles E. "Charles Edward Munroe 1849–1938." *Journal of the American Chemical Society* 61, no. 5 (May 1939): 1301–16.

Clegg, Brian. *The First Scientist: A Life of Roger Bacon*. New York: Carroll & Graf, 2003.

Coston, Martha J. *A Signal Success. The Life and Travels of Mrs. Martha J. Coston.* Philadelphia, PA: Lippincott Co., 1886.

Donovan, Arthur. *Antoine Lavoisier: Science, Administration, and Revolution.* Cambridge: Cambridge University Press, 1996.

Earle, James. *Commodore Squib: The Life, Times and Secretive Wars of England's First Rocket Man, Sir William Congreve, 1772–1828.* Newcastle upon Tyne: Cambridge Scholars Publishing, 2010.

Fant, Kenne. *Alfred Nobel. A Biography.* Translated by Marianne Ruuth. New York: Arcade Publishing, 2014.

Garcia, Carmelo. "Statue to Honour Canary Girls at Former Factory Site. September 26, 2024." *BBC News*. https://www.bbc.com/news/articles/c0r85lxzz95o [accessed November 16, 2024].

Hamilton, Peggy. *Three Years or the Duration. The Memoirs of a Munition Worker 1914–1918.* London: Peter Owens Publisher, 1978.

Kitz, Janet. *Shattered City: The Halifax Explosion and the Road to Recovery.* Halifax, Nova Scotia: Nimbus Publishing, Ltd., 1990.

Kurzer, Frederick. "The Life and Work of Edward Charles Howard FRS." *Annals of Science* 56, no. 2 (October 1999): 113–41.

Labrude, Pierre, and Corinne Becq. "Le pharmacien et chimiste Henri Braconnot (Commercy 1780-Nancy 1855)" ["Pharmacist and chemist Henri Braconnot (Commercy 1780-Nancy 1855)]." *Revue d'histoire de la pharmacie* 51, no. 337 (1st Quarter 2003): 61–78.

McCallum, Ian. *Blood Brothers: Hiram and Hudson Maxim: Pioneers of Modern Warfare*. London: Greenhill Books, 2006.

Médard, Louis. "L'œuvre scientifique de Paul Vieille (1854–1934)" ["The Scientific Work of Paul Vieille (1854–1934)]." *Revue d'histoire des sciences* 47, nos. 3–4 (March-April 1994): 381–404.

Messel, Rudolph. "Hermann Johann Philipp Sprengel." *Journal of the Chemical Society* 91 (1907): 661–3.

Minutaglio, Bill. *City on Fire: The Explosion that Devastated a Texas Town and Ignited a Historic Legal Battle.* Austin: University of Texas Press, 2014.

Norris, Megan. *Out of the Ashes: The Mother's Love that Healed the Scars of the Bali Bombings.* Newport, New South Wales: Big Sky Publishing, 2022.

Peterson, Larry. *Deadly Dust: The Imperial Sugar Inferno.* Castroville, TX: Black Rose Writing, 2016.

Pilato, Denise E. "Martha Coston: A Woman, a War, and a Signal to the World." *International Journal of Naval History* 1, no. 1 (April 2002): 115–27.

Robbins, Peggy. "The Confederacy's Bomb Brothers." [Gabriel and George Rains] *Journal of Mine Action* 6, no. 1 (April 2002): 1–7.

Ruston, Maureen. *The Canary Girls of Chilwell: the Story of No. 6 Shell Filling Factory, Chilwell, Nottinghamshire.* Beeston, Nottinghamshire: Beeston & District Local History Society, 2008.

Spiller, John. "Obituary Notices. Frederick August Abel." *Journal of the Chemical Society, Transactions* 87 (1905): 565–70.

Stoltzenberg, Dietrich. *Fritz Haber: Chemiker, Nobelpreisträger, Deutscher, Jude.* [in German] Weinheim: VCH, 1994.

Stoltzenberg, Dietrich. *Fritz Haber: Chemist, Nobel Laureate, German, Jew.* [abridged version of German original] Philadelphia, PA: Chemical Heritage Foundation Press, 2004.

Sztejnberg, Aleksander. "The Eminent French Chemist Claude-Louis Berthollet (1748–1822) in the Literature between the 19th and 21st Centuries." *Substantia* 4, no. 2 (September 2020): 139–49.

Thornton, Rosemary. *Penniman: Virginia's Own Ghost City.* Portsmouth, VA: Gentle Beam Publications: 2020.

Toobin, Jeffrey. *Homegrown: Timothy McVeigh and the Rise of Right-Wing Extremism.* New York: Simon & Schuster, 2023.

Wilkinson, Norman B. *Lammot du Pont and the American Explosives Industry 1850–1884.* Charlottesville, VA: University Press of Virginia, 1984.

Woollacott, Angela. *On Her Their Lives Depend: Munitions Workers in the Great War.* Oakland, CA: University of California Press, 1994.

B. Nuclear Explosives

Avery, Donald A. *The Science of War: Canadian Scientists and Allied Military Technology during the Second World War.* Toronto: University of Toronto Press, 1998.

Bird, Kai, and Martin J. Sherwin. *American Prometheus: The Triumph and Tragedy of J. Robert Oppenheimer.* New York: A. A. Knopf, 2005.

Dainton, Frederick Sydney. "George Bogdan Kistiakowsky. 18 November 1900—7 December 1982." *Biographical Memoirs of Fellows of the Royal Society* 31 (November 1985): 376–408.

Frisch, Otto R. *What Little I Remember.* Cambridge: Cambridge University Press, 1979.

Gaponov, Yuri V. "Igor Vasiliyevich Kurchatov: The Scientist and Doer. (January 12, 1903—February 7, 1960)." Translated from *Yadernaya Fizika. Physics of Atomic Nuclei* 66, no. 1 (2003): 3–7.

Goldschmidt, Bertrand. *Atomic Rivals. A Candid Memoir of Rivalries among the Allies over the Bomb.* Translated by Georges M. Temmer. New Brunswick, NJ: Rutgers University Press, 1990.

Goodchild, Peter. *Edward Teller: The Real Dr. Strangelove.* Cambridge, MA: Harvard University Press, 2004.

Gorelik, Gennady. "The Paternity of the H-Bombs: Soviet-American Perspectives." *Physics in Perspective* 11, no. 2 (June 2009): 169–97.

Ham, Paul. *Hiroshima Nagasaki: The Real Story of the Atomic Bombings and Their Aftermath.* New York: Picador, 2015.

Hecker, Siegfried S., and Elliot A. Serbin. *Hinge Points. An Inside Look at North Korea's Nuclear Program.* Stanford, CA: Stanford University Press, 2023.

Hersey, John. *Hiroshima.* New York: Alfred A. Knopf, Inc., 1946.

Holloway, David. *Stalin and the Bomb: The Soviet Union and Atomic Energy, 1939–1956.* New Haven, CT: Yale University Press, 1996.

Howes, Ruth H., and Caroline L. Herzenberg. *Their Day in the Sun. Women of the Manhattan Project.* Philadelphia, PA: Temple University Press, 1999.

Kelly, Cynthia C., ed. *The Manhattan Project: The Birth of the Atomic Bomb in the Words of Its Creators, Eyewitnesses, and Historians.* New York: Black Dog & Leventhal, 2020.

Lemmerich, Jost. *Science and Conscience: The Life of James Franck.* Translated by Ann M. Hentschel. Stanford, CA: Stanford University Press, 2011.

Lourie, Richard. *Sakharov. A Biography.* Waltham, MA: Brandeis University Press, 2002.

Kiernan, Denise. *The Girls of Atomic City: The Untold Story of the Women Who Helped Win World War II.* New York: Touchstone, 2013.

Kunetka, James. *The General and the Genius: Groves and Oppenheimer—The Unlikely Partnership that Built the Atomic Bomb.* Washington, DC: Regnery History, 2015.

Monk, Ray. *Robert Oppenheimer. A Life Inside the Center.* New York: Random House, 2012.

Oleynikov, Pavel V. "German Scientists in the Soviet Atomic Project." *Nonproliferation Review* 7, no. 2 (Summer 2000): 1–30.

Pertuiset, Simone. *Manet and the Lion Hunter.* [Biography of Eugène Pertuiset] Wilmington, DE: ECD—Tip of An Era Press, 2022.

Szilárd, Leo, Gertrude Weiss-Szilárd, and Spencer R. Weart. *Leo Szilárd: His Version of the Facts—Recollections and Correspondence.* Cambridge, MA: MIT Press, 1978.

Tsukerman, Veniamin Aronovich, and Zinaida Matveevna Azarkh. *Arzamas-16. Soviet Scientists in the Nuclear Age: A Memoir.* Translated by Timothy Segay. Bramcote, Nottinghamshire: Bramcote Press, 1998.

Ulam, Stanislaw M. *Adventures of a Mathematician.* Oakland, CA: University of California Press, 1991.

Waldof, Teresa Wilhelm. *Wilhelm's Way: The Inspiring Story of the Iowa Chemist Who Saved the Manhattan Project.* La Vergne, TN: Third Generation Publishing, 2022.

Weiner, Tim. "Abdul Qadeer Khan, 85, led Pakistan's Nuclear Program." *The Buffalo News.*October 12, 2021, B5.

About the Author

Benjamin C. Garrett retired in 2015 from the Federal Bureau of Investigation (FBI). In his last position, he served in the Laboratory Division, Quantico, Virginia, as the senior scientist for weapons of mass destruction. He holds an undergraduate degree from Davidson College, Davidson, North Carolina, and a doctorate in analytical chemistry from Emory University, Atlanta, Georgia. He served in the US Army and spent two decades working on defense, environmental, and intelligence programs prior to joining the FBI. He is the author of *Historical Dictionary of Nuclear, Biological, and Chemical Warfare* (2017).

www.ingramcontent.com/pod-product-compliance
Ingram Content Group UK Ltd.
Pitfield, Milton Keynes, MK11 3LW, UK
UKHW020220290926
15470UKWH00001B/301